ENERGY SCIENCE, ENGINEERING AND TECHNOLOGY

GEOTHERMAL ENERGY: THE RESOURCE UNDER OUR FEET

ENERGY SCIENCE, ENGINEERING AND TECHNOLOGY

Additional books in this series can be found on Nova's website at:

https://www.novapublishers.com/catalog/index.php?cPath=23_29&seriesp=Energy%20Science%2C%20
Engineering%20and%20Technology&sort=2a&page=1

Additional E-books in this series can be found on Nova's website at:

https://www.novapublishers.com/catalog/index.php?cPath=23_29&seriespe=Energy+Science%2C+
Engineering+and+Technology

ENERGY SCIENCE, ENGINEERING AND TECHNOLOGY

GEOTHERMAL ENERGY: THE RESOURCE UNDER OUR FEET

CHARLES T. MALLOY
EDITOR

Nova Science Publishers, Inc.

New York

For permission to use material from this book please contact us:
Telephone 631-231-7269; Fax 631-231-8175
Web Site: http://www.novapublishers.com

LIBRARY OF CONGRESS CATALOGING-IN-PUBLICATION DATA

Geothermal energy : the resource under our feet / editor, Charles T. Malloy.
p. cm.
Includes index.
ISBN 978-1-60741-502-2 (hardcover)
1. Geothermal resources. I. Malloy, Charles T.
GB1199.5.G456 2009
333.8'8--dc22
2010012220

Published by Nova Science Publishers, Inc. ✝ New York

CONTENTS

PREFACE

The Earth houses a vast energy supply in the form of geothermal resources. Domestic resources are equivalent to a 30,000 year energy supply at our current rate for the United States. Geothermal energy is used in all 50 U.S. states today. But, geothermal energy has not reached its full potential as a clean secure energy alternative because of issues with resources, technology, historically low natural gas prices, and public policies. These issues affect the economic competitiveness of geothermal energy. This book explores geothermal energy's viability, risk and cost analysis.

Chapter 1 - The Earth houses a vast energy supply in the form of geothermal resources. Domestic resources are equivalent to a 30,000-year energy supply at our current rate for the United States! In fact, geothermal energy is used in all 50 U.S. states today. But geothermal energy has not reached its full potential as a clean, secure energy alternative because of issues with resources, technology, historically low natural gas prices, and public policies. These issues affect the economic competitiveness of geothermal energy.

Chapter 2 - This is a testimony of Susan Petty, President, AltaRock Energy, Inc.before the United States Senate Committee.

Chapter 3 - Geothermal energy has been exploited for power generation since at least 1904.[1] However, the last few years have witnessed a conspicuous revival in interest in geothermal technologies both old and new. In fact, 2008 was a watershed year for the industry. The U.S. Department of Energy (DOE) revived its Geothermal Technologies Program (GTP) with new funding that made possible substantial new investments in geothermal research, development and technology demonstration. The U.S. Department of the Interior's (DOI) Bureau of Land Management (BLM) also significantly increased the amount of Federal land available for geothermal exploration and development and worked to streamline the complex permitting and leasing process. Installed geothermal capacities in the United States and abroad continued to increase as well.

Chapter 4 - Features a letter from the United States Department of Energy.

Chapter 5 - This document presents the results of an eight-month study by the Department of Energy (DOE) and its support staff at the national laboratories concerning the technological requirements to commercialize a new geothermal technology, Enhanced Geothermal Systems (EGS). EGS have been proposed as a viable means of extracting the earth's vast geothermal resources. Those who contributed to the study and authored portions of the report include: Allan Jelacic, Raymond Fortuna, Raymond LaSala and Jay Nathwani (DOE); Gerald Nix (retired), Charles Visser, and Bruce Green (National Renewable Energy

Laboratory); Joel Renner (Idaho National Laboratory); Douglas Blankenship (Sandia National Laboratories); Mack Kennedy (Lawrence Berkeley National Laboratory); and Carol Bruton (Lawrence Livermore National Laboratory). Richard Price (TMS Inc.) and Clifton Carwile (consultant) also made substantial contributions. Michael Reed, Jim McVeigh, Jihan Quail, and Christina Van Vleck (SENTECH, Inc.) and Raymond David (National Renewable Energy Laboratory) contributed to the design and production of this report.

Chapter 6 - Of the 15,400 MWt (4.38 x 10^6 tons) global installed base of GSHPs, about 56 percent of this capacity is installed in the U.S., corresponding to about 65 percent of the GSHP unit installations. Europe follows, with about 39 percent of the installed capacity, and Asia has about 5%. In Europe, Sweden is the dominant player in the GSHP market, with almost 2500 MWt (711,000 tons) installed— more than double any other European country.

The U.S. GSHP market is split roughly evenly between residential and commercial applications, with only a very small market for industrial applications.

GSHPs can provide significant primary unit energy savings compared to typical ASHPS or typical furnaces with air conditioners. Savings are often in the range of 30 to 60 percent of space-conditioning energy consumption, depending on GSHP efficiency, technology replaced, climate, and application.

In: Geothermal Energy: The Resource Under Our Feet
Editor: Charles T. Malloy

ISBN: 978-1-60741-502-2

Chapter 1

GEOTHERMAL - THE ENERGY UNDER OUR FEET, GEOTHERMAL RESOURCE ESTIMATES FOR THE UNITED STATES*

Bruce D. Green and R. Gerald Nix

EXECUTIVE SUMMARY

The Earth houses a vast energy supply in the form of geothermal resources. Domestic resources are **equivalent to a 30,000-year energy supply at our current rate for the United States!** In fact, geothermal energy is used in all 50 U.S. states today. But geothermal energy has not reached its full potential as a clean, secure energy alternative because of issues with resources, technology, historically low natural gas prices, and public policies. These issues affect the economic competitiveness of geothermal energy.

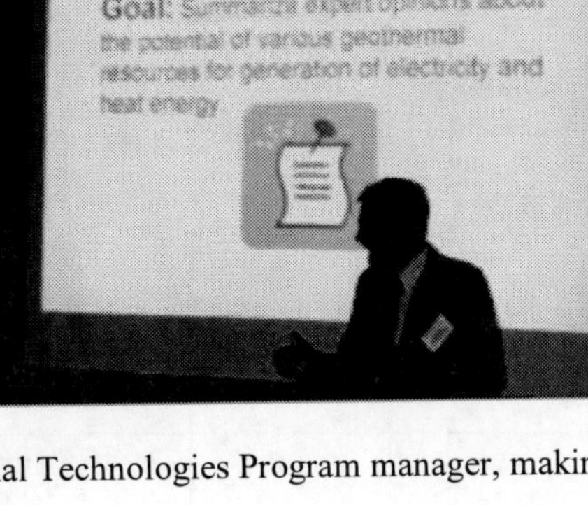

Dr. Roy Mink, DOE Geothermal Technologies Program manager, making opening comments.

* This is an edited, reformatted and augmented version of a National Renewable Energy Laboratory publication dated November 2006.

The geothermal heat pump working-group during their resource deliberations. John Geyer (l), Wael El-Sharif (c), and Jack DiEnna (r) made up this working group.

On May 16, 2006, the National Renewable Energy Laboratory (NREL) in Golden, Colorado hosted a geothermal resources workshop with experts from the geothermal community. The purpose of the workshop was to re-examine domestic geothermal resource estimates. The participating experts were organized into five working groups based on their primary area of expertise in the following types of geothermal resource or application: (1) Hydrothermal, (2) Deep Geothermal Systems, (3) Direct Use, (4) Geothermal Heat Pumps (GHPs), and (5) Co-Produced and Geopressured.

Geothermal resources are categorized in several layers of accessibility and feasibility, from broadest criteria (i.e., total physically available), to criteria that includes technical and economic considerations. The total resource base is scaled downward to accessible resource, and finally to a category called developable resource (see p. 4 for explanation).

The table below shows estimates for the different geothermal resource categories, as compiled by the workshop experts. **These estimates show the enormous potential of the U. S. geothermal energy resource.** New low-temperature electric generation technology may greatly expand the geothermal resources that can be developed economically today.

Findings by Resource Category

			Estimated Developable Resource*		
	Estimated Accessible Resource (MWe)	**2006 (Actual MWe)**	**2015 (MWe)**	**2025 (MWe)**	**2050 (MWe)**
Shallow Hydrothermal[1] (Identified) >90°C/1 94°F	30,000	2,800	10,000	20,000	30,000
Shallow Hydrothermal[1] (Unidentified) > 150°C/302°F	120,000		TBD	TBD	TBD
Co-Produced & Geopressured[2]	>1 00,000	2[3]	10,000 to 15,000	70,000	>1 00,000
Deep Geothermal[4]	1,300,000 to 13,000,000	0	1000	10,000	130,000

Thermal Uses	(MWt)	(MWt)	(MWt)	(MWt)	
Direct Uses[5]	>60,000	620	1600	4,200	45,000
Geothermal Heat Pumps[6]	>1,000,000	7,385	18,400	66,400	>1,000,000
GHP[6] Avoided Power	120,000	880	2,100	8,000	120,000

* Please note that these resource estimates represent a consensus of a group of experts who considered existing resource assessments (referenced on next page). There is considerable uncertainty in the estimates as many resources are hidden and exploration to date has been relatively limited. The figures shown above are *not* a resource assessment, but, even with uncertainty, clearly show that the U.S. geothermal resource is a very large and important domestic energy source.

1 *Assessment of Geothermal Resources of the United States – 1978*, USGS Circular 790 (p. 41 and 157). Includes identified and unidentified resources; 2015 and later estimates are a consensus of the experts at the workshop. Estimated accessible figure includes identified (~30,000 MW) and unidentified (~120,000 MW) (i.e., hidden or showing no surface manifestations) hydrothermal resources.

2 "Geothermal Electric Power Supply Possible from Gulf Coast, Midcontinent Oil Field Waters," *Oil and Gas Journal,* September 5, 2005, and SMU Geothermal Laboratory *Geothermal Energy Generation in Oil and Gas Settings Conference* findings, March 13 – 14, 2006, and USGS Circular 790.

3 Based on Mafi Trench Unit on offshore platform now in operation.

4 *Energy Recovery from Enhanced/Engineered Geothermal Systems* (EGS), Massachusetts Institute of Technology (MIT), September 2006.

5 OIT Geo-Heat Center, using analysis based on USGS Circulars 790 and 892.

6 Geothermal Heat Pump Consortium, based on Energy Information Administration data and projections. The 'avoided power' figure represents the peak power not required or offset through use of GHPs. Thus, GHPs act as a proven demand and growth management option for utilities.

INTRODUCTION

The United States possesses vast underground stores of heat whose full potential has yet to be realized. The Earth's interior reaches temperatures greater than 4,000°C (>7,200°F), and this geothermal energy flows continuously to the surface. The energy content of domestic geothermal resources to a depth of 3 km (~2 mile) is estimated to be 3 million quads, **equivalent to a 30,000-year supply of energy at our current rate for the United States!** While the entire resource base cannot be recovered, the recovery of even a very small percentage of this heat would make a large difference to the nation's energy supplies. New low- temperature electric generation technology may greatly expand the geothermal resources that can be developed economically today.

Geothermal resources could meet a substantial portion of the nation's energy needs in the 21st century. In fact, when including geothermal heat pumps (GHPs), geothermal energy is used in all 50 U.S. states today. The U.S. Department of Energy's (DOE) Geothermal Technologies Program seeks to make geothermal energy the nation's environmentally preferred baseload energy alternative. The Program's mission is to work in partnership with U.S. industry to establish geothermal energy as an economically competitive contributor to the nation's energy supply.

The purpose of the workshop was to re-examine domestic geothermal resource estimates. There were two guiding questions: what is the total potential accessible resource in the United States, and, given favorable circumstances and using existing practices with improved technology, and with institutional issues solved, how much of the resource is developable by 2015, 2025, and 2050? Resource types include hydrothermal, deep geothermal systems, co-produced, geopressured, direct use, and GHPs.

The goal of the workshop was to *gather and summarize* expert opinions about the potential of various geothermal resources for generation of electricity and utilization of heat energy. The workshop was not a formal assessment, but a recorded discussion by a group of experts who collectively state their opinions based on their experiences, knowledge, and interpretations of various detailed assessments (e.g., USGS Circular 790).

Resource Definitions

The *total resource base* in the United States, both renewable and non-renewable, is very large, with an energy content of over 657,000 billion barrels of oil equivalent (BBOE), or nearly 50,000 times the annual current rate of national energy consumption. Figure 1 shows graphically what the *total resource base* looks like, and in descending order, the values for the estimated accessible geothermal resource and estimated developable resource.

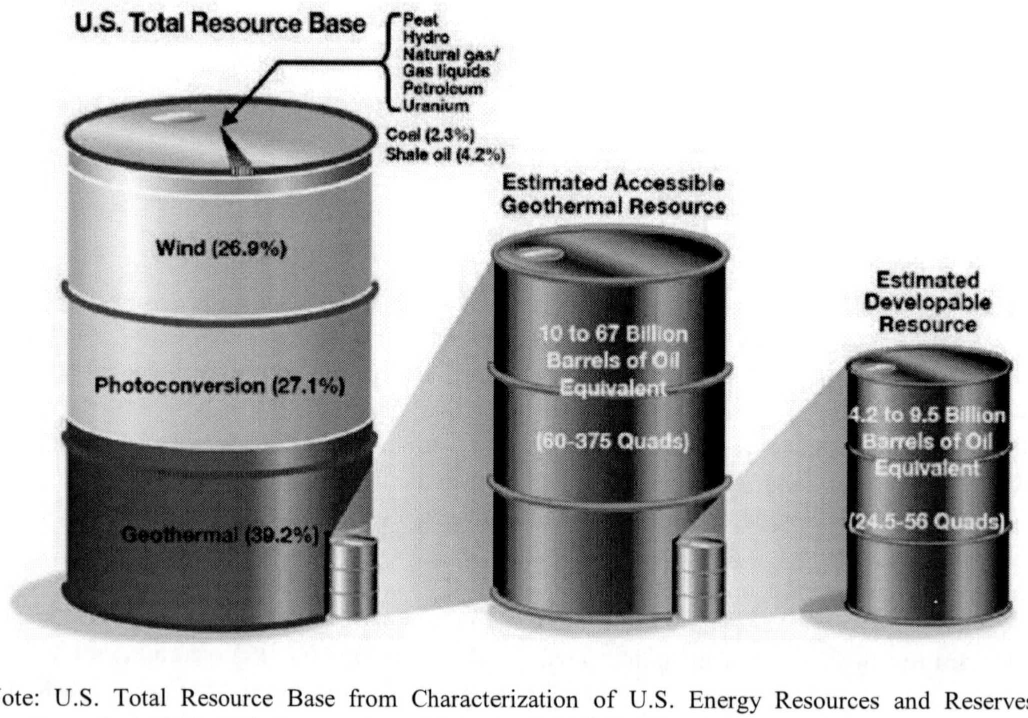

Note: U.S. Total Resource Base from Characterization of U.S. Energy Resources and Reserves, December 1989, U.S. Department of Energy, DOE/CE-0279. Data for "Estimated Accessible Geothermal Resource" and "Estimated Developable Resource" are from Table 4 of this report.

Figure 1. U.S. Energy and Geothermal Resources

Energy resources are traditionally classified according to the degree of certainty and the economic feasibility of exploiting the particular resource. The U.S. Geological Survey (USGS) and the U.S. Department of Energy (DOE) have used such identifying terms for resource classification. Following are simple definitions of these classification terms:

U.S. Total Resource Base – Resource base is all of a given material in the Earth's crust, whether its existence is known or unknown and regardless of cost considerations[a].

Estimated Accessible Geothermal Resource– The accessible resource base for geothermal energy is that part of the resource base shallow enough to be reached by production drilling in the foreseeable future[a].

Estimated Developable Resource – This category is the subset of the accessible resource base that the workshop experts believe likely to be developed in future years.

[a]**Source:** Assessment of Geothermal Resources of the United States – 1978, USGS Circular 790 (p. 4).

DESCRIPTION OF GEOTHERMAL RESOURCE CATEGORIES

Hydrothermal

A hydrothermal system is defined as a subterranean geothermal reservoir that transfers heat energy upward by vertical circulation of fluids driven by differences in fluid density that correspond to differences in temperature (see Figure 2). Hydrothermal systems can be classified into two types—vapor-dominated and hot water— depending on whether the fluid is steam or liquid water, respectively.

Geothermal resources are available throughout the entire U.S.

Most high-temperature geothermal resources occur where magma (molten rock) has penetrated the upper crust of the Earth. The magma heats the surrounding rock, and when the rock is permeable enough to allow the circulation of water, the resulting hot water or steam is referred to as a hydrothermal resource. Such resources are used today for the commercial production of geothermal power. They benefit from continuous recharge of energy as heat flows into the reservoir from greater depths.

Deep Geothermal Systems

Deep geothermal systems (a.k.a. enhanced geothermal systems or EGS) are defined as engineered reservoirs that have been created to extract heat from economically unproductive geothermal resources. The deep geothermal/EGS concept is to extract heat by creating a subsurface fracture system to which water can be added through injection wells. The water is heated by contact with the rock and returns to the surface through production wells, just as in naturally occurring hydrothermal systems.

Hydrofracturing and stimulation techniques are used widely in the oil and gas industry to extend production, and can be used to greatly extend and expand use of geothermal resources. Figure 3 gives a graphic idea of the domestic scope of geothermal resources at just 6 kilometers (3.7 miles), a nominal drilling depth in the oil and gas industry.

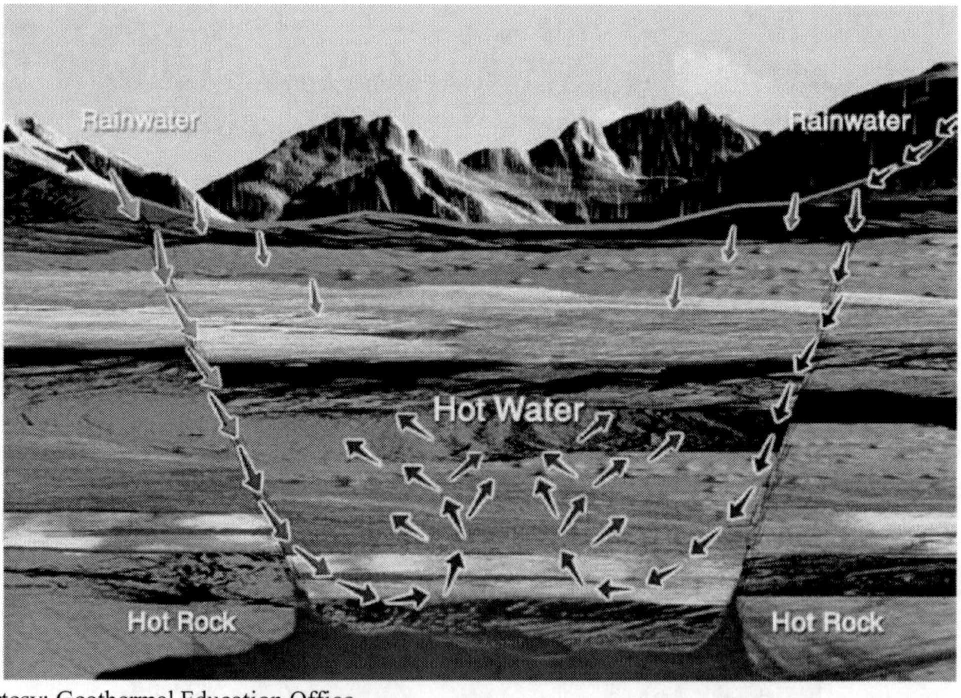

Courtesy: Geothermal Education Office

Figure 2. llustration of a hydrothermal reservoir, showing the natural recharge, fractures, and heat source

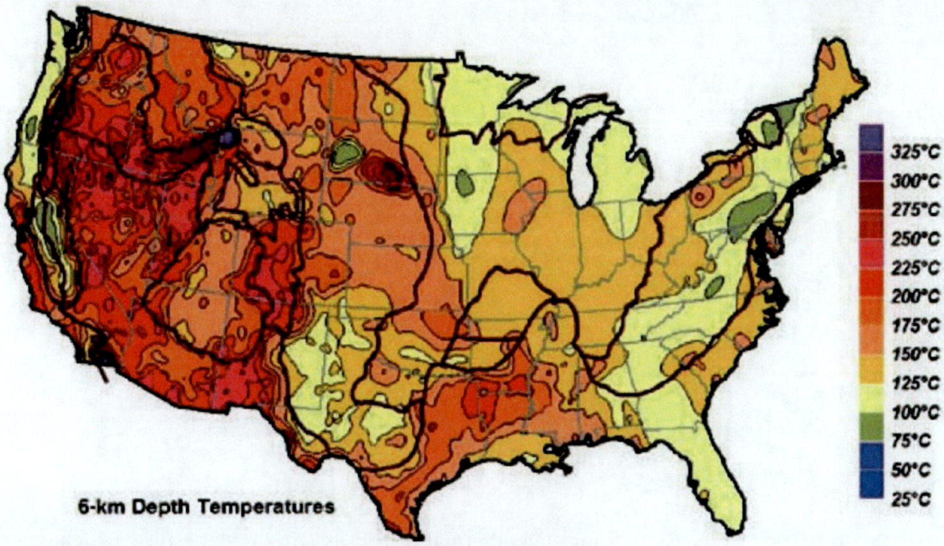

Courtesy: Southern Methodist University Geothermal Laboratory.

Figure 3. Estimated Earth temperature at 6-km (3.7-mile) depth.

Direct Use

Hot water from geothermal resources is used directly to provide heat for buildings, crop and lumber drying, industrial process heat needs, aquaculture, horticulture, ice melting on sidewalks, roads, and bridges, and district heating systems. In direct use applications, a well (or series of wells) brings hot water to the surface; a mechanical system— piping, heat exchanger, pumps, and controls—delivers the heat to the space or process.

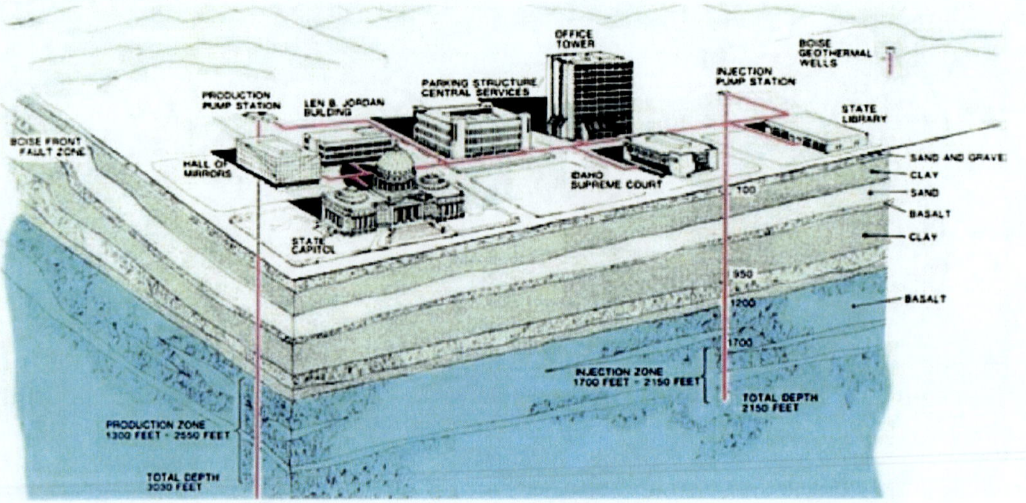

Figure 4. An illustration of the State Capitol geothermal district-heating system, Boise, Idaho

Often, direct use applications use geothermal fluids not hot enough for electricity generation. To improve efficiencies, used water from geothermal power plants can be 'cascaded' down for lower temperature uses, such as in greenhouses or aquaculture. Flowers, vegatables, and various fish species and alligators are examples of products from greenhouse and aquaculture systems.

Geothermal Heat Pumps

Geothermal heat pumps (GHPs) use the Earth's huge energy storage capability to heat and cool buildings, and to provide hot water. GHPs use conventional vapor compression (refrigerant- based) heat pumps to extract the low-grade heat from the Earth for space heating. In summer, the process reverses and the Earth becomes a heat sink while providing space cooling (see Figure 5). GHPs are used in all 50 U.S. states today, with great potential for near-term market growth and savings.

Geopressured Resources

The geopressured resource consists of deeply buried reservoirs of hot brine, under abnormally high pressure, that contain dissolved methane. Geopressured brine reservoirs with pressures approaching lithostatic load are known to occur both onshore and offshore beneath the Gulf of Mexico coast, along the Pacific west coast, in Appalachia, and in deep sedimentary basins elsewhere in the United States.

The resource contains three forms of energy: methane, heat, and hydraulic pressure. In the past, DOE conducted research on geopressured reservoirs in the northern Gulf of Mexico sedimentary basin, and operated a 1-megawatt (MW) power plant using the heat and methane from the resource (i.e., Pleasant Bayou, TX, 1989 – 1990).

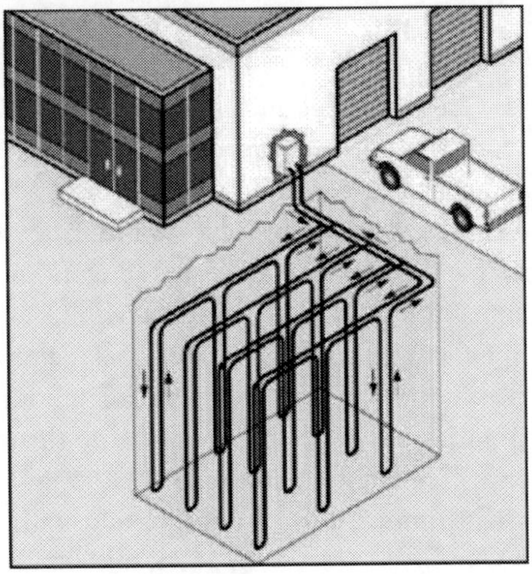

Figure 5 - Geothermal heat pump (GHP) illustration for a commercial application

Co-Produced Geothermal Fluids

Sometimes referred to as the 'produced water cut' or 'produced water' from oil and gas wells, co- produced geothermal fluids are hot and are often found in waterflood fields in a number of U.S oil and gas production regions (See Table 1). This water is typically considered a nuisance to the oil and gas industry (and industry is accountable for proper disposal), but could be used to produce electricity for internal use or sale to the grid.

Like geopressured resources, co-produced geothermal resources can deliver near-term energy savings, diminish greenhouse gas emissions, and extend the economical use of an oil or gas field. New low-temperature electric generation technology may greatly expand the geothermal resources that can be developed economically today.

Table 1. Co-Produced Geothermal Fluids

Estimated equivalent geothermal power from processed water associated with existing hydrocarbon production, using 140°C (285°F) as a nominal fluid temperature.		
State	**Total Processed Water, 2004 (bbl)**	**Power, MW @ 140°C (285)°F)**
Alabama	203,223,404	47
Arkansas	258,095,372	59
California	5,080,065,058	1169
Florida	160,412,148	37
Louisiana	2,136,572,640	492
Mississippi	592,517,602	136
Oklahoma	12,423,264,300	2860
Texas	12,097,990,120	2785
Total	32,952,140,644 bbl	7,585 MW

Courtesy: Dr. David Blackwel, Southern Methodist University

Process Used

The participating experts were organized into five working groups based on their primary area of expertise in the following types of geothermal resource or application: (1) Hydrothermal, (2) Deep Geothermal Systems, (3) Direct Use, (4) Geothermal Heat Pumps, and (5) Co-Produced and Geopressured. Because of similarities (e.g., oil and gas industry involvement), the co- produced and geopressured resource working groups were combined.

Through opening comments and guidance, Dr. Roy Mink of DOE and Dr. Gerry Nix of NREL introduced the participants to the task, purpose, goals, and approach of the workshop. The facilitator provided each working group with a set of guiding questions to help obtain results consistent with stated purpose, goals, and approach to this workshop. These included: what is the total potential accessible resource in the U.S., and, given favorable circumstances and using existing practices with improved technology, and with institutional issues solved, how much of the resource is developable by 2015, 2025, and 2050?

The Gulf Coast region of the U.S. is rich in geopressured and co-produced energy resources

Workshop participants during data integration discussions – Dr. John Lund makes a point

The working groups answered the guiding questions and formulated their responses, then presented their findings to the other participants. During the final stage of the workshop, the participants discussed data integration and related issues. See Table 2 for the findings of the working groups.

Table 2. Findings by Resource Category

			Estimated Developable Resource*		
	Estimated Accessible Resource (MWe)	**2006 (Actual MWe)**	**2015 MWe)**	**2025 (MWe)**	**2050 (MWe)**
ShallowHydrothermal[1] (Identified) >90°C/194°F	30,000	2,800	10,000	20,000	30,000
Shallow Hydrothermal[1] (Unidentified) >150°C/302°F	120,000		TBD	TBD	TBD
Co-Produced & Geopress[2]	>100,000	23	10,000 to 15,000	70,000	>100,000
Deep Geothermal[4]	1,300,000 to 13,000,000	0	1000	10,000	130,000
Thermal Uses	**(MWt)**	**(MWt)**	**(MWt)**	**(MWt)**	
Direct Uses[5]	>60,000	620	1600	4,200	45,000
Geothermal Heat Pumps[6]	>1,000,000	7,385	18,400	66,400	>1,000,000
GHP[6]Avoided Power	120,000	880	2,100	8,000	120,000

* Please note that these resource estimates represent a consensus of a group of experts who considered existing resource assessments (referenced below). There is considerable uncertainty in the estimates as many resources are hidden and exploration to date has been relatively limited. The figures shown above are **not** a resource assessment, but, even with uncertainty, ***clearly show that the U.S. geothermal resource is a very large and important domestic energy source.***

Chena Hot Springs Resort (Alaska) geothermal power plant dedication in August 2006. New technology such as that being used at Chena could greatly extend the feasibility of using geothermal resources.

Table 3. Co-Produced and Geopressured Resources (Estimated Developable Resources)[2]

	2015	2025	2050
Co-Produced	5,000 MWe	10,000 to 20,000 MWe	30,000 to 40,000 MWe
Geopressured	5,000 to 10,000 MWe	50,000 to 60,000 MWe	70,000 to 80,000 MWe

Table 4. Energy Equivalents by Resource Category (see appendices for calculation method)

Resource Category	Estimated Accessible Resource	Estimated Developable Resource – 2050
Shallow Hydrothermal[1] (Identified) >90˚C/194˚F	0.81 Quads 135 million BOE	0.81 Quads 135 million BOE
Shallow Hydrothermal[1] (Unidentified) >150˚C/302˚F	3.2 Quads 540 million BOE	–
Co-Produced & Geopressured[2]	2.7 Quads 450 million BOE	2.7 Quads 450 million BOE
Deep Geothermal[4]	35.1 to 351 Quads 5.8 to 58.5 BBOE	3.5 tp 35 Quads 0.58 to 5.8 BBOE
Direct Use[5]	0.88 Quads 150 million BOE	15 Quads 112.5 million BOE
Geothermal Heat Pumps (GHP)[6]	15 Quads 2.5 billion BOE	15 Quads 2.5 BBOE
GHP[6] Avoided Power	1.8 Quads 300 million BOE	1.8 Quads 300 million BOE

1. USGS Circular 790 (p. 157), includes identified and unidentified resources; 2015 and later estimates are a consensus of the experts at the workshop. Estimated accessible figure includes identified (~30,000 MW) and unidentified (~120,000 MW) (i.e., hidden or showing no surface manifestations) hydrothermal resources.
2. USGS Circular 790, *Geothermal Electric Power Supply Possible from Gulf Coast, Midcontinent Oil Field Waters,* "Oil and Gas Journal," September 5, 2005, and SMU Geothermal Laboratory *Geothermal Energy Generation in Oil and Gas Settings Conference* findings, March 13 – 14, 2006.
3. Based on Mafi Trench Unit on offshore platform now in operation.
4. *Energy Recovery from Enhanced/Engineered Geothermal Systems* (EGS), Massachusetts Institute of Technology (MIT), September 2006.
5. OIT Geo-Heat Center, using analysis based on USGS Circulars 790 and 892.
6. Geothermal Heat Pump Consortium, based on Energy Information Administration data and projections. The 'avoided power' figure represents the peak power not required or offset through use of GHPs. Thus, GHPs act as a proven demand and growth management option for utilities.

Energy Comparison Information

U.S annual energy consumption equals about 100 quads (EIA, 2004).
U.S annual electricity production equals about 40 quads (EIA, 2004).
U.S. petroleum demand equals about 21 million bbl/day, 7.67 billion bbl/yr (EIA, 2004).
World petroleum demand equals about 84 million bbl/day, 30 billion bbl/yr (EIA, 2004).

Energy Equivalents

1 Quad = 0.170 billion barrels of oil, 170 million barrels of oil
1 Quad = 45 million short tons of coal
1 Quad = 1 trillion cubic feet of dry natural gas

1000 KWh = 0.59 barrels of crude oil
1000 KWh = 0.15 short tons (300 pounds) of coal
1000 KWh = 3,300 cubic feet of dry natural gas

1 barrel of crude oil = 1,700 kWh
1 barrel of crude oil = 5,600 cubic feet of dry natural gas
1 barrel of crude oil = 0.26 short tons (520 pounds) of coal

APPENDICES

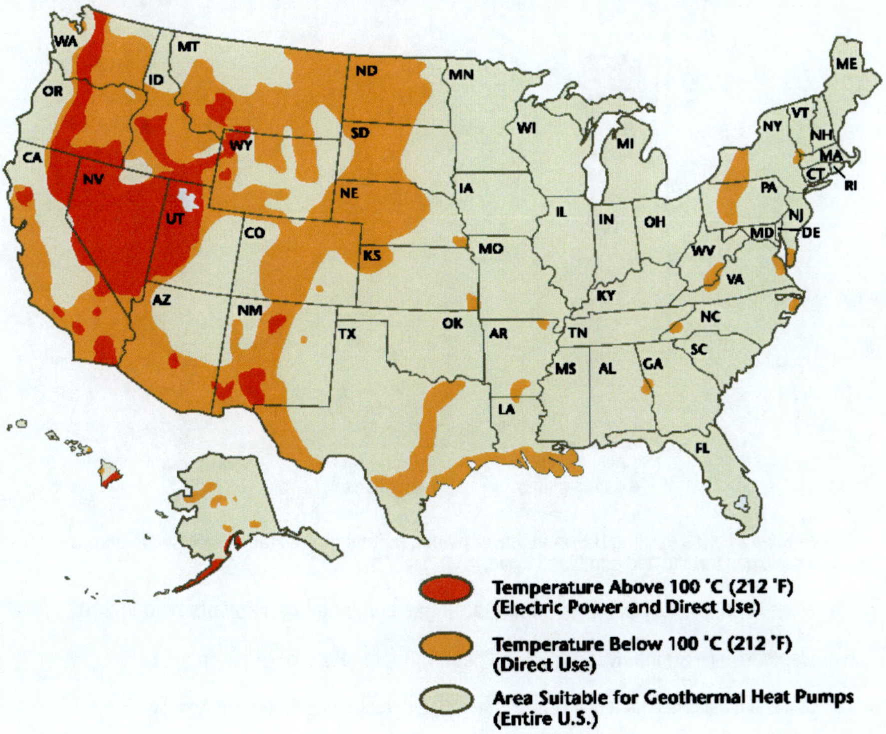

The Nation's geothermal resources represent a huge and viable energy resource, providing the U.S. with various ways to use them and enhance national security, and economic and environmental health.

Present Development

Today's U.S. geothermal industry is a $2-billion-per-year enterprise involving over 2800 MW of electricity generation capacity, about 620 MW of thermal energy capacity in direct-use applications such as indoor heating, greenhouses, food drying, and aquaculture, and over 7,300 MW of thermal energy capacity from geothermal heat pumps. The international market for geothermal power development could exceed a total of $25 billion for the next 10 to 15 years. At the present time, U.S. technology and industry stand at the forefront of this international market.

Chart 1. Carbon Dioxide (CO_2) for U.S. Power Plants
(Pounds per Megawatt-hour)

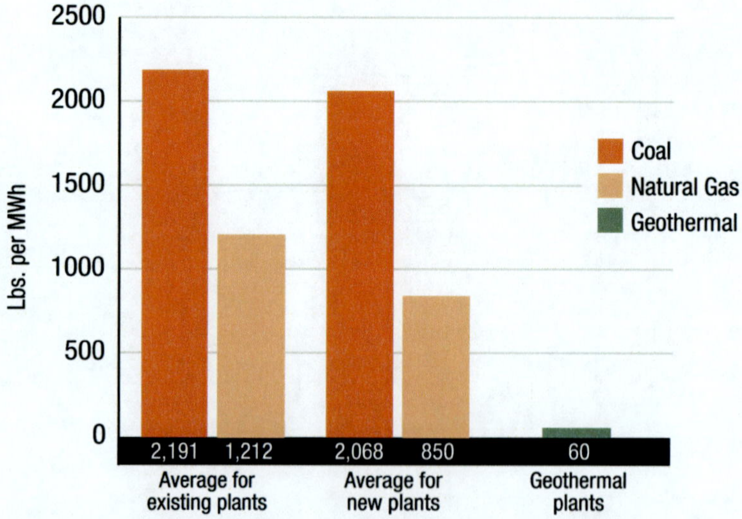

Existing - For all existing U.S. coal power plants; natural gas averages include steam cycle, simple gas turbine and combined cycle.

New - Coal plants built in 1990s; natural gas combined cycle plants built in 2002.

Geothermal - 60 lbs./MWh for flash plants; 0 lbs./MWh for binary plants.

CO_2 is not classified as a pollutant by the U.S. Environmental Protection Agency.

Chart 2. Sulfur Dioxide (SO_2) for U.S. Power Plants

(Pounds per Megawatt-hour)

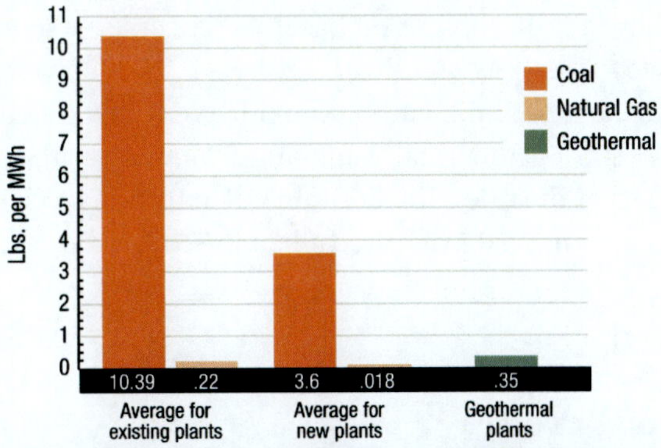

Existing - For all existing U.S. coal power plants; natural gas averages include steam cycle, simple gas turbine and combined cycle.

New - Coal plants built in 1990s; natural gas combined cycle plants built in 2002.

Geothermal - .35 lbs./MWh for flash plants, 0 lbs./MWh for binary plants. SO_2 from geothermal plants is from hydrogen sulfide contained in geothermal fluids. Modern systems return 90 percent of the hydrogen sulfide to the reservoir.

Chart 3. Nitrogen Oxide (NO_X) for U.S. Power Plants

(Pounds per Megawatt-hour)

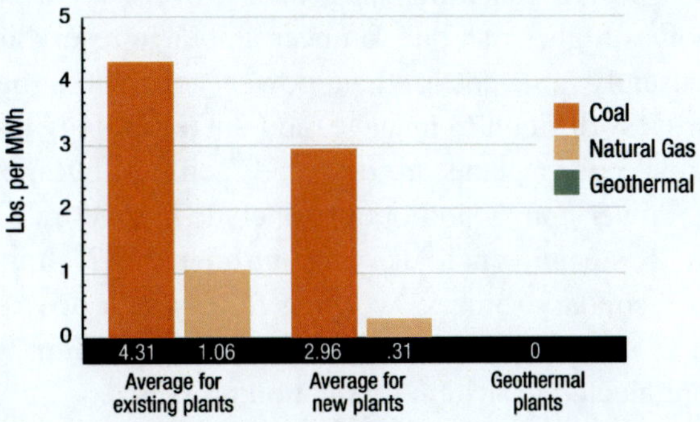

Existing - For all existing U.S. coal power plants; natural gas averages include steam cycle, simple gas turbine and combined cycle.

New - Coal plants built in 1990s; natural gas combined cycle plants built in 2002.

Source: Coal and natural gas emissions information from Platts Research and Consulting, based on data from the Environmental Protection Agency's Continuous Emissions Monitoring Systems, 2003; geothermal information from U.S. Department of Energy, 2000

Benefits

Geothermal energy benefits the nation by helping to solve three problems— energy reliability and security, economic development, and air quality. Geothermal resources can help address the shortage of new electricity generating capacity in the United States cited in the National Energy Policy Act of 2005. As a baseload generation source, geothermal energy is well proven and reliable.

Geothermal power plants emit little carbon dioxide, very low quantities of sulfur dioxide, and no nitrogen oxides. U.S. geothermal generation annually offsets the emission of 22 million metric tons of carbon dioxide, 200,000 tons of nitrogen oxides, and 110,000 tons of particulate matter from conventional coal-fired plants. See Charts 1, 2, and 3 for details on comparative emissions.

Profile of a Geothermal Project

Geothermal power and direct use development begins with exploration to locate an economic reservoir, using a variety of techniques. Wells are drilled to measure subsurface temperatures and flow rates and to produce and inject the hydrothermal fluid. Once the reservoir has been proven, the site is developed either for power generation or a direct use application.

Geothermal projects are capital-intensive, and the major expenses are incurred before the project produces revenue. Exploration represents only about 10% of the total cost of a successful project, but many projects can fail at this stage. A high degree of risk evolves from the need for success on the first wells drilled into the reservoir. The extent to which these wells produce hot fluids influence subsequent investment decisions. Although the most expensive element of a power generation project is surface plant construction, drilling to create a well field involves higher risk due to uncertainties in reservoir characteristics. Direct use applications are usually less costly than power generation, because the resource is shallower, the fluids are less difficult to manage, and the technology less complex.

Typically, geothermal power plants are baseload facilities, but they may be operated in a load- following mode. Power conversion options include (1) the transformation (flashing) of hot geothermal fluids to steam which drives a turbine or (2) transfer of heat from the geothermal fluids to a secondary (binary) working fluid which drives a turbine. Geothermal plants have very high availabilities and capacity factors, exceeding 90%. Liquids produced from the reservoir are reinjected to sustain production pressures.

Geothermal heat pump (GHP) systems use highly durable heat exchangers that are placed into the ground, a water source, or a well, and deliver heating and cooling very efficiently, as well as providing hot water. Commercial and institutional GHP installations are often cost-competitive today, achieving substantial energy savings for building owners and tenants. Once connected to the ground (or water), the GHP system is typically integrated with traditional HVAC equipment. Only accredited designers and installers (see: www.igshpa.okstate.edu) should be used for GHP installations. There are federal tax benefits for use of GHPs, and some states also offer incentives (see: www.geoexchange.org).

Strategic Value

Use of domestic geothermal energy resources has substantial strategic value for the nation. Geothermal resources can contribute to:

◀ Clean electricity generation

◀ Baseload power production, having established high capacity factors

▶ Coproduction and enhanced oil recovery, thus gaining more oil

◀ Distributed energy systems with modular and shorter development timeframe advantages

▶ Direct-use for building energy needs

▼ Ethanol and biodiesel production – thermal energy requirements

▲ Hydrogen production – via off-peak electrolysis

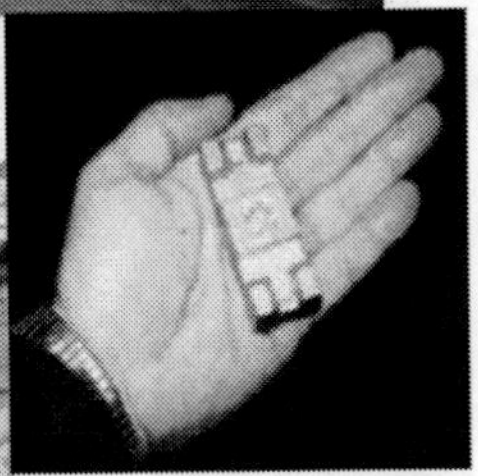

▲ Mineral recovery, such as silica and zinc – both strategic minerals in short supply.

▲ Rural economic development – aquaculture and horticulture, and lumber drying

◀ Climate change mitigation – by offsetting needs for fossil-fueled power plants

Clean, baseload, distributed electricity is being produced from geothermal resources today. Big Geysers, a 75-MW geothermal power plant located in northern California, is shown above.

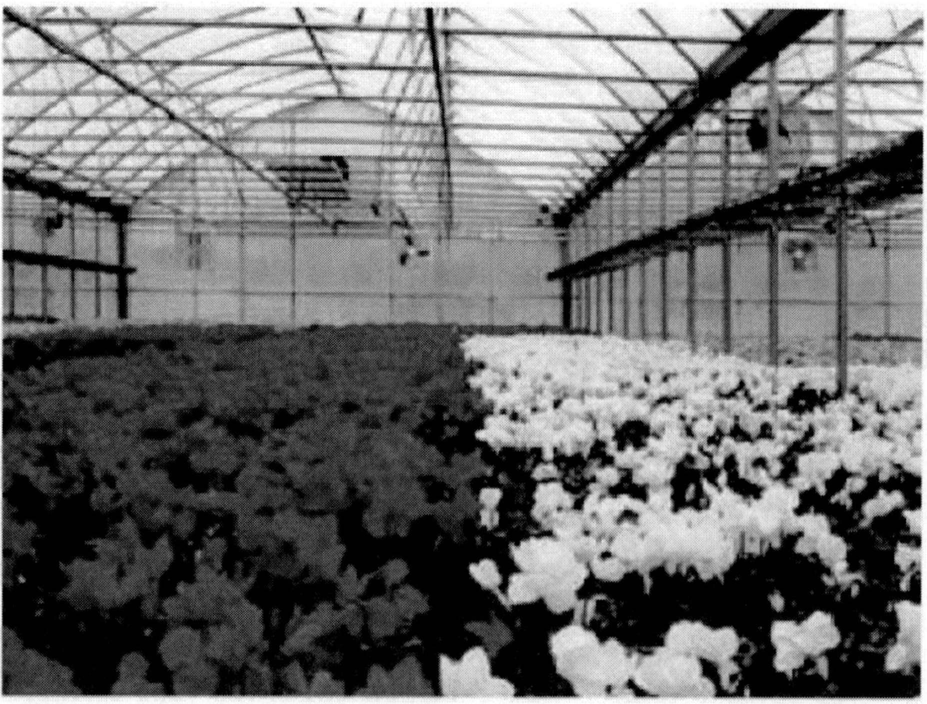

Bruce Green, NREL/PIX13079

Horticulture represents a rapidly growing domestic industry that lends itself well to rural economic development.

Residential application of a geothermal heat pump.

List of Workshop Participants and Staff

Hydrothermal

Jim Lovekin, GeothermEx, Inc., CA
Al Waibel, Columbia Geoscience, OR
Matthew Sares, Colorado Geologic Survey, CO

EGS

Susan Petty, Black Mountain Technology, WA

Direct Use

John Lund, Oregon Institute of Technology,
Geo-Heat Center, OR

Co-Produced (combined w/Geo-Press.)

David Blackwell, Southern Methodist University,
Geothermal Laboratory, TX

Richard Erdlac, University of Texas
Permian Basin, TX
Mark Milliken, Rocky Mountain Oil Testing Center, WY

Geopressured

Chip Groat, University of Texas – Austin, TX

Geothermal Heat Pumps

Wael El-Sharif, Geothermal Heat Pump Consortium, D.C.
Jack DiEnna, Geothermal Heat Pump Consortium, PA
John Geyer, JG&A, Inc., WA

DOE, INL, and NREL Staff

Roy Mink, DOE
Ray Fortuna, DOE
Jay Nathwani, DOE
Sandy Glatt, DOE
Joel Renner, INL
Gerry Nix, NREL
Bruce Green, NREL

Observers

Colin Williams, USGS
Marshall Reed, USGS
Kermit Witherbee, BLM

Facilitators

Kathleen Rutherford, The Keystone Center
Jody Erikson, The Keystone Center

Several geothermal power plants are visible at The Geysers in northern California.

Members of the entire workshop group during discussions about resource data integration.

How Energy Equivalent Calculations Were Made

Electricity as the equivalent energy content of the produced electricity –

1. Barrels (bbls) of oil equivalent (BOE)
   [(electricity production potential/MWe) (1000 kW/MW) (8760 hrs/yr.) (0.9 capacity factor) (3413 BTU/kW-hr)] 1 bbl oil equivalent/6 x 10^6 BTU = BOE/yr. for a minimum of 30 years.
2. Quads (Q)
   [(electricity production potential/MWe) (1000 kW/MW) (8760 hrs/yr.) (0.9 capacity factor) (3413 BTU/kW-hr)] 1/1 0^{15} BTU/Q

NOTE: If properly designed, plants are sustainable.
Thermal calculations assumed a duty cycle of 50% –

1. BOE
   [(MWt) (1000 kW/MW) (8760 hrs/yr.) (0.5 duty cycle) (3413 BTU/kW-hr)]
   (1 bbl oil equivalent/6 x 10^6 BTU)
2. Quads (Q)
   [(MWt) (1000 kW/MW) (8760 hrs/yr.) (0.5 duty cycle) (3413 BTU/kW-hr)] 1/10^{15} BTU/Q

NOTE: Actual fuel displacement for alternative production sources is somewhat larger.

Supporting Documents

1. *Assessment of Geothermal Resources of the United States* – 1978, USGS Circular 790 (p. 157), includes identified and unidentified resources; 2015 and later estimates are a consensus of the experts at the workshop.
2. *Geothermal Electric Power Supply Possible from Gulf Coast, Midcontinent Oil Field Waters*, "Oil and Gas Journal," September 5, 2005, and *Geothermal Energy Generation in Oil and Gas Settings Conference* findings, March 13 – 14, 2006, and SMU Geothermal Laboratory. Available at: http://www.smu. edu/geothermal/ publications/Oil&GASJ2005_ McKenna.pdf
3. *Energy Recovery from Enhanced/Engineered Geothermal Systems* (EGS), Massachusetts Institute of Technology (MIT), September 2006. Available at: http://www1 .eere.energy. gov/geothermal/egs_technology.html *Characterization of U.S. Energy Resources and Reserves*, December 1989, U.S. Department of Energy, DOE/CE-0279.
4. *Characterization of U.S. Energy Resources and Reserves*, December 1989, U.S. Department of Energy, DOE/CE-0279.

NOTICE

This report was prepared as an account of work sponsored by an agency of the United States government. Neither the United States government nor any agency thereof, nor any of their employees, makes any warranty, express or implied, or assumes any legal liability or responsibility for the accuracy, completeness, or usefulness of any information, apparatus, product, or process disclosed, or represents that its use would not infringe privately owned rights. Reference herein to any specific commercial product, process, or service by trade name, trademark, manufacturer, or otherwise does not necessarily constitute or imply its endorsement, recommendation, or favoring by the United States government or any agency thereof. The views and opinions of authors expressed herein do not necessarily state or reflect those of the United States government or any agency thereof.

REPORT DOCUMENTATION PAGE

Form Approved
OMB No. 0704-0188

The public reporting burden for this collection of information is estimated to average 1 hour per response, including the time for reviewing instructions, searching existing data sources, gathering and maintaining the data needed, and completing and reviewing the collection of information. Send comments regarding this burden estimate or any other aspect of this collection of information, including suggestions for reducing the burden, to Department of Defense, Executive Services and Communications Directorate (0704-0188). Respondents should be aware that notwithstanding any other provision of law, no person shall be subject to any penalty for failing to comply with a collection of information if it does not display a currently valid OMB control number.

PLEASE DO NOT RETURN YOUR FORM TO THE ABOVE ORGANIZATION.

1. REPORT DATE *(DD-MM-YYYY)* 10/2006	2. REPORT TYPE Technical Report	3. DATES COVERED *(From - To)* 2006

4. TITLE AND SUBTITLE
Geothermal -- The Energy Under Our Feet

5a. CONTRACT NUMBER
DE-AC36-99-GO10337

5b. GRANT NUMBER

5c. PROGRAM ELEMENT NUMBER

6. AUTHOR(S)
Bruce Green and Gerry Nix, National Renewable Energy Laboratory

5d. PROJECT NUMBER
NREL/TP-840-40665

5e. TASK NUMBER
GT06.1000

5f. WORK UNIT NUMBER

7. PERFORMING ORGANIZATION NAME(S) AND ADDRESS(ES)
National Renewable Energy Laboratory
1617 Cole Blvd.
Golden, CO 80401-3393

8. PERFORMING ORGANIZATION REPORT NUMBER
NREL/TP-840-40665

9. SPONSORING/MONITORING AGENCY NAME(S) AND ADDRESS(ES)

10. SPONSOR/MONITOR'S ACRONYM(S)
NREL

11. SPONSORING/MONITORING AGENCY REPORT NUMBER

12. DISTRIBUTION AVAILABILITY STATEMENT
National Technical Information Service
U.S. Department of Commerce
5285 Port Royal Road
Springfield, VA 22161

13. SUPPLEMENTARY NOTES
NREL Technical Monitor: Gerry Nix

14. ABSTRACT *(Maximum 200 Words)*
On May 16, 2006, the National Renewable Energy Laboratory (NREL) in Golden, Colorado hosted a geothermal resources workshop with experts from the geothermal community. The purpose of the workshop was to re-examine domestic geothermal resource estimates. The participating experts were organized into five working groups based on their primary area of expertise in the following types of geothermal resource or application: (1) Hydrothermal, (2) Deep Geothermal Systems, (3) Direct Use, (4) Geothermal Heat Pumps (GHPs), and (5) Co-Produced and Geopressured. The workshop found that the domestic geothermal resource is very large, with significant benefits.

15. SUBJECT TERMS
Geothermal; resources; hydrothermal; geopressured; co-produced, geothermal heat pump; GHPs; earth energy; deep geothermal systems; EGS; enhanced geothermal systems; direct use; and energy resources.

16. SECURITY CLASSIFICATION OF:			17. LIMITATION OF ABSTRACT	18. NUMBER OF PAGES	19a. NAME OF RESPONSIBLE PERSON
a. REPORT	b. ABSTRACT	c. THIS PAGE			
Unclassified	Unclassified	Unclassified	UL		19b. TELEPHONE NUMBER *(Include area code)*

In: Geothermal Energy: The Resource Under Our Feet
Editor: Charles T. Malloy
ISBN: 978-1-60741-502-2

Chapter 2

STATEMENT OF SUSAN PETTY, PRESIDENT-ALTAROCK ENERGY, INC., BEFORE THE SENATE COMMITTEE ON ENERGY AND NATURAL RESOURCES, REGARDING SENATE BILL 1543, NATIONAL GEOTHERMAL INITIATIVE ACT OF 2007

Susan Petty

Mr. Chairman and Members of the Committee, I am honored to have the opportunity to speak to you regarding Senate Bill 1543, the "National Geothermal Initiative Act of 2007," which was introduced to the Senate on June 5, 2007, by Senator Bingaman to encourage increased production of energy from geothermal resources.

One of the goals of S. 1543 is to achieve 20% of electric power generation from geothermal energy by 2050. You may be asking yourself if this a realistic goal? In the fall of 2004, I was included in a 12 member panel led by Dr. Jefferson Tester of the Massachusetts Institute of Technology that looked at the Future of Geothermal Energy. Our group consisted of members from both industry and academia. While some of us started the study convinced that it was possible to engineer or enhance geothermal systems (EGS) with today's technology, many of us, including myself, were skeptical. As we reviewed data, and listened to experts who were actively researching new methods, testing them in the field, and starting commercial enterprises to develop power projects from geothermal energy using this emerging technology, I believe all of us became convinced that a way had been found to tap into the vast geothermal resource under our feet.

Everywhere on Earth, the deeper you go, the hotter it gets. In some places, high temperatures are closer to the surface than others. We have all heard of the "Ring of Fire, " characterized by volcanoes, hot springs and fumaroles around the rim of the Pacific Ocean, including the Cascades, the Aleutian Islands, Japan, the Philippines and Indonesia. We know that along the tectonic rifts such as the Mid-Atlantic Ridge including Iceland and the Azores, the East African Rift Valley, the East Pacific Rise, the Rio Grande Rift running up through New Mexico and Colorado and the Juan de Fuca Ridge the earth's heat is right at the surface.

But other geologic settings allow high temperatures to occur at shallow depths, such as the faulted mountains and valleys of the Basin and Range, the deep faults in the Rocky Mountains and the Colorado Plateau. In addition, the sedimentary basins that insulate granites heated by radioactive decay along the Gulf Coast, in the Midwest, along the Chesapeake Bay and just west of the Appalachians can not only provide oil and gas, but hot water as well. (See Figure 1)

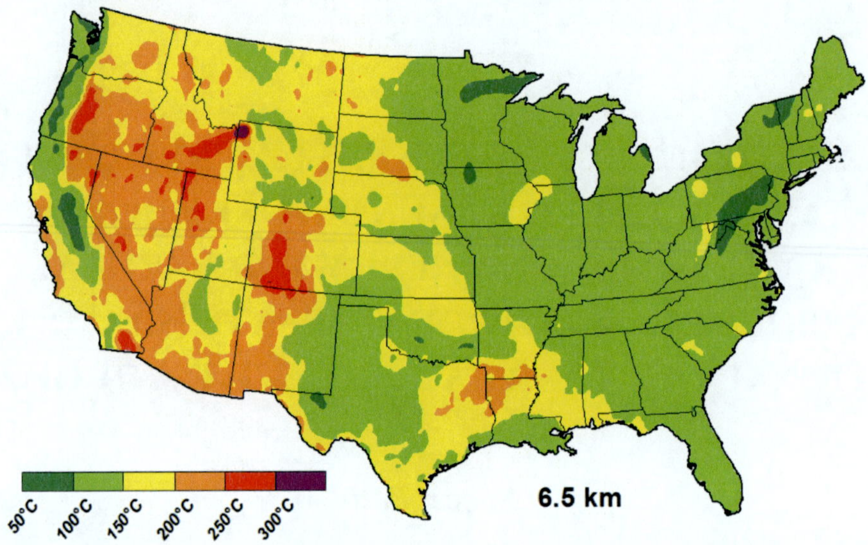

Figure 1. Temperature at a depth of 6.5 km – Continental United States

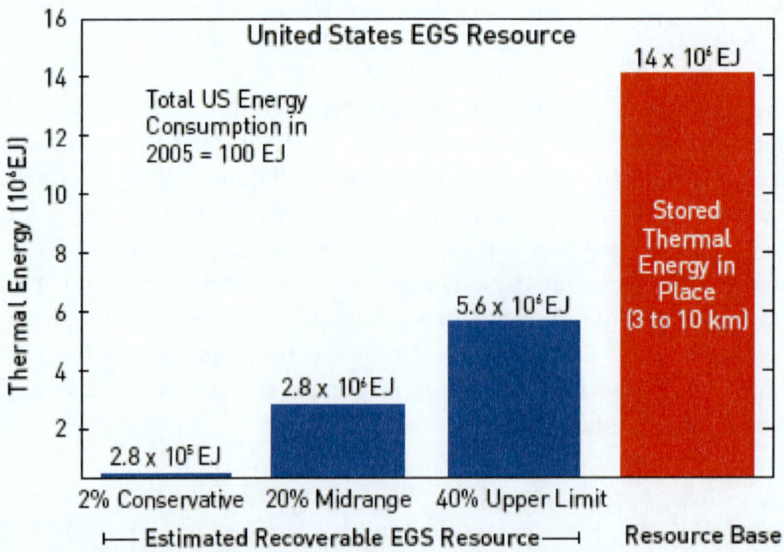

Figure 2. Comparison of Stored thermal energy in place with potential recoverable energy between 3 km and 10 km.

The heat contained in this vast resource is so large that it is really difficult to contemplate. Even with very conservative calculations, the MIT study panel found that the amount of heat that could be realistically recovered in the US from rocks at depths of 3 km to 10 km (about 2 miles to 6 miles) is almost 3,000 times the current energy consumption of the country. (See Figure 2) Listening to the experience of those developing the Soultz project in France, the Rosemanowes project in the UK and the Cooper Basin project in Australia, the panel members began to understand that the technology to recover this heat was here today. We can drill wells into high temperature rocks at depths greater than 3 km. We can fracture large volumes of hot rock. We can target wells into these man-made fractures and intersect them. We can circulate water through these created fractures, picking up heat and produce it at the other side heated to the temperature of reservoir rocks. We can produce what we inject without having to add more water. Long term tests have been conducted at fairly modest flow rates on these created reservoirs without change in temperature over time. No power plants have yet been built, but several are in progress in Europe.

Does this mean that we can build economic geothermal power plants based on EGS technology right now? At the best sites, where high temperatures occur at shallow depths in large rock masses with similar properties, geothermal power production from EGS technology is economic today. But to bring on line the huge resource stretching across the country from coast to coast, we need to do some work.

I'd like to talk about the economics of geothermal power production so you can better understand what needs to happen to enable widespread development of power projects using EGS.

At some places in the Earth's crust, faults and fractures allow water to circulate in contact with hot rock naturally. These are hydrothermal systems where natural fractures and high permeability allow high production rates. Even low temperature systems can be economic if the flow rates produced are high enough. The capital cost for the wells and wellfield-related equipment generally is between 25% - 50% of the total capital cost of the power project. The capital cost for hydrothermal projects can range from around $2,500/installed kW to over $5,000/kW, largely depending on the flow rate per well and the depth of the wells. The levelized break-even cost of energy for commercially viable hydrothermal projects currently ranges from $35/MWh to over $80/MWh. Of this, about $15-25/MWh is operating cost. The rest is the cost to amortize the power generation equipment and the wellfield.

Hydrothermal power is a good deal: Clean, small foot print, cost-effective. So why isn't more power from hydrothermal sources on line? The issue for hydrothermal power is risk. Because the risk related to finding the resource and successfully drilling and completing wells into the resource is high, development by utilities is unlikely. In order to accept this risk, independent power producers need a long-term contract at a guaranteed price and a high return on their investment. Utilities are loath to give a long-term contract because the payments to the generator will be treated as debt in determining their debt-to-equity ratio for credit and bond ratings.

Hydrothermal projects also tend to be small in size. While some of the potential future hydrothermal projects might be large, many of these are associated with scenic volcanic features protected as national parks or revered by Native Americans. A large scale project might mitigate the risk by spreading it over a much larger number of MW. In addition, there is a true economy of scale for geothermal power projects. For instance, the same number of people are needed to operate a 10 MW geothermal project as operate a 120 MW, or even a 250 MW, project.

Most of the really good (i.e. economic) hydrothermal systems are in the arid West. Not only is cooling water--which improves project economics by improving plant efficiency--an issue in this part of the country, but also the wide open spaces mean high-potential sites are often far from transmission, operators, supplies and large population centers with a high demand for power. Little potential for producing power from conventional geothermal, i.e. hydrothermal, sources exists in the Midwest, Southeast or East Coast.

Still, hydrothermal power has the potential to supply the country with more than 20,000 MW, or about 2% of our current installed capacity. However, the very high reliability of geothermal power means that this would be about 4% of our current annual generation. And this power is baseload or power that is available night and day.

Over the years, the cost of generating electricity from hydrothermal sources has dropped from around $130/MWh to less than $50/MWh. This was facilitated by incentives provided both by the market during the mid-1980s oil crisis, and by the government in the form of tax subsidies encourage the construction of over 2,000 MW of geothermal power that went on line from 1986-1995. Some of this drop in cost is due to research conducted by the US Department of Energy (DOE). For instance, in 1980 the DOE completed the first demonstration binary power plant at Raft River. This plant enabled the use of fluids at temperatures much lower than had been developed in the past. Industry commercialized this technology, and now most of the new geothermal power plants being built today are binary plants. DOE research, together with industry, developed high-temperature tools that are now essential to the evaluation of geothermal wells. A combination of DOE-supported research and industry effort as improved binary power plant efficiency by almost 50% from the earliest commercial plants in the 1980s, and flash power-plant efficiency by almost 35% over the same time period. This translates directly into reduction in overall project cost and power prices because fewer wells and less equipment is needed to generate the same amount of energy.

The MIT study started with the current state of the geothermal industry. The first task we realized we needed to undertake was a realistic look at the size and potential cost of developing geothermal power across the continent. It has long been realized by scientists that a vast geothermal resource exists everywhere as long as technology allows us to drill deep enough, develop a reservoir by creating fractures or enhancing natural fractures, and connect wells to circulate fluid through that reservoir. The US Geological Survey has been tasked with a detailed evaluation of the US geothermal resource, but this could not be finished in time for our study. The MIT panel, therefore, undertook a preliminary assessment of the geothermal resource in the US.

Using data collected over the years with DOE support, maps of the temperature at depth were developed by Dr. David Blackwell's group at SMU. Temperature at the midpoint of 1 km thick slices was projected at 1 km intervals starting at a depth of 3 km and extending down to 10 km, a reasonable limit for drilling using today's technology. The heat resource contained in each cubic kilometer of rock at these temperatures at each depth was then calculated. The amount of energy stored in this volume of rock is so enormous that it is really impossible to comprehend. (See Figure 1) We then looked at the studies that had estimated what fraction of this heat might be recovered, and at what efficiency this recovered heat might be turned into electric power. Studies showed that for economic systems, 40% or more of the total heat stored in the rock is recoverable. We also considered the more conservative recoverable estimates of 2% and 20%. Even at 2%, the amount of energy that could be realistically recovered, leaving economics and cost considerations aside, is more than 3,000 times the current total energy consumption of the US, including transportation uses.

In order to understand the technology needed to recover this energy, we turned to the published literature on the experiments done in the past at Fenton Hill, Rosemanowes, Hijiori, Ogachi and Soultz. We also brought in experts who are currently working on the Soultz project and on commercial engineered and enhanced geothermal projects in Europe and in Australia to tell us about the status of their work and their future efforts and needs. By the end of the study, we had concluded that EGS technology is technically feasible today. We can:

- Drill wells deep enough and successfully using standard geothermal and oil-and-gas drilling technology with existing infrastructure to tap the geothermal resource across the US, including areas in the Midwest, East and Southeast
- Consistently fracture large rock volumes of rock
- Monitor and map these created or enhanced fractures
- Drill production wells into the fractured rock
- Circulate cold water into the injection well and produce heated water from the production wells
- Operate the system without having to add significant amounts of water over time
- Operate the circulation system over extended test periods without measurable drop in temperature.
- Generate power from the circulating water at Fenton Hill and Ogachi

In addition, EGS power projects are scalable. Once the first demonstration unit has been tested at a site, the potential exists to develop a really large scale project of 250 to 1000 MW. Combined with the fact that good EGS sites where large bodies of hot rock with fairly uniform properties can be found across the US, that the sites are so many that they can be selected to avoid places with no transmission capacity or those located near areas of scenic beauty or environmental sensitivity, generating power from EGS technology looks like a winning proposition.

The real question then becomes, not is it realistic to anticipate generating 20% of our nation's electric power from geothermal energy, but can we make it cost effective?

The MIT panel included members from industry and research who are experts in the economics of power generation. The panel developed a list of key technologies that could help reduce the cost of generating power from EGS. They considered the changes in the cost of power generation from hydrothermal systems over the last 20 years, and the current state of EGS technology. They also considered research currently underway, not only that sponsored by DOE through universities and the national laboratories, but that being done by industry. Using models developed by both DOE and MIT, the cost of power and the impact on that cost of these possible technology improvements was examined. In addition, the panel looked at the impact of "learning by doing" on the cost of power.

We concluded that at the best sites, those with very high temperatures at depths of around 3-4 km in areas with low permeability natural fractures, EGS is economic today. Figure 3 shows the relative cost of power from a 300°C site at a depth of 3 km. With current technology power from this site could be generated for a levelized cost of power of about $74/MWh. This isn't the price that power could be sold for, since it doesn't include profit. It does, however, include financing charges at higher than utility rates, operating costs and the cost of amortizing the capital investment in the welfield and power plant. At deeper depths and lower temperatures, the cost of generating power using EGS technology is much higher, about $192/MWh. (Figure 4)

300°C at 3 km

- With current technology ~7.8¢/kWh
- With improved technology 5.4¢/kWh
- Areas for technology improvement
  - Conversion cycle efficiency
  - Drilling cost reduction/risk reduction
    - Fewer casing strings
    - Higher hard rock ROP
    - Better measurement while drilling for HT (risk?)
  - Improved stimulation technology
    - Better zone isolation
    - Better reservoir understanding
      - Stress measurement
      - Fracture ID
      - Higher flow per producer
      - Single well test methods

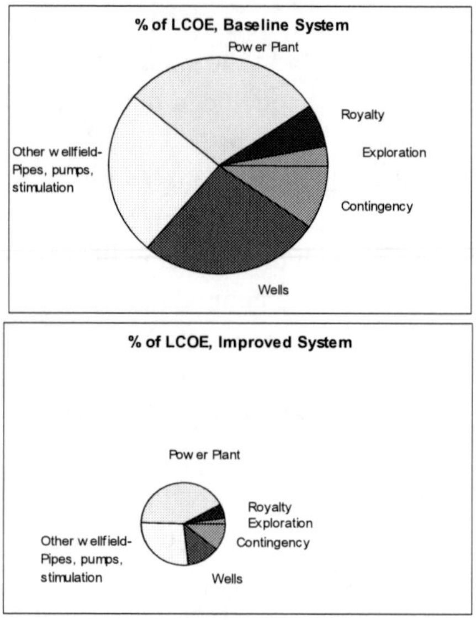

Figure 3. Levelized breakeven cost of EGS power from a high temperature site at 3 km

150°C at 5 km

- With current technology ~19.2¢/kWh
- With improved technology 7.4¢/kWh
- Areas for technology improvement
  - Conversion cycle efficiency
    - Improved HT pumping
    - More efficiency binary cycle
  - Drilling reduction/risk reduction
    - Fewer casing strings
    - Higher hard rock ROP
    - Better measurement while drilling for HT (risk?)
  - Improved stimulation technology
    - Higher flow per producer!
    - Better zone isolation
    - Better reservoir understanding
      - Stress measurement
      - Fracture ID
      - Single well test methods

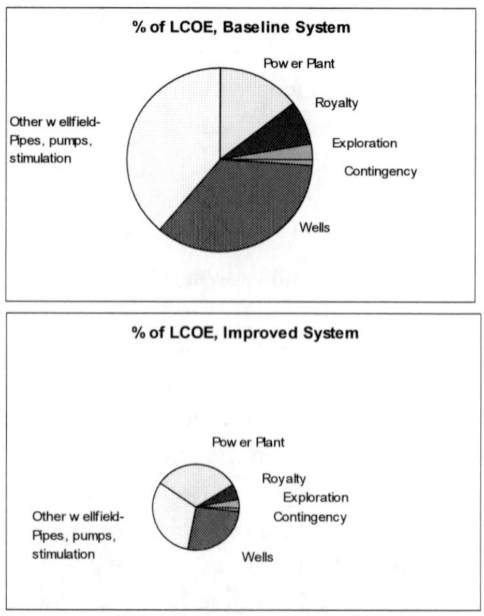

Figure 4. Levelized breakeven cost of EGS power from a moderate temperature site

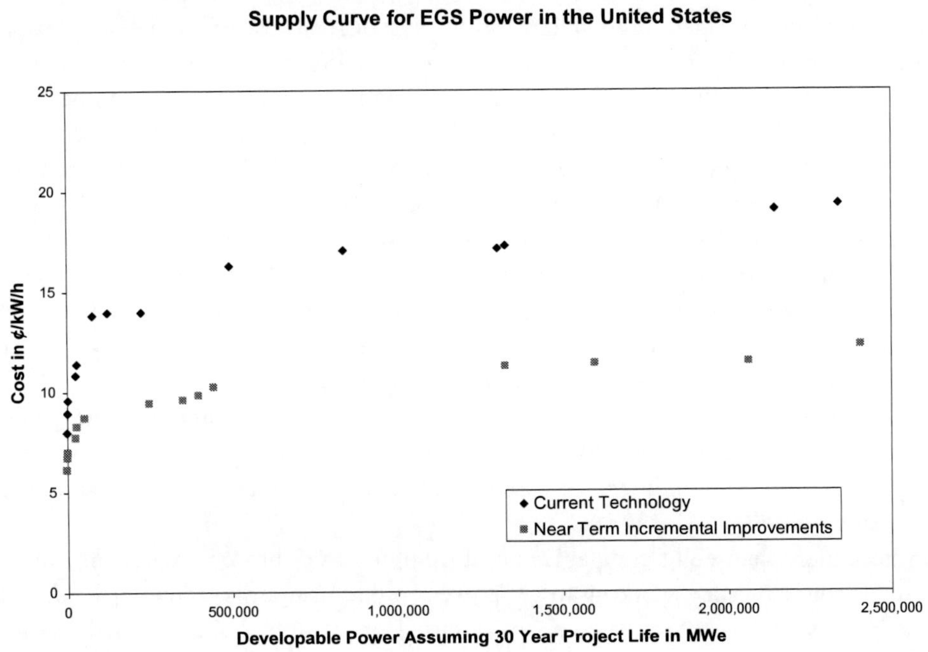

Figure 5. Supply curve for EGS Power in the United States – plots the levelized breakeven cost of power against the developable power assuming a 30 year project life

With incremental technology improvement, the cost of power could be cut in half or more, particularly for the deeper high temperature systems. These incremental technology improvements include things like improving conversion cycle efficiency, being able to isolate the part of the wellbore that has been treated so that untreated parts can be fractured, redesigning wells to reduce the number of casing strings and improved understanding of rock/fluid interaction to prevent or repair short circuiting through the reservoir. None of these technology improvements require game changing strategies, just the kind of advancement that comes from persisting in extending our knowledge to the next level. Looking at the high temperature example in Figure 3, the levelized cost of power could be cut to $54/MWh or about 27% with these technology improvements implemented. The moderate temperature site could see a much larger reduction of over 60% to $74/MWh.

Figure 5 shows a supply curve for EGS based geothermal power for the entire US. This curve shows the amount of power available at a certain cost. However, this is cost of power not price. In other words, this is not the price that an independent power producer would charge a utility for this power if they were selling it to them. However, it does give an idea of what could be economic in the future. The two sets of dots are calculated using current technology and the projected cost using future incrementally improved technology. Once the cost of power increases to around $100/MWh, it is clear that more than 400,000 MW would be available or development. This means that the amount of power we could develop is not limited by the resource available, but by the cost. And the cost is limited by the technology and the fact that we aren't doing this here in the US.

We concluded that at the best sites, those with very high temperatures at depths of around 3-4 km in areas with low-permeability natural fractures, EGS is economic today. With incremental technology improvement, the cost of power could be cut in half or more, particularly for the deeper high temperature systems. These incremental technology improvements include things such as improving conversion cycle efficiency, being able to isolate the part of the wellbore that has been treated so that untreated parts can be fractured, redesigning wells to reduce the number of casing strings and improved understanding of rock/fluid interaction to prevent or repair short circuiting through the reservoir. None of these technology improvements require game-changing or revolutionary strategies, just the kind of advancement that comes from persisting in extending our knowledge to the next level.

The cost of this type of technology improvement is not high. The panel felt that an investment of ~$368,000,000 over a period of about 8-10 years combined with industry involvement could result in 100,000 MW on line by 2030. This would be 10% of the current installed capacity and over 20% of the current electric generation of the country. Combined with the hydrothermal resource, it is a very realistic goal to have geothermal energy provide 20% of the nation's electricity by 2030. However, the effort would require federal support, university, laboratory and industry research, and development and a real commitment to renewable energy use.

Currently more than eight companies are developing EGS power projects in Europe and more than 20 companies are working to get power on line using this technology in Australia. AltaRock Energy Inc. is the only company focused on commercializing power generation from EGS technology in the US. In Europe, price subsidies and European Union-sponsored research are helping to start more than 50 EGS projects. In Australia, government grants, help with transmission access, research, and legislation requiring generation from renewable energy sources are driving EGS technology to commercialization. Other countries with fewer economic geothermal resources are planning to include geothermal energy in their generation portfolio. The US needs to commit to this clean, baseload, renewable power source for our own energy future.

SUMMARY

- The Future of Geothermal Energy: Impact of Enhanced Geothermal Systems (EGS) on the United States in the 21st Century
    - http://geothermal.inel.gov/publications/future_of_geothermal_energy.pdf
    - 12 member panel lead by Dr. Jefferson Tester through MIT
- Conclusions:
    - EGS power is technically feasible today
    - Potentially 100,000 MW can be on line by 2030 with federal investment of ~$350,000,000
    - Resource extends across US
    - Best resources economic today at high temperature, shallow sites
    - With incremental technology improvement, cost can be cut in half
    - With learning by doing and innovative technology improvement cost can be reduced for deep resources to ¼ cost with current technology
- Hydrothermal Systems
    - Natural permeability

- High flow rates
- Few big systems
- Located in Western US
- Exploration drilling is needed and remains risky
- Economic now even for low temperatures
- >2800 MW on line growing by about 300 MW/yr
- Potential for as much as 20,000 MW at economic costs over next 40 yrs
- >95% average availability
- Technology improvement reduced cost (not price) -13¢ per kWh in 1986 to about 5¢ per kWh in 2006

- Enhanced Geothermal Systems (EGS)
  - Resource is vast
  - Distributed across the US, but best sites in West
  - Low or no natural permeability
  - Reservoir must be engineered to:
    - Obtain high flow rates
    - Develop good heat exchange area
  - Exploration risk reduced
    - Temperature only needed
    - Drill deeper to get greater temperature
  - Large systems can be developed
  - Uses proven state-of-the-art drilling technology
  - Fracturing technology developing
  - MIT study identified key areas of technology improvement needed to reduce cost
  - Potential for CO_2 sequestration
  - 8 companies in Europe; ~20 companies in Australia working to commercialize
  - AltaRock Energy – first US company focused on EGS technology development

In: Geothermal Energy: The Resource Under Our Feet
Editor: Charles T. Malloy
ISBN: 978-1-60741-502-2

Chapter 3

2008 GEOTHERMAL TECHNOLOGIES MARKET REPORT*

Jonathan Cross and Jeremiah Freeman

ACKNOWLEDGMENTS

This report was made possible by the U.S. Department of Energy (DOE) Geothermal Technologies Program (GTP). The authors thank Jørn Aabakken, Alison Wise, Rachel Gelman (National Renewable Energy Laboratory) and Alexandra Pressman (Sentech) for technical input; and Lauren Boyd, Nicole Reed, Mike Murphy (DOE GTP) and Agatha Wein (New West Technologies) for comments on drafts throughout the process. The authors also thank Christina Van Vleck for graphic design. Of course, any remaining errors or omissions are the fault of the authors.

EXECUTIVE SUMMARY

Geothermal energy has been exploited for power generation since at least 1904.[1] However, the last few years have witnessed a conspicuous revival in interest in geothermal technologies both old and new. In fact, 2008 was a watershed year for the industry. The U.S. Department of Energy (DOE) revived its Geothermal Technologies Program (GTP) with new funding that made possible substantial new investments in geothermal research, development and technology demonstration. The U.S. Department of the Interior's (DOI) Bureau of Land Management (BLM) also significantly increased the amount of Federal land available for geothermal exploration and development and worked to streamline the complex permitting and leasing process. Installed geothermal capacities in the United States and abroad continued to increase as well.

* This is an edited, reformatted and augmented version of a U. S. Department of Energy publication dated July 2009.

Despite the positive advances for geothermal in recent years, strains from the global economic downturn that started late in 2008 are beginning to have an effect on financing in the industry. Geothermal power developers rely heavily on the equity markets and financing based on the monetization of production tax credits (PTCs), and these sources of capital are no longer readily accessible. Geothermal development also has a steep, front-loaded risk profile that makes projects very difficult to finance; exploratory drilling is an extremely expensive step early in the development process that carries the greatest risk.

Geothermal markets are also being affected by the downturn of the Icelandic economy. A particularly poignant example is the nationalization of Glitnir Bank, now Í slandsbanki, which was adept at providing geothermal developers with funding necessary to support risky exploration and drilling activities until they were able to secure financing from traditional sources. When it was nationalized by the Icelandic government in September of 2008, Glitnir largely disappeared from the pool of potential geothermal financing sources. Unfortunately, they were not the only financier of geothermal development to fall victim to the economic downturn. Only half of the 14 large financial companies that funded renewable energy projects over the past few years are still active today.[2]

In contrast to the economic arena, the policy environment in 2008 was favorable to continued geothermal power development. In the United States, the Emergency Economic Stabilization Act (EESA) of 2008, signed by President Bush on October 3, 2008, extended PTCs for geothermal energy production until January 1, 2011. The legislation also reinstituted a 30% individual tax credit for qualifying geothermal heat pumps (GHPs), capped at $2,000.[3] Additionally, state renewable portfolio standards (RPS) remained an effective driver for investments in a variety of renewable energy technologies, including geothermal.

At the non-electricity generating end of the geothermal technology spectrum, the market for GHPs continued to experience rapid growth despite the downturn in financial and real estate sectors.

The Air-Conditioning, Heating and Refrigeration Institute (AHRI) reported 2008 shipments of more than 71,000 units, indicating continued strong demand. The heat pump market still faces significant barriers, however, including: high installation and capital costs; a pervasive lack of consumer awareness; and insufficient market delivery infrastructure. In order for heat pumps to reach their full market potential, these barriers must be addressed through effective market conditioning strategies.

Low-temperature geothermal direct use applications typically include spas, district space heating, aquaculture, agricultural drying, and snow melting.* Though these applications remain only a small portion of total geothermal resource use in the United States, it is still noteworthy that their installed base has doubled in the past 15 years.[4] Direct-use geothermal energy is widely used internationally, including in Iceland, China and Japan. In Japan, geothermal power developers are competing with spa, hotel, and bath projects to access the direct-use energy resources.

Geothermal co-production with oil and gas is another exciting and likely possibility for the near future. These developments, along with the enormous potential of enhanced geothermal systems (EGS) projects, will transform geothermal energy in the United States from a western state-focused energy source into a ubiquitous source of baseload power.

---

* Some authorities include GHPs in the direct-use category but they are treated separately in this paper.

In conclusion, this is a particularly exciting time for the geothermal energy industry. Even in the face of a troubled economic climate, it seems likely that the next few years will see a marked increase in the use geothermal energy to meet the nation's growing electricity demand requirements.

Major 2008 Highlights

- 110 additional MW of geothermal power came online in the United States. (100 MW from binary plants and 10 MW from steam plants).
- The GTP made 21 awards totaling $43.1 million over four years.[5]
- Google.org, Google's philanthrophic arm, gave over $10 million in grants to two gethermal companies and one research university to support their work on EGS. Google's name-brand support thrust geothermal into the public spotlight and improved its standing as a viable alternative energy source, alongside wind and solar.[6]
- BLM leased 301,588 acres of land for geothermal power development, a substantial addition to the 244,000 acres leased for this purpose since July of 2007.[7]
- The United States signed the International Partnership for Geothermal Technology (IPGT) with Iceland and Australia. The IPGT will lead to joint technology development projects with partner countries, reducing the cost of advanced geothermal technology development for each country and increasing the available expertise for specific projects.[†]
- The economy-wide credit crunch dried up equity markets, making it extremely difficult for geothermal developers to locate financing for their projects.
- Glitnir Bank collapsed, taking with it an important source of geothermal financing.
- Price drops in the market for PTCs decreased their efficacy.
- Investments in geothermal continued to increase.
- Small, low-temperature power generation units began to account for a significant portion of the overall geothermal market, a trend expected to continue for at least the next several years.
- Modular low-temperature electricity generation units gained popularity. These units have the potential to become a major contributor to the national geothermal energy portfolio over the next few years.

1. INTRODUCTION

While geothermal energy technology has been in development in the United States for over 100 years[*], national interest in geothermal recently gained momentum as the result of new analysis that suggests massive electricity producing potential. The geothermal industry

[†] The International Partnership for Geothermal Technology (www.internationalgeothermal.org)

[*] While geothermal energy has been in use for over 100 years within the United States, its use for electrical production dates back to 1922. The first large-scale geothermal power plant began operation in 1960. See http://www1.eere.energy.gov/geothermal/history.html.

has also seen unprecedented investment growth following the transition to a new administration and its response to the economic climate through the American Recovery and Reinvestment Act of 2009 (the Recovery Act). While it tends to have a lower profile among the nation's renewable energy resources, geothermal is currently in the midst of a renaissance. In such a rapidly changing market, this report bears particular significance.

Geothermal energy technologies can be broken into four major categories: conventional hydrothermal, low-temperature, EGS, and direct use, including geothermal heat pumps (GHPs). The first three categories generate electricity, while the fourth is used primarily for heating and cooling and hot water production. This report will consider electricity generation technologies separately from direct use technologies due to differences in technology maturity and market characteristics.

This report describes market-wide trends for the geothermal industry throughout 2008 and the beginning of 2009. It begins with an overview of the GTP's involvement with the geothermal industry and recent investment trends for electric generation technologies. The report next describes the current state of geothermal power generation and activity within the United States, costs associated with development, financing trends, an analysis of the levelized cost of energy (LCOE), and a look at the current policy environment. The report also highlights trends regarding direct use of geothermal energy, including GHPs.[†] The final sections of the report focus on international perspectives, employment and economic benefits from geothermal energy development, and potential incentives in pending national legislation.

2. INVESTMENT

The 2008 Geothermal Technologies Program: $44M for EGS RD&D

Combined with rising energy prices and climate change concerns, significant renewed interest in geothermal energy came in 2007 with the release of Massachusetts Institute of Technology (MIT) report, "The Future of Geothermal Energy: Impact of Enhanced Geothermal Systems (EGS) on the United States in the 21st Century." This report presented exciting new research that has already had a profound effect on overall energy investment in the United States, suggesting that given appropriate funding, 100,000 MWe of geothermal could be developed through EGS technologies within 50 years. After the release of the MIT report, Congress directed the GTP to refocus its program onto the development and eventual deployment of EGS technology due to its potential as a nationwide energy resource. The GTP received an infusion of funding during the 2008 Fiscal Year of approximately $20 million (see Table 1).

Investments in Geothermal Energy on the Rise

Though 2008 presented enormous economic challenges, private investments in geothermal energy actually increased over prior years. Public market investment, project

[†] GHPs are also commonly referred to as ground source heat pumps.

acquisitions, and venture capital (VC)/private equity (PE) have shown a marked increase from 2005-2008 (Figure 1), with U.S. projects receiving the majority of worldwide investment in geothermal development in 2007 (Figure 2).

Table 1. GTP Budget Request FY 2007-2009

	Funding ($ in thousands)				
	FY 2007 Approp.	FY 2008 Request	FY 2008 Approp.	FY 2009 Request	FY 2009 Approp.
Enhanced Geothermal Systems	2,000	0	19,818	30,000	44,000
Oil and Gas Well Co-Production and Resource Assessment	3,000	0	0	0	0
TOTAL	**5,000**	**0**	**19,818**	**30,000**	**44,000**

Source: DOE, "EERE Fiscal-Year 2009: Budget-in-Brief"

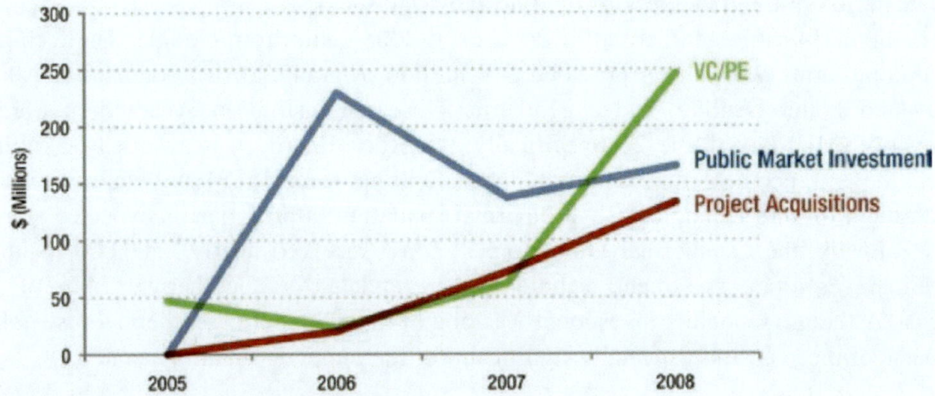

Source: New Energy Finance, January, 2009

Figure 1. Trends in U.S. Geothermal Investments (2005-2008)

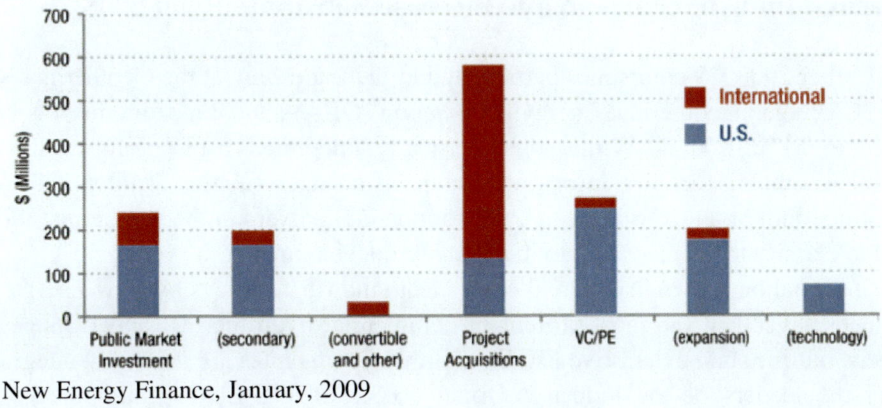

Source: New Energy Finance, January, 2009

Figure 2. U.S. and International Geothermal Investments

Table 2. Google.org Funding for Geothermal Research

Awardees	Funding
AltaRock Energy	$6,000,000
Potter Drilling	$4,000,000
SMU Geothermal Lab	$489,521

Source: Google.org, 2008

In 2007 and 2008, as the number of geothermal industry players grew, so did total investments in the sector. Many of these new developers are relatively small companies with few assets that are particularly vulnerable as the result of shrinking equity markets. Adding to the challenge, geothermal projects are notoriously difficult to finance because of large up-front capital costs, high risk, and long lead-time (see Financing section for more detail).[8]

Íslandsbanki, formerly known as Glitnir, has played an integral part in geothermal project financing, particularly during the early, high-risk stages of development. The nationalization of the Icelandic bank in 2008 ended its involvement in the U.S. geothermal industry. This blow to the industry did not result in a major setback because the crash of credit markets in this time period resulted in a lack of funding across all sectors.[9]

The highest-profile geothermal investment of 2008 came from Google. The tech giant's philanthropic arm, *Google.org*, provided $10 million in grants to two companies, AltaRock Energy and Potter Drilling, and a geothermal research institution at Southern Methodist University (SMU) (see Table 2). Specifically, AltaRock Energy was awarded $6 million to support the advancement of EGS, and Potter Drilling received $4 million to develop its breakthrough drilling technology, hydrothermal spallation; a prototype is expected sometime in 2009. Lastly, the Geothermal Laboratory at SMU received nearly $500,000 to improve geothermal resource assessment techniques and update the Geothermal Map of North America. Although Google's investment was one of many made in geothermal research over the course of the year, it is especially significant for the publicity that it generated.

3. State of Power Generation & Current Activity in the U.S.

Geothermal Industry Participants Increase Substantially in 2008

In October 2008, 79 companies participated in the tradeshow at the Geothermal Resource Council (GRC) and Geothermal Energy Association (GEA) annual meeting in Reno, Nevada, compared to 51 in 2007.[10] While some vertically integrated firms perform all stages of development, others specialize in one or two specific stages such as drilling or engineering and construction. For an overview of all the commercial players in the geothermal industry, it is useful to classify them according to their stage of development.

The information shown in Figure 3 comes from industry surveys by New Energy Finance and includes several of the most prominent commercial hydrothermal and EGS geothermal developers, but is not an exhaustive list. Five of these companies are vertically integrated, and represent the leaders of the industry: Ormat (U.S.), PNOC-EDC (Philippines), Chevron (U.S.), Enel (Italy), and Calpine (U.S.). Two firms perform all stages except research and development (R&D): PT Pertamina (India) and Reykjavik Energy (Iceland). Six companies in the United States are dedicated to drilling and confirmation: Baker Drilling, Parker Drilling, ThermaSource, and Geothermex; along with Boart Longyear, and Halliburton who also perform exploration.

GTP 2008 Funding Opportunity Announcement Receives the Largest Number of Applicants in the Program's History

In October of 2008, DOE awarded $43.1 to 21 applicants over four years for research, development and demonstration (RD&D) associated with EGS.* This is the greatest number of award recipients and of first-time recipients, 13 of the 21, in the history of the program (See Table 3). Specifically, for the 2008 fiscal year, $8.7 million was awarded to fund 17 component technologies research and development projects, while roughly $11.1 million was provided for the four demonstrative projects.

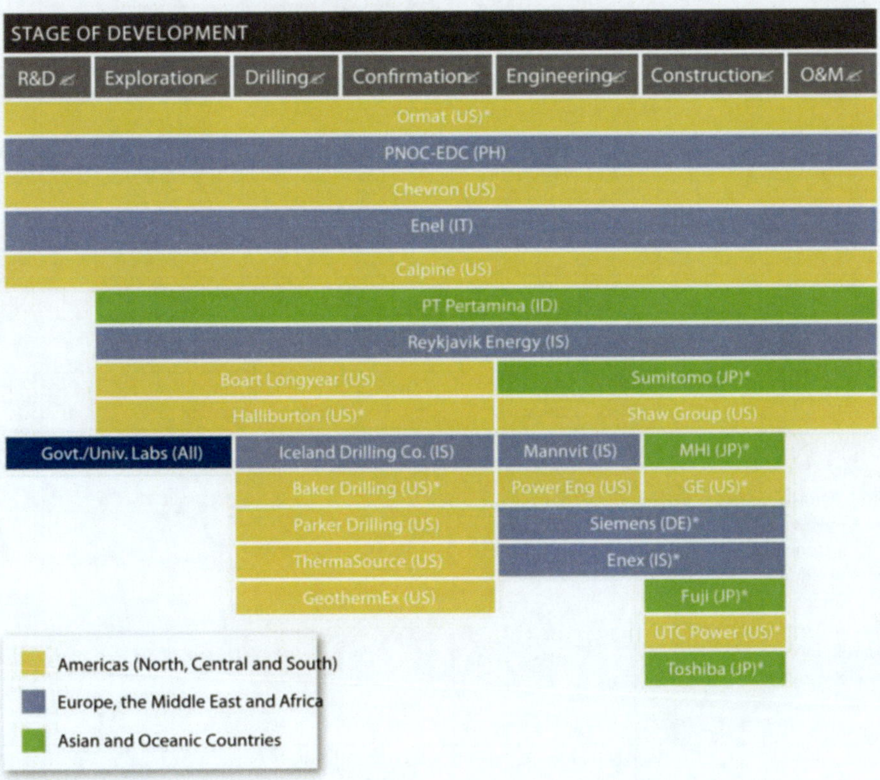

*The U.S. Department of Energy does not endorse any company listed in this report.
Source: New Energy Finance, 2008

Figure 3. Companies in the Geothermal Value Chain *(not comprehensive)* *

---

* DOE's commitment of $43.1 million is subject to annual appropriations.

Table 3. GTP FOA Awardees: October 2008

COMPONENT TECHNOLOGIES R&D			
Awardees	**Location**	**Project Description**	**Funding**
• Baker-Hughes, Inc.	Houston, Texas	Develop an ultrasonic borehole televiewer	$3,139,364
• Colorado School of Mines • Boise State University • Flint, LLC • Mt. Princeton Geothermal, LLC	Golden, Colorado	Geophysical characterization of geothermal systems using joint inversion of electrical and seismic data	$867,564
• Composite Technology • Wood Group ESP • New England Wire Technology	Lafayette, Colorado	Develop high temperature motor windings for electric submersible pumps	$987,739
• Foulger Consulting • Geosystem with WesternGeco • US Navy • Magma Energy US Corporation • Lawrence Berkeley National Laboratory	Menlo Park, California	Develop tools and methods suited to monitoring EGS-induced micro-earthquakes	$561,729
• GE Global Research • Auburn University • GE Energy	Niskayuna, New York	Develop high temperature electronics platform and temperature sensor	$1,599,934
• Hattenbrug, Dilley, and Linnell, LLC • University of Utah	Anchorage, Alaska	Use of Fluid Inclusion Stratigraphy (FIS) chemical signature to identify open fracture systems	$313,858
• Hi-Q Geophysical Inc. • Ormat Technologies Inc. • Lawrence Berkeley National Laboratory	Ponca City, Oklahoma	Develop surface and borehole seismic methodologies	$817,757
• MIT • Chevron • Los Alamos National Laboratory	Cambridge, Massachusetts	Develop geomechanical model of reservoir fluid flow	$508,633
• MIT • New England Research	Cambridge, Massachusetts	Combine geophysical methods with a rock physics model for fracture characterization	$1,019,769
• Perma Works and Frequency • Management International • ElectroChemical Systems Inc • Draka Cableteq • Pacific Systems Inc • Tiger Wireline Inc • Viking Engineering • Electronic Workmanship Standards, Inc. • Eclipse NanoMed • Kuster Company • Honeywell SSEC	Albuquerque, New Mexico	Develop high-temperature well monitoring tools	$2,200,000
• Schlumberger	Sugar Land, Texas	Extend temperature operating range of electric submersible pumps	$1,245,751
• Schlumberger	Sugar Land, Texas	Develop downhole monitoring system for electric submersible pumps	$1,253,959
• Stanford University	Stanford, California	Develop wellbore tools and reservoir engineering approaches	$967,541
• Texas A&M University • Sandia National Laboratory • University of Mississippi	College Station, Texas	Develop improved seismicity-based reservoir characterization technology techniques	$820,198

Table 3. (Continued)

Awardees	Location	Project Description	Funding
• Texas A&M University • Sandia National Laboratory • University of Mississippi	College Station, Texas	Develop three-dimensional numerical model to predict reservoir stimulation	$690,953
• University of Utah	Salt Lake City, Utah	Demonstrate absorbing tracers and develop fluorimeter to measure tracer concentration	$1,091,039
• University of Utah	Salt Lake City, Utah	Investigate fracture stability	$978,180
SYSTEM DEMONSTRATION			
• AltaRock Energy Inc • Northern California Power Agency • University of Utah • Texas A&M University • SAIC • Temple University	Seattle, Washington	Demonstrate innovative stimulation process to create EGS reservoir by drilling below permeable zone and stimulating low permeability zone	$6,014,351
• Geysers Power Co. LLC • Lawrence Berkeley National Laboratory	Middletown, California	Demonstrate deepening of wells into high-temperature zones	$5,697,700
• Ormat Nevada, Inc • Lawrence Berkeley National Laboratory • University of Utah • Pinnacle Technologies • GeoMechanics International • University of Nevada – Reno • TerraTex/Schlumberger	Reno, Nevada	Demonstrate ability to stimulate multiple wells at Brady Field, Nevada	$3,374,430
• University of Utah • APEX Petroleum Engineering Services • HiPoint Reservoir Imaging • Chevron	Salt Lake City, Utah	Demonstrate monitored hydraulic stimulation of existing injection well at Raft River Idaho	$8,928,999
		System Demonstrations Total	$24,015,480
		Total Department of Energy Funding	$43,079,448

Source: DOE EE/RE

These RD&D projects target GTP's goal of reaching EGS technology readiness by 2015. Though successful EGS development will provide long-term nationwide benefits, near-term gains in geothermal expansion will likely come from conventional high-temperature hydrothermal, co-produced fluids, and low-temperature resources once considered uneconomical for commercial electricity generation.

U.S. Geothermal Capacity Increases by 3.8% in 2008

In 2008, an estimated 110 MWe of nameplate capacity was installed within the United States, bringing the cumulative total to 3,040 MWe (see Table 4). Of this total, 100 MWe was sourced from binary plants and 10 from steam plants.

Electricity generated from geothermal sources reached 15 billion kWh in 2008, representing approximately 0.36% of the total U.S. electrical production and 12.13% of electricity generated from renewable resources, excluding hydropower (see Figure 4).[11]

Table 4. New Geothermal Power Plants Online in 2008

Start Year	State	Power Plant	Nameplate Capacity (MWe)
2008	Idaho	Raft River	15.8
2008	Nevada	Galena	20.0
2008	New Mexico	Lightning Dock	0.24
2008	Utah	Hatch	14.0
2008	Wyoming	NPR3	0.25
2008	California	Herber South	10.0
2008	California	North Brawley	50.0
		TOTAL	**110.29**

Source: New Energy Finance, 2009.

Though growth has been modest, the United States remained the leader in installed geothermal capacity in 2007 (see Figure 5).[12] Because the majority of electricity production is currently from hydrothermal sources, geothermal power generation in 2008 remained limited to western states that contain these resources (see Figure 6). As more low-temperature and co-produced resources are exploited, geothermal energy is expected to expand eastward (see Figure 7). Additionally, temperatures viable for EGS production are available throughout the U.S. at a depth of 10 km (see Figure 8)*, which is reachable with current drilling technology.

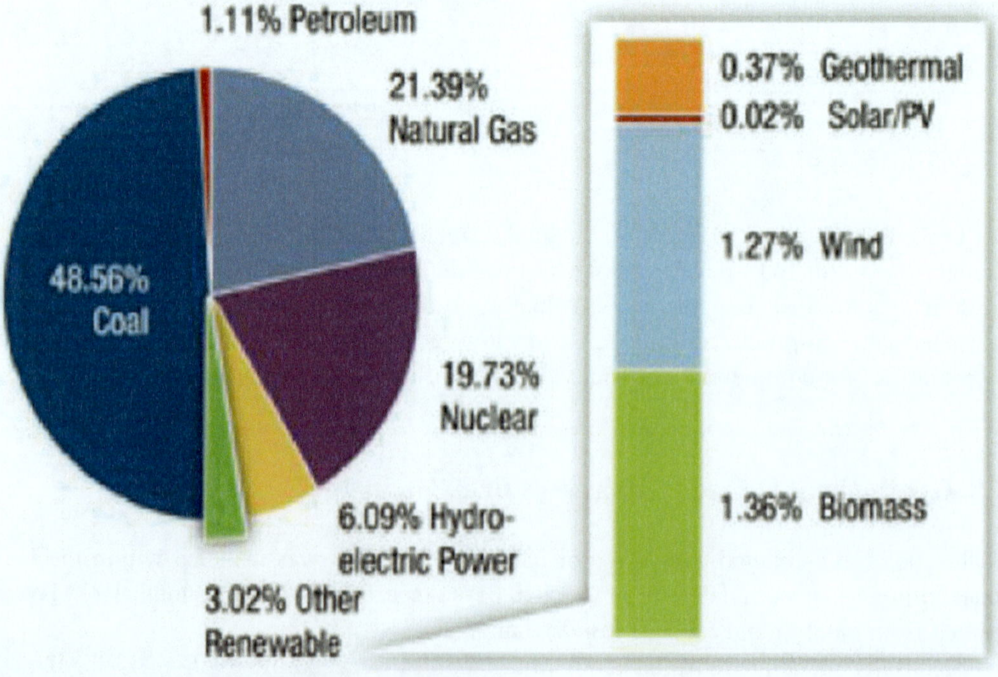

Source: EIA, "Electric Power Monthly" (March 2009)

Figure 4. U.S. electricity generation by type

* Geothermal electricity production can come from resources as low as 74°C (165°F).

Source: Bertani, R. "World Geothermal Generation in 2007" (September 2007)

Figure 5. Top Ten Countries with Geothermal Power Generation (2007)

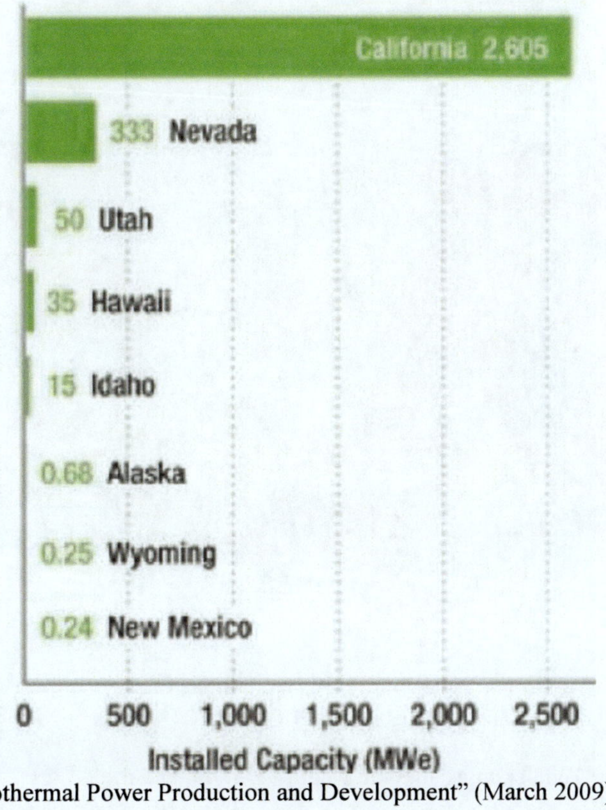

Source: GEA, "U.S. Geothermal Power Production and Development" (March 2009)

Figure 6. Installed U.S. Geothermal Capacity in 2008

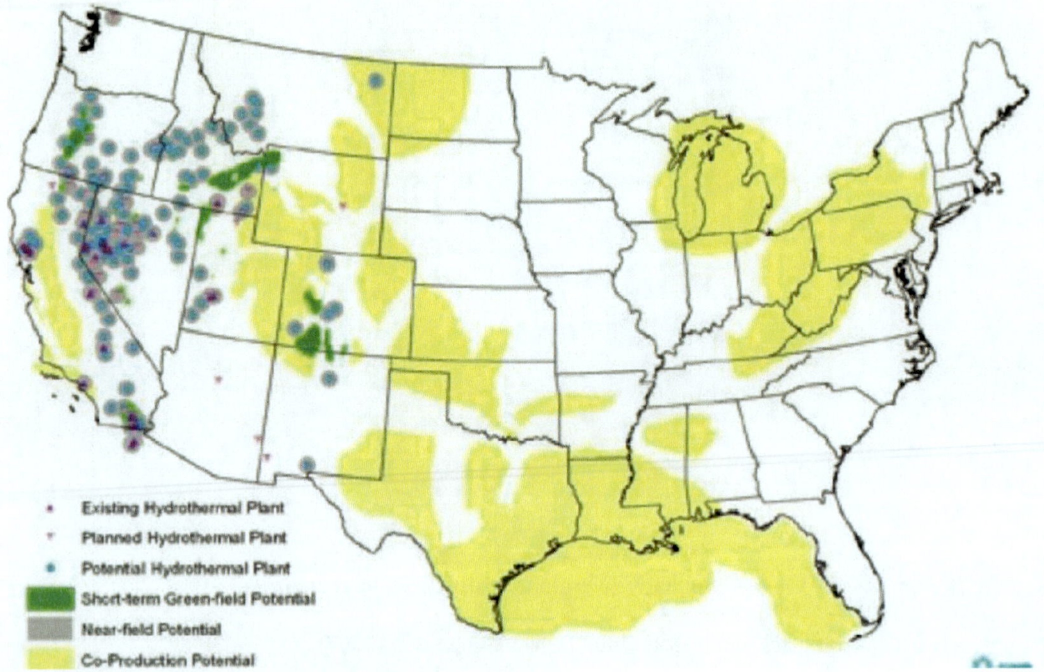

Source: DOE, National Renewable Energy Laboratory

Figure 7. Short-Term Geothermal Energy Potential

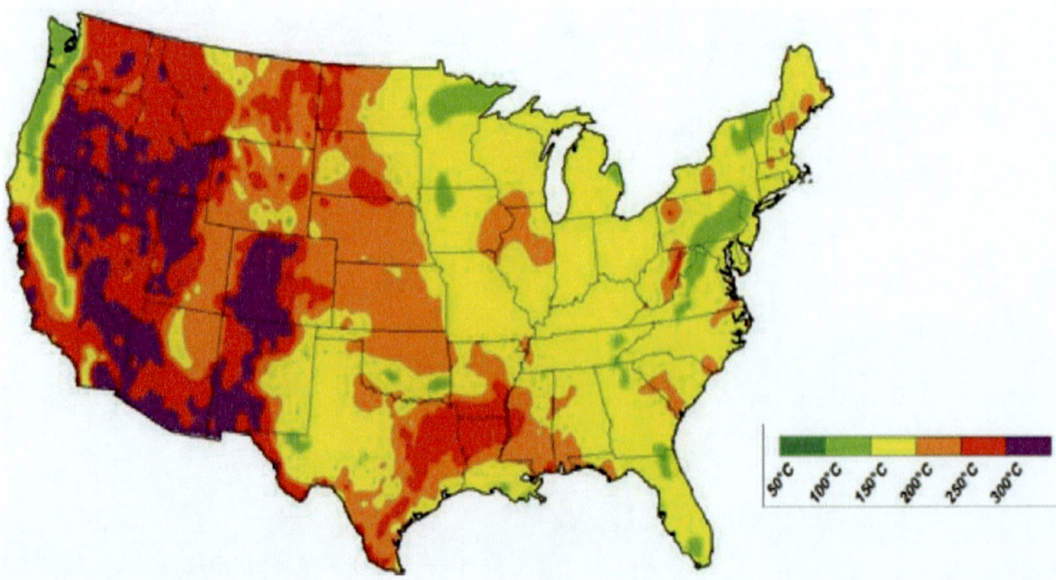

Source: Tester, J., et al. 2006. "The Future of Geothermal Energy: Impact of Enhanced Geothermal Systems (EGS) on the United States in the 21 Century"

Figure 8. Subsurface Temperatures at 10km Depth - EGS Potential

USGS Releases the First National Geothermal Assessment in More than 30 Years (September 2008)

With funding support from Congress and the DOE, the United States Geological Survey (USGS) released an assessment of domestic geothermal electricity production potential in September of 2008.[13] This assessment focused on electric generation potential in 13 western states* and estimated 39,090 MWe of potential from conventional hydrothermal reservoirs. This figure includes 9,057 MWe from discovered sources, and a mean estimated power production potential from 'undiscovered' geothermal resources of 30,033 MWe.† These figures suggest that only 23% of sources capable of producing geothermal electricity with today's technology have been discovered in the United States. The undiscovered source estimates are based on analysis of the local geology and the calculated potential of current discovered sources in the states examined. The assessment also predicts an additional 517,800 MWe of generation could come from implementing EGS technologies in high temperature, low permeability rock formations (see Figure 9).

The Geothermal Development Pipeline in 2008: 126 Projects with 3,638-5,650 MWe of Capacity

In August of 2008, the GEA reported that the 103 projects in development ranged from 2,805 MWe to 3,979 MWe in capacity.[14] By March 2009, the number of projects in development had increased to 126 and an additional 752-1,670 MWe of geothermal generating capacity had been added to the pipeline (see Figure 10).‡ According to the GEA, in addition to the eight current western states producing geothermal power, projects exist at various stages in five additional states: Arizona, Colorado, Oregon, Washington and Florida (see Figure 11).

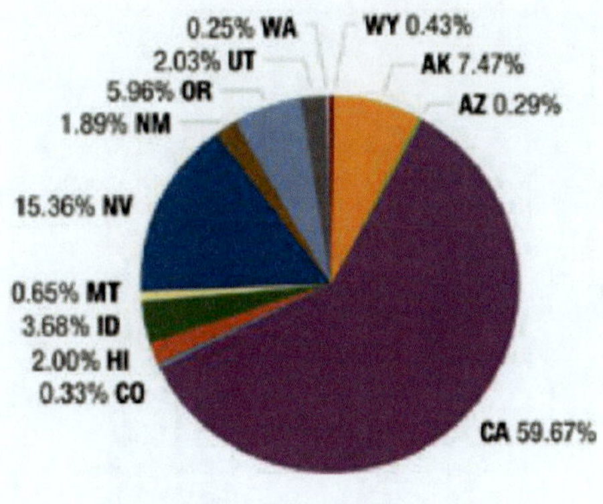

(A) Identified

* The 13 states assessed were; Alaska, Arizona, California, Colorado, Hawaii, Idaho, Montana, Nevada, New Mexico, Oregon, Utah, Washington, and Wyoming.

† Figure may be as high as 73,286 MWe at a 5% probability.

‡ It is important to note that while the overall number of development projects increased, this change in number also accounts for projects that have been completed and removed from the total.

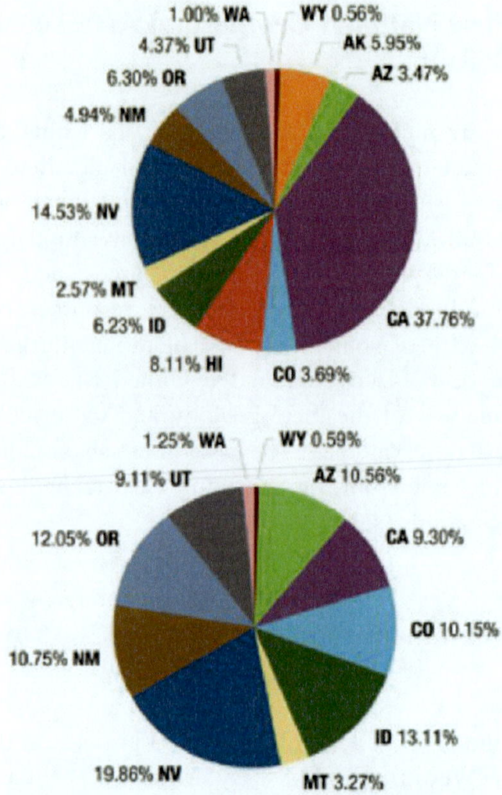

(B) Undiscovered (C) Enhanced

Source: Department of the Interior's BLM, "Assessment of Moderate- and High-Temperature Geothermal Resources of the United States" 2008

Figure 9. Distribution of Identified, Undiscovered, and EGS Resources

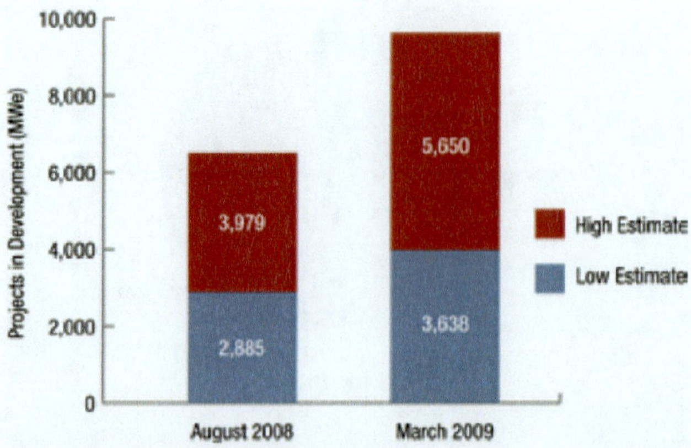

Source: Geothermal Energy Association, "U.S. Geothermal Power Production and Development" (August 2008 and March 2009).

Figure 10. The Geothermal Project Pipeline (2008-2009)

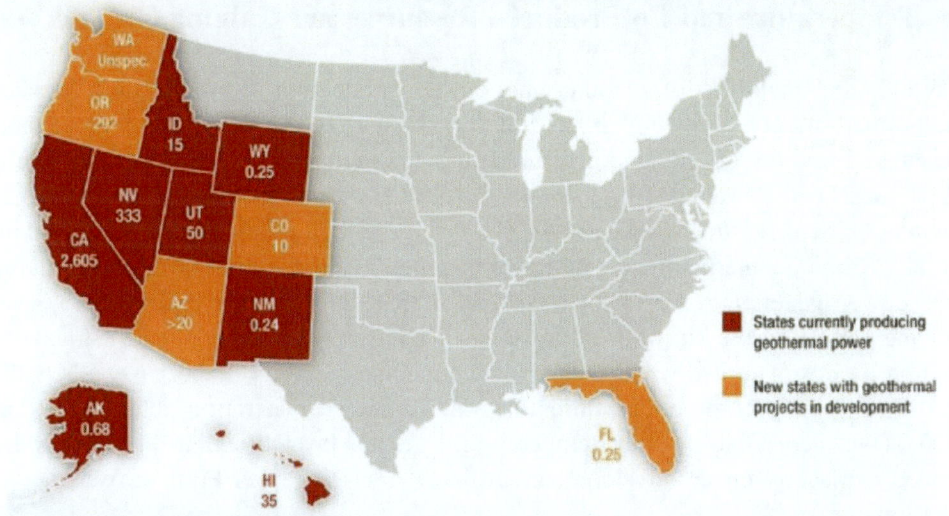

Source: Geothermal Energy Association, "U.S. Geothermal Power Production and Development" (August 2008 and March 2009).

Figure 11. States with Geothermal Projects under Development

Of the 126 projects in development, ten are currently in the final stages and will add roughly 329-457 MWe of capacity.[15] As the number of projects under development continues to grow and see completion, the overall installed capacity is expected to bounce back from its 2000 decline (see Figure 12). The decline resulted from a reduction in output from the U.S.'s largest production site, The Geysers Geothermal Field in California. A number of plants were closed due to overproduction of geothermal resources. As the result of recovery measures, some of these plants are now beginning to reopen.*

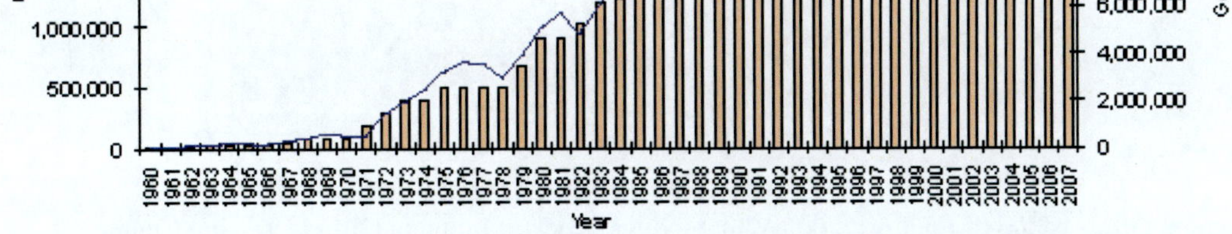

Source: Energy Information Administration (EIA), "Annual Energy Review 2007, " June 2008.

Figure 12. Installed Capacity and Generation, 1960-2007

* For example, the Bottle Rock Geothermal Power Plant in Cobb, California began operation in 1985, with a 55 MWe capacity. However, the steam field (resource) only allowed for 15 MWe of production. As a result, the operation of the Bottle Rock Power plant was suspended in 1990. Seventeen years later, in 2007, the plant was re-opened and began delivering power to the grid.

Low–Temperature and Co-Produced Resources are Gaining Ground

While the majority of geothermal power production comes from conventional hydrothermal sources, the geothermal industry is starting to tap the enormous potential represented by co-produced, geo-pressured and low-temperature resources. In September 2008 at the Naval Petroleum Reserve No. 3 (NPR3), Ormat Technologies and the Rocky Mountain Oilfield Testing Center (RMOTC) achieved the first successful generation of electricity from geothermal technologies integrated with existing oil infrastructure. The Ormat power generating unit known as the Ormat Energy Converter (OEC) has been producing 150-250 gross kilowatts of power since its inception (see Figure 13). An average of 40 billion barrels of heated water is co-produced annually from oil and gas wells within the United States; these co-produced fluids have an estimated generation potential of 3,000 to 14,000 MWe, depending on their temperature.[16] At the Jay Oilfield in Florida another co-produced project is under development, utilizing a UTC Power/Pratt & Whitney binary generation unit.

Binary units have expanded the resource base for geothermal power by allowing for the exploitation of lower temperature geothermal fluids.[*] Until recently, only temperatures over 93ºC (200ºF) were deemed commercially viable for successful electric generation from geothermal resources. In 2006 at Chena Hot Springs in Alaska, successful power generation occurred at a temperature of 74ºC (165ºF).[†]

Nameplate capacity for binary plants ranges from 200-280 kW to more than 100 MW. The major manufacturers of binary cycle units in use in the United States are UTC Power/ Pratt & Whitney, which sold approximately 100 of its PureCycle units in 2008, and Ormat Technologies, which sold around 12 of its OEC units in 2008. Other companies that produce binary cycle generators include:

Source: Office of Fossil Energy, "2009 Winter News: Rocky Mountain Oilfield Testing Center

Figure 13. Ormat's OEC Producing Power From Co-Produced fluids in Wyoming

[*] In a binary cycle, the heat from a geothermal fluid is transferred to another fluid that vaporizes at a lower temperature and higher pressure than water. The vapor from this second fluid then drives a turbine generator.

[†] The Chena Hot Springs resort facility used a UTC Power/Pratt & Whitney PureCycle system. The lowest temperature previously used for commercial energy conversion was 208°F.

Source: Raser Technologies

Figure 14. Raser's Hatch Power Plant in Beaver Creek, Utah

- Barber-Nichols (Organic Rankine Cycle/ORC)
- Mafia-Trench (ORC)
- Turboden (ORC)
- Enex (ORC)
- GE
- Siemens (Kalina Cycle)
- Exorka (Kalina Cycle)
- Gulf Coast Geothermal ("Green Machine") (ORC)
- Deluge Inc.
- Linear Power Ltd.

NEW BINARY PLANT DESIGNS REDUCE CONSTRUCTION LEAD TIME

Recently introduced binary-cycle plant designs have allowed power developers to substantially reduce plant construction lead times. One notable example is Raser Technology's Hatch Power Plant in Utah*, completed during November 2008. The plant consists of 50 UTC Power/Pratt & Whitney PureCycle binary units capable of producing at least 10 MW of net electricity (see Figure 14). The entire project was built and put online in less than one year, with construction completed in just six months rather than the typical three year timeframe.

The project is remarkable not only because of the rapid construction, but also because of the flexibility of its modular approach. Employing small, off-the-shelf UTC Power/Pratt & Whitney units, a plant can be scaled to the local geothermal resource, energy demand and available financing. Raser has subsequently confirmed that the geothermal resource at Hatch may have the potential to generate more than 200 MW. The company plans to add ten more units in 2009.

* The Hatch Power Plant was formerly known as Thermo.

DOE-FUNDED PROJECTS TARGET EGS DEPLOYMENT

The USGS Assessment of Geothermal Resources revealed that the majority of future power generation potential lies with EGS (see Figure 15). However, the technology necessary to exploit EGS resources is not yet commercial-ready. The GTP refocused its long-term technology development goals to address this state of affairs. The Program selected four field demonstration projects in 2008 focused on EGS reservoir creation, development, management and successful power production. These projects are located on the fringes of pre-existing conventional geothermal fields with active power generating capabilities in order to share infrastructure. Two field projects are located at the Geysers in northern California, run by AltaRock Energy and Calpine. A third project is located at Brady's Hot Springs, Nevada, and the fourth is at Raft River, Utah.[17]

The four field projects link steam production lines to current power plant facilities on site; no new facilities are under construction. AltaRock will utilize Northern California Power Agency (NCPA) power plants, and Geysers Power Company will utilize their own power plants. The University of Utah will use preexisting plants at the Raft River geothermal field operated by U.S. Geothermal, and Ormat will use their existing power facilities at Brady's Hot Springs.[18]

In addition to AltaRock Energy, Geysers Power Company (Calpine), University of Utah, and Ormat Technologies, other major entities involved in EGS development in the United States are U.S. Geothermal, Inc., APEX Petroleum, Engineering Services, and HiPoint Reservoir Imaging.[19]

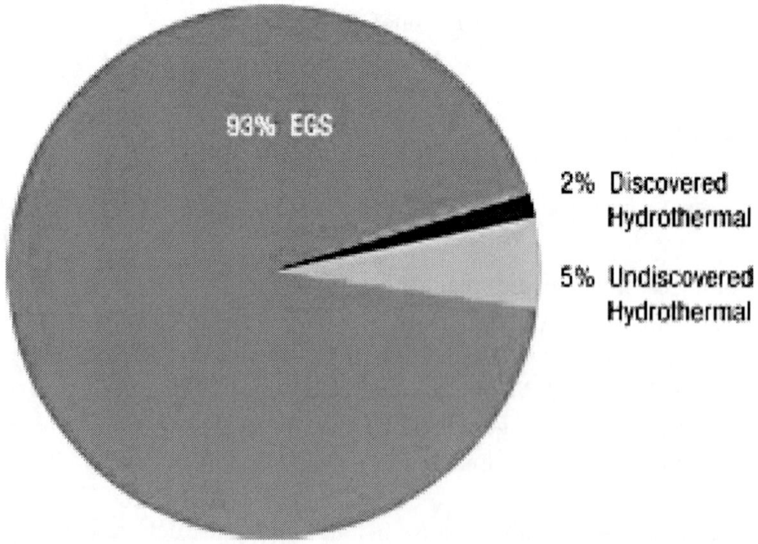

Source: USGS, "Assessment of Moderate- and High-Temperature Geothermal Resources of the United States"

Figure 15. Future Geothermal Potential by Resource Category

Developmental Stage	Cost ($ per kW installed)
Exploration	14
Permitting	50
Steam Gathering	250
Exploratory Drilling	169
Production Drilling	1,367
Plant & Construction	1,700
Transmission	100
Total	3,650

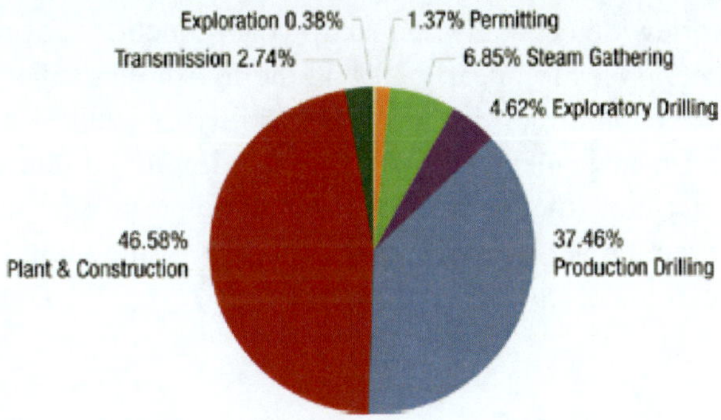

Source: Taylor, M. New Energy Finance, 2009

Figure 16. Estimated Developmental Costs for a Typical 50 MWe Geothermal Power Plant

4. Cost of Development, Operation and Maintenance

Conventional Hydrothermal Plants Typically Cost $3,000 to $4,000 per Installed KW[20]

The development of geothermal energy requires the consideration and evaluation of a number of factors, such as site (geography), geology, reservoir size, geothermal temperature, and plant type. In 2008, New Energy Finance published a breakdown of estimated costs for each developmental stage (see Figure 16). The majority of the overall cost is typically attributed to construction of the power plant, due to the high cost of raw materials including steel. The second highest cost intensive processes are the exploratory and production drilling stages, which together comprise 42.1% of the total cost.

Though geothermal power production is very capital-intensive with high first-cost and risk, it boasts fairly low operating costs and a high capacity factor*, making it one of the most economical baseload power generation options available. As previously noted, a number of factors contribute to the cost of developing a geothermal power plant. The power conversion technology (plant type) in use also has an effect on cost. Low-temperature reservoirs typically use binary power plants, while moderate- to high-temperature reservoirs employ dry steam or flash steam plants, based on whether the production wells produce primarily steam or water, respectively. Recent cost comparisons between flash, dry steam and binary plants do not demonstrate a clear winner.[21]

Power Plant Construction Costs Decline in 2008

After years of steady increases in plant construction costs, 2008 saw a 5% decline, according to Cambridge Energy Research Associates (CERA) (see Figure 17). Additional

* Capacity factor measures the amount of real time a facility is utilized to generate power.

cost reductions of approximately 7%-10% are expected for 2009 due to the declining worldwide economy and sharp cost reductions for raw materials, including steel and copper. Steel prices fell nearly 30% in the fourth quarter of 2008, a backlash from steep increases in the beginning of the year. Availability of equipment such as drilling rigs, labor, engineering and management services, has also improved due to delays and cancellations of other new plant construction projects. While some power purchase agreements were renegotiated to reflect higher overhead prices in 2008, some utilities may postpone negotiations and wait for costs to decrease.[22]

Geothermal Development Cycle and Risk Profiles

Sizable up-front capital requirements, pervasive resource and development uncertainty, and long project lead times lead to risk-related mark-up over other renewable and traditional energy alternatives. These factors, combined with current economic conditions, mean private firms seeking to develop geothermal projects may face greater difficulties in obtaining the requisite capital for exploration and development. Industry analysts suggest that although financing is still available, the terms will be less attractive to investors and developers.[23] Nevertheless, equity investors see real opportunity in the sector.

Development of Geothermal Power: Project Cycle

The primary stages of the geothermal development cycle are exploration, resource confirmation, drilling and reservoir development, plant construction and power production. Each of these steps carries with it different varieties and levels of risk. As Figure 18 shows, the risks associated with each stage call for different types of equity investors, who will expect a reward commensurate with the level of risk they assume.

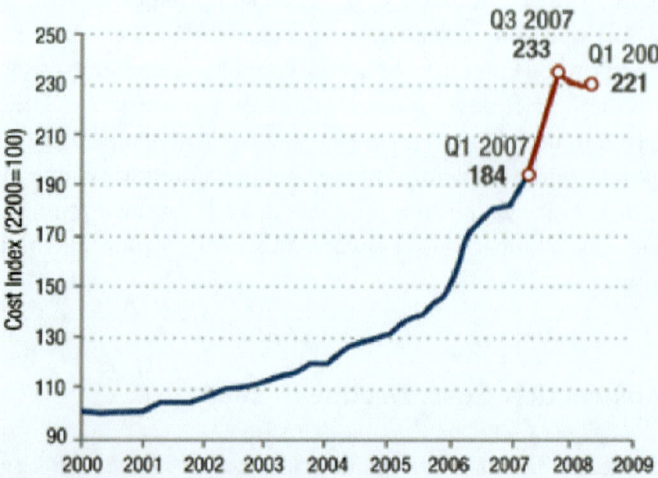

Source: Cambridge Energy Research Associates

Figure 17. IHS/CERA Power Capital Costs Index (PCCI)

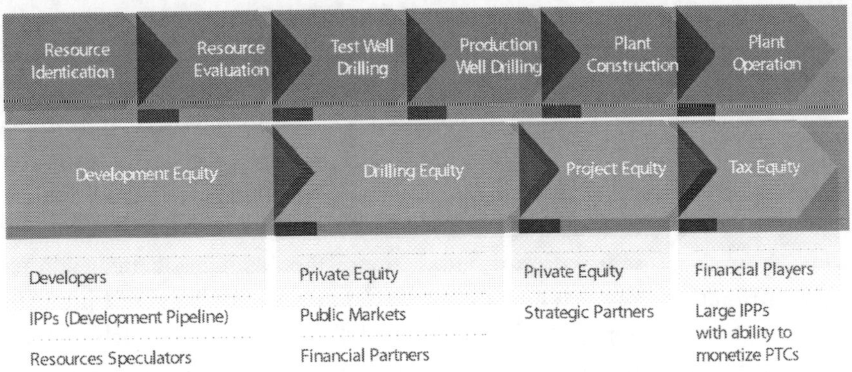

Source: Geothermal Investment: An Equity Provider's Perspective, Geothermal Investors' Forum, October 2007

Figure 18. The Geothermal Development Cycle

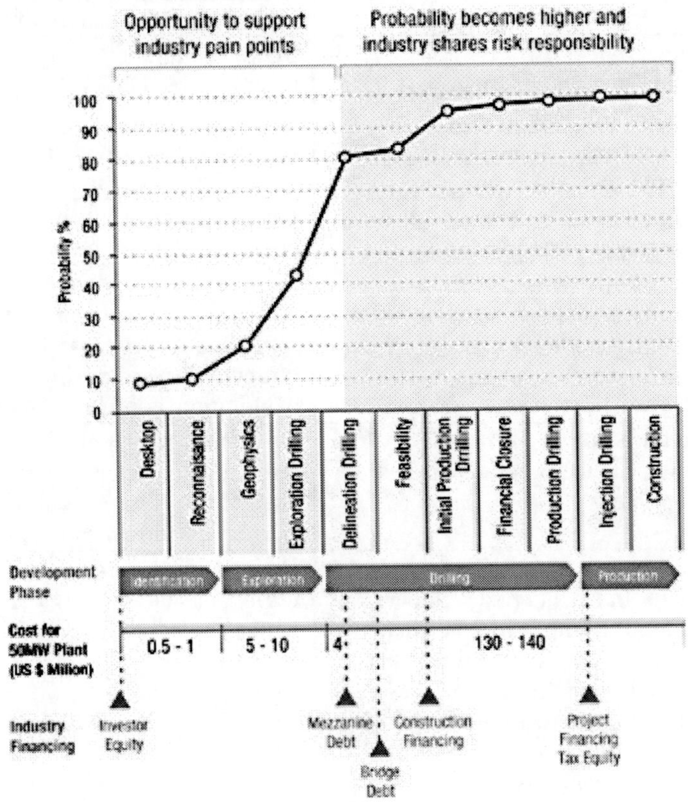

Source: Geothermal Energy Association, Update on U.S. Geothermal Power Production and Development, January 16, 2008.

Figure 19. Risk and Financing for Each Phase of Project Development

Though geothermal projects vary widely in terms of technical elements, location, and economic and political environments, financial models employed are relatively consistent. The greatest risk is associated with the initial stages of development, prior to the verification of the geothermal resource (see Figure 19). Activities such as the drilling of exploratory wells may prove unsuccessful even if geological data are favorable. Additionally, cost and risk increase proportionately with drilling depth. As the project moves toward the production phase, this risk begins to decline and financing options are more readily available.

In the exploration stage, prior to the validation of the geothermal resource, equity financing predominates, usually in the form of seed or venture capital. The project developer may also fund a portion out of its own budget. Debt financing, i.e., bank loans, typically enters the investment cycle following the successful demonstration of the geothermal resource, when the risk is greatly diminished. Though the costs associated with exploration and resource confirmation only account for approximately 10% of overall project costs, the risk associated with these activities is still too high for traditional debt lenders. Power developers have identified strategies to address the risk inherent at each development stage (see Table 5).

Table 5. Geothermal Project Risk Mitigation Strategies

Project risk	Mitigation strategy
Exploration Stage Lack of heat or fluid for heat extraction. A 25% success rate.	Make maximum use of surface technologies, Go-No Go exploration steps
Resource Capacity Risk 70% drilling success risk	Drill and test deep wells, develop a rigorous resource model.
Regulatory Risk Minimal with the proper planning	Utilize an experienced permitting consultant, begin the process early
Drilling Risks Risk of drilling a dry well, approximately 70% success risk	Prepare geological model and drill with blow out protectors and control of well insurance. Create a "risk fund" that can mitigate investor drilling risk during the exploration, confirmation project stages.
Plant Construction Risk There is minimal risk if the previous items are completed appropriately.	Use a credible supplier/contractor, get turnkey fixed price/date certain contract, use field-proven technology supplier, get start-up performance guarantee.
Financing Risk Financing issues for independent developers include: exploration financing (investor may want returns equal to multiples of investment), require an investment-grade power purchaser, construction financing (interest rates may be up to 10% or more, construction lender requires "take out" guarantee at commissioning), term financing usually based on 30% equity/70% debt, IRR in the high teens, interest 7% or more for 15 years.	Execute financeable take or pay PPA with utility, execute binding commitment with lender

Source: Getting Geothermal Electricity Projects Online, Presentation by Daniel J. Fleischmann, ORMAT Nevada, Inc., July 23, 2007.

Tax Equity Financing - Special Purpose Entities and the Partnership Flip

The third major source of capital available to the geothermal project developer is tax equity financing, monetized through the creation of a special purpose vehicle (SPV)* and what is commonly known as a partnership flip. Federal tax subsidies amount to a large share of development financing for a variety of types of renewable energy plants, including solar, wind and geothermal. Geothermal projects qualify for the Federal PTC under Section 45 of the Internal Revenue Code (IRC), a 2.1 cent per kWh credit claimed on the electricity generated by the geothermal plant for up to 10 years. The geothermal developer may also depreciate the geothermal property over five years, allowed under the Modified Accelerated Cost-Recovery System (MACRS). In addition, intangible drilling costs can be deducted immediately, either amortized over the five-year period or folded into basis in the geothermal wells and reservoirs, and depletion can be claimed on the investment in the reservoir.[24] In sum, a substantial portion of the cost of developing the geothermal project can be covered by these tax benefits.

To claim the credits under Section 45 of the IRC, the taxpayer must be the owner and operator of the renewable energy property, but most developers cannot directly utilize these tax subsidies to build their projects. The trick is to monetize them or convert them into capital through a partnership flip. Under this arrangement, the geothermal developer brings in an outside entity, typically a large, institutional investor that can take advantage of the available tax credits, forming a special purpose entity or vehicle. The developer enters into a disproportionate allocation partnership with this investor, an arrangement made for tax purposes wherein the party attempting to monetize the tax credits is allocated the majority share of project income and loss. The tax-oriented investor is allocated 99% of the geothermal plant's economic returns, i.e., income and tax credits, until they reach a target yield--an after-tax return on investment previously agreed upon by both parties. This is typically designed to occur towards the end of the 10-year tax credit period, once the project is completed and in operation, providing a revenue stream. On this flip date the investor's percentage interest is reduced to 5% and the developer has the option to buy the remaining interest. That is, the project flips back to the developer.[25]

The total tax equity generated against the project at the outset of such a partnership arrangement is essentially the present value of cash income, PTCs, depreciation-related tax savings, depletion interest and investor-paid intangible drilling cost deductions and taxes. The amount of tax equity depends upon the overall capital cost of the project, the quantity of electricity generated, negotiated prices under the power purchase agreement with the end user, and the tax equity yield. For every 50 basis point increases in yield, the portion of plant cost covered by the tax credits is reduced by approximately 10%. That is, with each hike in yield the equity investor is providing less value in return for the tax credits. At some point, it may be beneficial for the project developer to retain the tax benefits for its own future use.[26] To qualify for the production tax credits, geothermal projects must be placed into service by December 2010, though plants that come online after this date are still eligible for a 10% investment tax credit (ITC). The developer may utilize tax equity financing at the initiation of the project or, if it has access to sufficient debt financing during the plant construction phase,

* SPE/SPVs are legal entities created to accomplish specific objectives, in this case financing of a specific asset, the geothermal plant.

it may opt to sell the tax-oriented investor an interest in the project after it has started operating.

Availability of Geothermal Project Financing Declines in 2008

The current global economic crisis impacts the ability to obtain tax equity for such partnership flip scenarios, which require the resources of large institutional investors, such as Morgan Stanley, GE Financial Services, and the now-bankrupt Lehman Brothers. This sector has contracted rapidly over the last six months. According to industry analysts, just half of the 14 large financial companies that funded renewable energy projects over the past two years are still active in this market, resulting in a $2-3 billion tax equity shortfall at the close of 2008. Tax equity yields are 150-170 basis points higher than just one year ago.[27] A representative of GE Financial Services was recently quoted as saying that the firm simply did not have the resources to take on any new tax-monetization investments in renewable energy projects.[28] Finally, the increased borrowing costs and tax equity yields may even require developers to renegotiate power purchase agreements with their utility customers.[29]

As fewer financiers are active in the markets, there are limited financing options available for geothermal companies, more stringent financing terms and added deal-making complexity.[30] Development capital that is still available will tend to gravitate to renewable energy projects that demonstrate the greatest potential for project returns for a given level of risk.[31] Resource uncertainty, high up-front capital cost and attendant risk associated with geothermal energy production could potentially handicap new projects currently in the pipeline for the sector. Developers may be required to drill additional boreholes to secure up to 50% of capacity rather than the 30-35% that was previously sufficient, increasing up-front costs as much as $20 million.[32] Debt providers generally require that 25% of the resource capacity is proven and a long-term PPA in place prior to lending.[33] There are numerous reports of geothermal projects on hold for want of financing in early 2009.[34]

Geothermal companies that have advanced to the later project stages, with available cash and liquidity, are surer bets for investors because they have greater flexibility to develop their projects. They might be able to rely on the strength of their balance sheets to finance projects outright or use them to obtain better deals from investment partners. Ormat, despite a significant decline in its stock from a 2008 high of $53.54 to $38.18 (quoted May 27, 2009), remains in this category of developers. On the other hand, small or inexperienced developers, with limited project portfolios or projects in the early exploration or drilling stages, will be severely impacted. For example, Sierra Geothermal Power (SGP), currently active with five exploration-stage projects in Nevada, has run low on available working capital and such a weak hand to show to investors hinders its ability to attract new capital.[35]

Geothermal Costs Less Than Other Renewables and Some Conventional Sources

Geothermal power production boasts fairly low operating costs and high capacity factor, making it one of the most attractive baseload generation options available among renewables.

On a levelized cost of energy (LCOE) basis*, which provides an apples-to-apples comparison of generation options, geothermal is very competitive with other renewable and conventional technologies. Most recently, the financial advisory and asset management firm, Lazard, calculated LCOE for various alternative and conventional electric generating technologies. With tax incentives included, it estimated geothermal LCOE between $0.042 and $0.069 per kWh depending on technology employed (See Figures 20). An earlier 2005 study conducted by the California Energy Commission estimated geothermal LCOE between $0.04 and $0.09 per kWh with PTCs added (See Table 6). Despite the high upfront cost and risk, geothermal installation costs are lower than nuclear, solar, small hydro, and selected biomass technologies.[36]

The costs to develop a given geothermal plant can vary tremendously depending on resource characteristics, the conversion technology utilized by the plant, and other factors, such as raw materials, drilling, and financing costs. It should also be noted that PTCs play a major role in making geothermal more competitive (see Figure 21). Without Federal tax incentives, costs can soar to between $0.078 and $0.11 6 per kWh, highlighting the importance of these incentives (see Table 8 and Figure 21).[37] In 2008, the PTC for geothermal was $0.021 per kWh.[38]

5. National Policy, Geothermal Leasing and Permitting

EESA of 2008 Extends Geothermal Tax Incentives

On October 3, 2008, President Bush signed the EESA of 2008 (H.R. 1424), which included the Energy Improvement and Extension Act of 2008. The bill extended PTCs for electricity produced by geothermal facilities (as well as other renewable energy sources) by two years, bringing the sunsets of these credits to the end of 2010. This brought a renewed sense of certainty to the investment market, as these tax credits were set to expire at the end of 2008. This act also created a 30% tax credit for GHPs, with a cap of $2,000.[39]

Table 6. LCOE for Various Geothermal Generation Technologies ($/kWh)

Technology	Without PTC	With PTC
Dry steam	$0.0781	$0.0691
Dual flash steam	$0.0563 to $0.0979	$0.0473 to $0.0889
Binary	$0.049 to $0.1021	$0.040 to $0.0931

Source: California Energy Commission, "Geothermal Strategic Value Analysis, June 2005.

* LCOE includes a more complete set of cost variables, including fixed and variable costs, financing and fuel costs. It is defined as a constant annual costs that equivalent on a present-value basis to the annual costs, which may be variable. It may include capital and financing costs, insurance costs, ad valorem/property tax costs, fixed and variable operations and maintenance costs, corporate taxes, and costs of fuel. (*Comparative Costs of California Central Station Electricity Generation Technologies*, California Energy Commission Report CEC-200-2007-011 SF, December 2007)

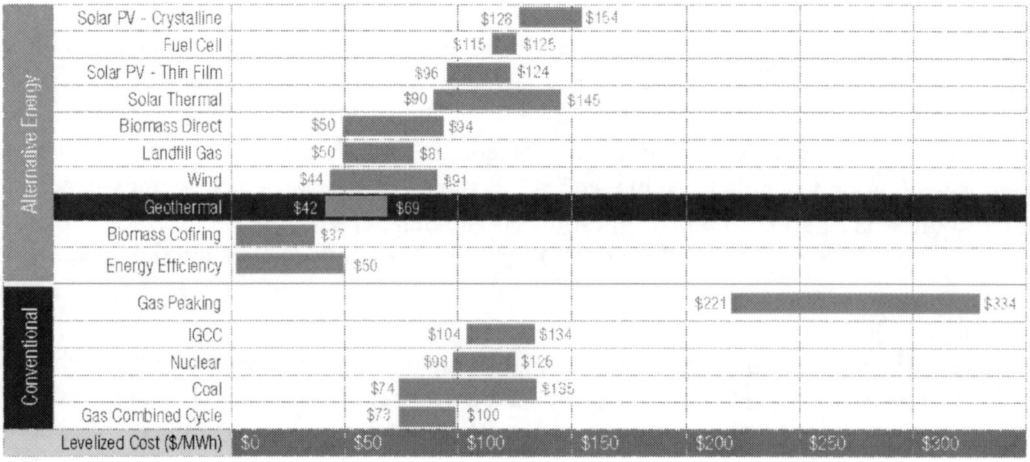

Source: Lazard, June 2008

Figure 20. Levelized Cost of Energy per MWh of various power technologies

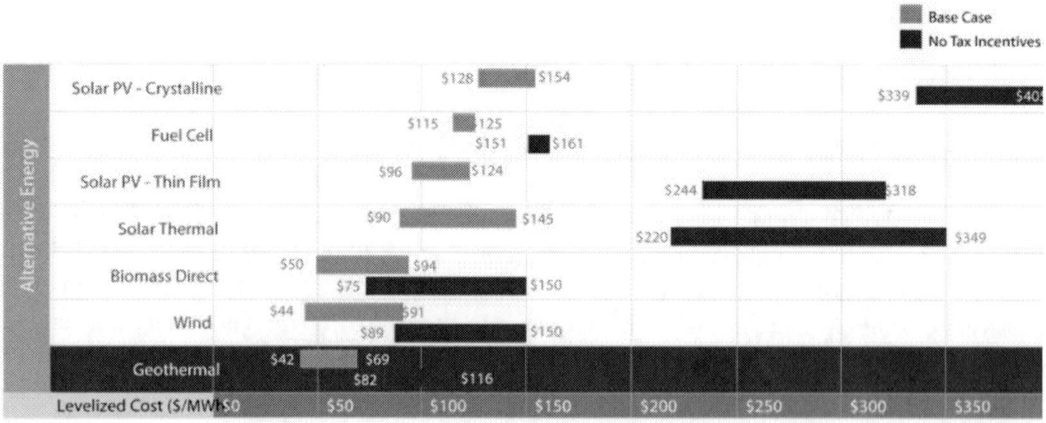

Source: Lazard, June 2008

Figure 21. Levelized Cost of Renewable Technologies With and Without Tax Incentives

Renewable Portfolio Standards Drive Renewable Energy Development

Renewable Portfolio Standards (RPSs) are widely considered to be an essential driver for development of geothermal and other renewable energy technologies. Currently, RPSs exist only at the state level in the United States. The diverse set of authoring entities has resulted in a disparate set of policies governing geothermal technologies. Many of the state RPSs target small-scale or residential geothermal projects, but do not provide adequate incentives for large-scale exploration or plant development. As of May 2009, 32 states and Washington, D.C. have implemented RPS guidelines that are either mandatory or goal-oriented (see Figure 22). A national RPS is currently under consideration in Congress (see "Looking Ahead" section for more detail).

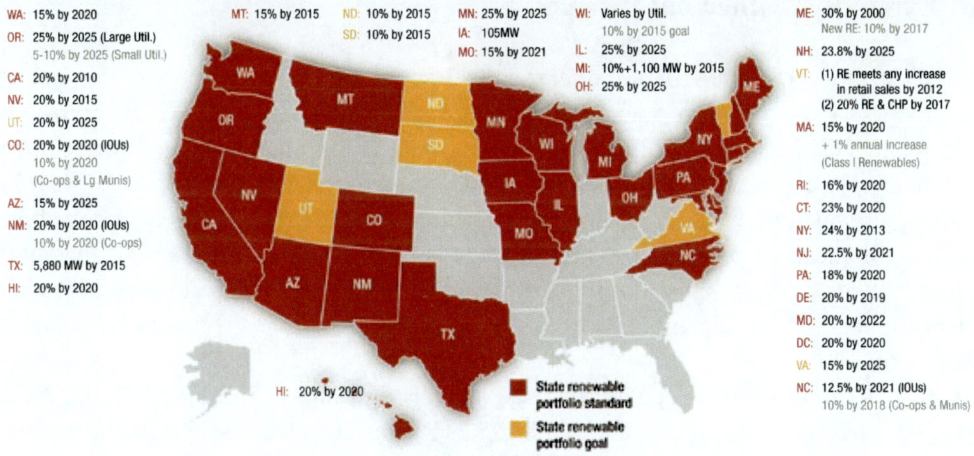

Source: Database of State Incentives for Renewable Energy (DSIRE), May 2009.

Figure 22. States with Renewable Portfolio Goals and Policies

Western Renewable Energy Zone to Expedite Renewable Energy Development and Delivery

The Western Renewable Energy Zone (WREZ), an initiative launched in May 2008 by the Western Governors' Association and DOE, seeks to identify the most cost-effective and environmentally sustainable areas within the western United States to develop renewable energy resources and facilitate their delivery to major load centers. The project promotes stakeholder collaboration and information exchange between state and Federal governments and non-governmental organizations with a regional approach to energy development. Eleven states, two Canadian provinces, and areas in Mexico that are part of the Western Interconnection are currently participating in the project.

EPAct 2005: New Procedures for Federal Geothermal Leases

BLM manages over 700 million acres of subsurface mineral estate and through 480 leases has made just 700,000 acres available for geothermal development, highlighting the vast potential for development of domestic geothermal energy.[40/41] The Energy Policy Act (EPAct) of 2005 addressed the growing backlog of lease applications by fostering greater cooperation among the Federal agencies involved in the leasing process. The BLM and the Forest Service signed a memorandum of understanding (MOU) in 2006 that lead to the completion of the Programmatic Environmental Impact Statement (PEIS), which amends federal resource management plans and land use plans. Site-specific analysis of leasing nominations, permit applications, and operations plans can refer back to the PEIS, reducing the processing time for leasing and permitting.

BLM Expands Geothermal Leasing

On December 17, 2008, BLM released its Record of Decision for Geothermal PEIS signed by the Department of the Interior's Assistant Secretary for Land and Minerals Management. This decision (1) allocates BLM lands as open to be considered for geothermal leasing or closed for geothermal leasing, and identifies those National Forest System lands that are legally open or closed to leasing; (2) develops a reasonably foreseeable development scenario that indicates a potential for 12,210 megawatts of electrical generating capacity from 244 power plants by 2025, plus additional direct uses of geothermal resources; and (3) adopts stipulations, best management practices, and procedures for geothermal leasing and development.

BLM held a competitive auction of lease parcels on August 5, 2008 in Reno, Nevada, offering 35 parcels encompassing a total of 105,211 acres. The lease sale brought in a record $28.2 million in bids for geothermal energy development. A second lease sale was held in December 2008 offering 61 parcels totaling 196,377 acres in the states of Utah, Oregon, and Idaho. Cumulatively the two sales totaled 301,588 acres and generated more than $34.5 million in revenue.

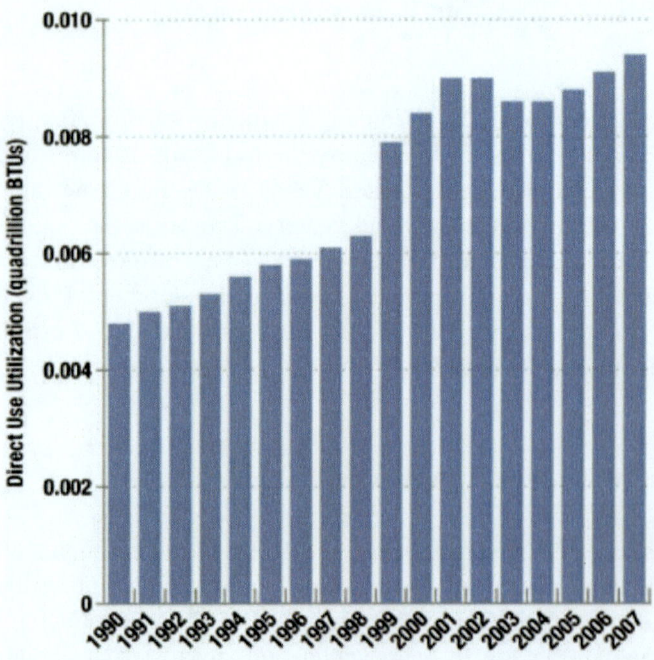

Year	Direct Use Utilizatio (Quadrillion BtU
1990	0.0048
1991	0.0050
1992	0.0051
1993	0.0053
1994	0.0056
1995	0.0058
1996	0.0059
1997	0.0061
1998	0/0063
1999	0.0079
2000	0.0084
2001	0.0090
2002	0.0090
2003	0.0086
2004	0.0086
2005	0.0088
2006	0.0091
2007	0.0094

Source: Lund, J., Oregon Institute of Technology, Geo-Heat Center, March 2008.

Figure 23. Geothermal direct use of energy (1990-2007)

6. DIRECT-USE AND GHPS

Direct-Use & GHPs: Strong Market Growth in 2008

Direct-use applications typically include aquaculture, greenhouses, industrial and agricultural processes, pools and spas, and space and district heating. Direct use of geothermal energy consumed 0.0094 quadrillion BTUs (quads) in 2007 (see Figure 23).[*] In 2008, the installed capacity for direct uses, excluding heat pumps, was estimated to be 704 MWt with an annual consumption of 10,332 TJ/yr (2,869 GWh/yr) using an overall escalation of 4%.[†] All non-heat pump direct uses had a calculated capacity factor of 46 percent, identical to past-calculated values.[42] While a projection has been made for 2008, direct-use estimates are difficult to determine because there are a wide array of uses, locations are geographically diverse, and temperature and flow-rates are unknown.[43]

The U.S. GHP Installed Base is World's Largest - More than 1 Million Units[*]

Installed GHP capacity in the United States in 2007 was equivalent to 10,839 MWt with a capacity factor of 10 percent. The thermal energy consumed totaled 33,445 TJ/yr (9,287 GWh/yr), roughly 0.03 17 quads (see Figure 24). In 2008, the geothermal heat pump capacity was estimated to be 12,031 MWt with an annual consumption of 37,124 TJ/yr (10,309 GWh/yr). This estimate was produced using Lund's annual escalation factor for geothermal heat pumps, which was 11 percent for 2008.[44]

Based on the latest Energy Information Administration (EIA) Form EIA-902, "Annual Geothermal Heat Pump Manufacturers Survey", GHP manufacturers shipped 86,396 GHPs in 2007, a 36% increase over the 2006 total of 63,682. The total rated capacity of GHPs shipped in 2007 was 291,300 tons, which represents almost a 19% increase over the 245,603 tons shipped in 2006 (see Figure 25). AHRI reported 2008 shipments of more than 71,000 units, indicating continued strong demand despite worsening economic conditions (see Table 7).

WaterFurnace, the Canadian GHP company and a market leader in the United States and Canada, witnessed a doubling of sales between 2003 and 2007, with a 26% year-over-year sales increase from 2007 through 3Q 2008.[45]

Table 7. Geothermal Heat Pump Shipments (1999-2008)

1999	2000	2001	2002	2003	2004	2005	2006	2007	2008^
41,679	35,581	N/A	37,139	36,439	43,806	47,830	63,682	86,396	71,000

Source: EIA, AHRI, 2009.

^ Advance data from the AHRI.

[*] In February 2009 the EIA released its annual, Geothermal Heat Pump Manufacturing Activities. The data within the report were only applicable for the 2007. 2008 data were estimated given annual escalation factors used by John Lund, OIT Geo-Heat Center.

[†] The annual escalation factor of 4% = average percent increase since 1990.

[*] In 2007, it was estimated that the number of heat pumps installed totaled over 800,000. One geothermal heat pump has an assumed average size of 12 kW. Therefore, the assumed 12,031 MWt installed for 2008 allows for an estimate of 1,002,583 heat pumps installed.

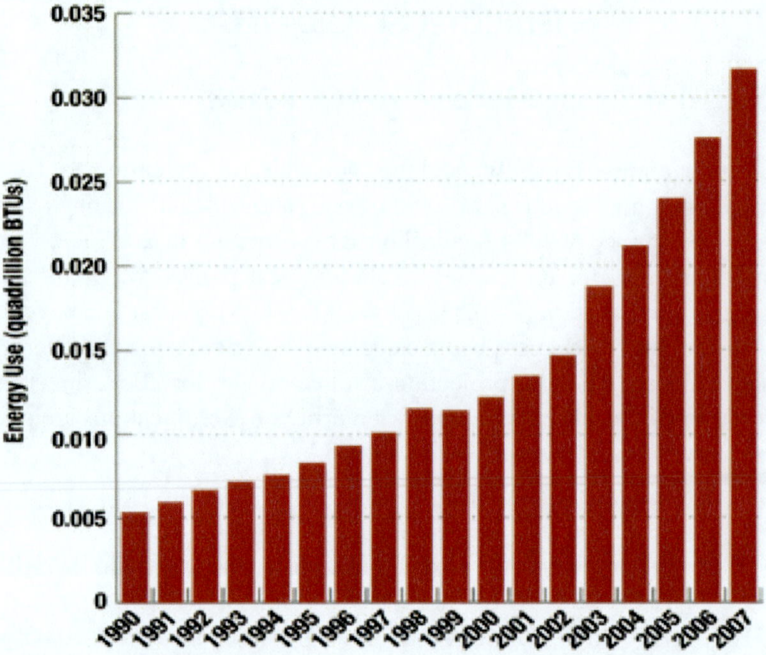

Source: EIA, "Geothermal Heat Pump Manufacturing Activities 2007" (Released February 2009)

Figure 24. GHP Primary Energy Consumption

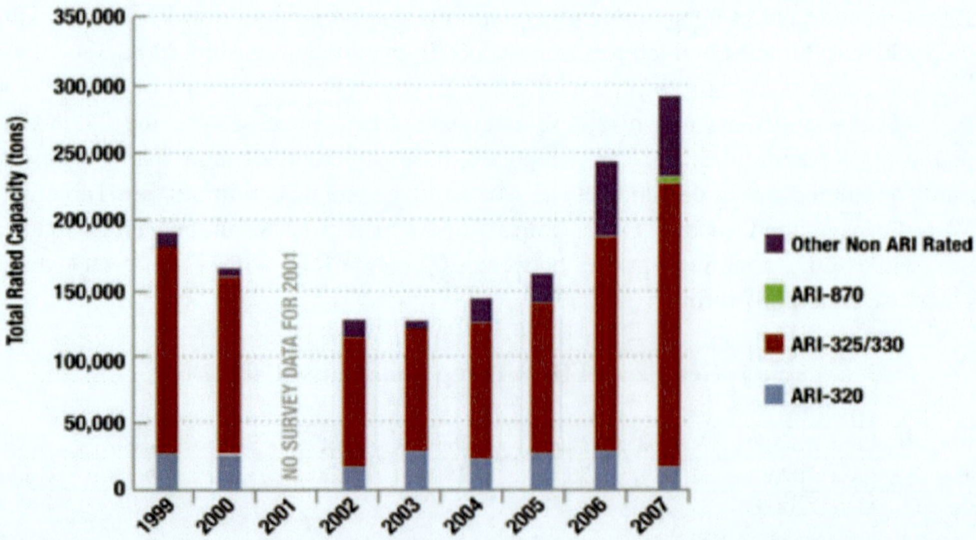

*ARI 320 refers to ARI rated water source heat pumps, ARI 325 to ARI rated ground water source heat pumps (open loop), and ARI 330 to ARI ground source heat pumps (closed loop).
Source: EIA, "Geothermal Heat Pump Manufacturing Activities 2007" (Released February 2009).

Figure 25. Capacity of GHP Shipments by Model Type (2009)*

U.S. GHP Market Segmentation–Evenly Divided between Residential and Commercial Applications

GHPs can be used in a wide variety of applications, including residential, commercial, institutional and multifamily buildings. Currently, GHP shipments are fairly evenly divided between residential and commercial building applications (see Figure 26). According to ClimateMater, GHPs were installed in 1 out of every 38 new U.S. homes in 2008. This represents a 2.6% market share for the segment.[46] The retrofit market for schools has grown substantially in recent years; there are currently more than 600 schools with GHP systems.

As shown in Figure 27, GHPs have a presence in all census regions, although the market has historically been dominated by the Midwestern and southern states, which are home to the major GHP manufacturers and have more personnel trained in GHP installation and maintenance than other regions.

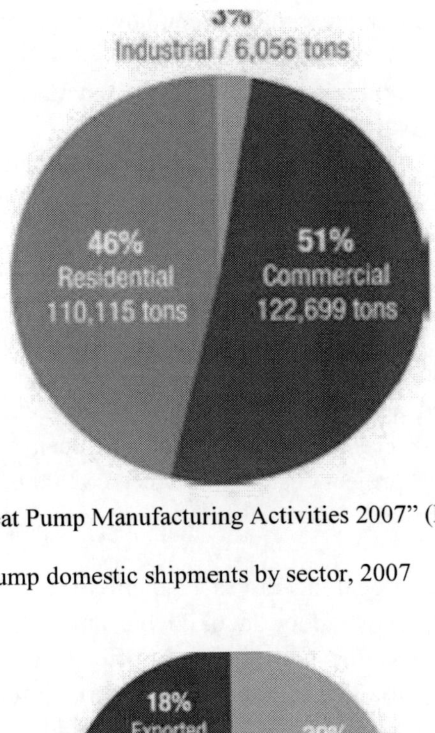

Source: EIA, "Geothermal Heat Pump Manufacturing Activities 2007" (Released February 2009)

Figure 26. Geothermal heat pump domestic shipments by sector, 2007

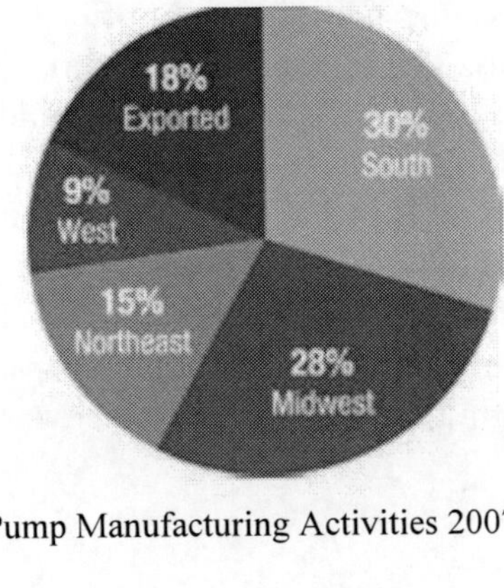

Source: EIA, "Geothermal Heat Pump Manufacturing Activities 2007" (Released February 2009)

Figure 27. GSHP Shipments by Census Region in Tons (2007)

Four Companies Hold 80 Percent of the U.S. GHP Market

Although 40 firms respond to the EIA's GHP survey,"[47] just four companies account for over 80% of annual sales.[48] The top four manufacturers are ClimateMaster (a unit of LSB Industries), Florida Heat Pump (a unit of Bosch), WaterFurnace International, Inc., and Trane (a business unit of Ingersoll Rand.) An additional 10-15 firms account for the remainder of the U.S. market. Some serve the entire nation while others cater to specific market niches. In addition, certain GHPs are rebranded and resold under different names. Major firms within this group include McQuay International (a unit of Daikin), Mammoth and several regional manufacturers. Carrier markets water-source heat pump and GHP systems designed by other manufacturers under their own label.[49]

Tax Credits and Incentives Set to Increase GHP Deployment

The GHP market still faces significant barriers, however, including: high installation and capital costs; a pervasive lack of consumer awareness; and insufficient market delivery infrastructure. In order for heat pumps to reach their full market potential, these barriers must be addressed through effective market conditioning strategies.[50]

First and foremost, GHP systems are generally more expensive than conventional heating and cooling systems due to the costs associated with installation of the ground connection. The remaining components, the "balance of system," cost roughly the same as the equipment that comprise common air-source heat pumps. Average installed costs for 2008 are roughly \$5,000–\$6,000/ton* for the residential market and \$6,000 to \$10,000/ton for commercial applications.[51] Though GHPs have higher initial capital costs their operation and maintenance costs tend to be lower than some conventional alternatives.

To encourage continued adoption of GHPs in the residential market, EESA of 2008 renewed the EPAct 2005 tax credit for GHPs that had been allowed to lapse. The credit covers 30 percent of the GHP project cost, not to exceed \$2,000. The recent Recovery Act of 2009 greatly enhanced the scope of the tax credit by removing the \$2,000 cap (see Looking Ahead section).

There are a variety of additional incentives available from Federal and state governments and utilities to encourage greater adoption of GHPs among residential, institutional and commercial consumers. Thirty-four states currently offer GHP incentives, generally in the form of rebates, loan programs, tax exemptions (property, sales and use), renewable and green building requirements for public buildings, stricter building codes that require specified energy efficiency requirements, public benefits funds created through utility surcharges, and green building incentives, such as expedited permitting.[52] Other possible incentive options include design assistance programs, innovative loop leasing and financing strategies, low interest loans, consumer rebates, GHP utility rate tariffs, contractor training programs, to nurture the delivery infrastructure, and public support by Federal and state agencies, including program marketing and funding for demonstration and showcase facilities.[53]

* One ton is equivalent to 12,000 BTU/hr.

7. International Activities

Worldwide Geothermal Capacity Continues to Grow

The United States remained the leader in installed geothermal capacity in 2008, followed in order by the Philippines, Indonesia, Mexico and Italy (see Table 8). While international figures are not yet known for 2008, New Energy Finance estimates 335 MWe of capacity has been added outside the United States (see Table 9). Some important international geothermal developers include; Geodynamics (Australia), Petratherm (Australia), Green Rock Energy (Australia), Chevron Geothermal & Power (USA), and Enel Green Power (Italy).

Table 8. Installed geothermal capacity of the top ten countries in 2007

Country	Installed Capacity (MWe)
USA	2,687
Philippines	1,970
Indonesia	992
Mexico	953
Italy	811
Japan	535
New Zealand	472
Iceland	421
El Salvador	204
Costa Rica	163

Source: Bertani, R. "World Geothermal Generation in 2007" (September 2007)

Table 9. Total 2008 MWe increases by country

Country	2008 Increases in Capacity (MWe)
France	1.5
Germany	6.9
Iceland	90.0
Indonesia	60.0
Kenya	35.0
New Zealand	121.6
Philippines	20.0
TOTAL	**335.0**

Source: New Energy Finance, 2008

International Direct-Use Geothermal is Widespread

Direct-use geothermal is widely utilized abroad. Iceland and Turkey both employ a tremendous amount of geothermal energy to serve their heating and cooling requirements (see Table 10). Iceland satisfies 89% of its heating and cooling needs with geothermal, whereas Turkey has increased its installed base of district heating systems from 820 MWt to 1,495 MWt, almost 50% in just 5 years. Japan has over 2,000 hot spring resorts, 5,000 public bath houses, and 15,000 hotels with natural hot springs. Switzerland makes extensive use of its geothermal resources as well, with more than 30,000 GHPs, and uses drain water from tunnels to heat homes and melt roadway ice and snow.[54]

International installed GHP capacity has experienced strong growth in recent years. Annual growth rates exceed 10% over the last 10 years. Most of this activity occurred in the North American and European markets.[55] The European GHP market is expected to experience continued strong growth due to a variety of energy efficiency and climate protection goals and policies by the European Union (EU) countries and stakeholder organizations.[56] These include the EU Proposal for a Directive of the European Parliament and of the Council on the Promotion of Renewable Energy[57]; the Ground Reach Initiative, a collaborative effort to utilize GHPs to meet Kyoto Treaty climate targets[58]; and the European Geothermal Energy Council's recent strategy document.[59]

As noted earlier, direct-use applications are quite diverse and include everything from agricultural to resorts and spas. Table 11 below contains a complete breakdown of direct-use categories.*

Table 10. Top direct-use countries

Country	GWh/yr	MWt	Main Applications
China	12,605	3,687	Bathing
Sweden	12,000	4,200	GHP
USA	8,678	7,817	GHP
Turkey	6,900	1,495	District Heating
Iceland	6,806	1,844	District Heating
Japan	2,862	822	Bathing
Hungary	2,206	694	Spas/Greenhouse
Italy	2,098	607	Spas/Space heating
New Zealand	1,969	30	Industrial Uses

Source: Lund, John (2007). Characteristics, Development and Utilization of Geothermal Resources. Geo-Heat Center, Oregon Institute of Technology, GHC Bulletin, p.6.

Table 11. Direct use application breakdown by installed capacity and annual energy use

Application	Installed Capacity	Energy Use
Geothermal Heat Pumps	56.5%	33.2%
Bathing/swimming/spas	17.7%	28.8%
Space heating (with district heating)	14.9%	20.2%

* Dr. John Lund includes GHPs with other direct use geothermal applications.

Table 11. (Continued)

Application	Installed Capacity	Energy Use
Greenhouse heating	4.8%	7.5%
Aquaculture	2.2%	4.2%
Industrial	1.8%	4.2%
Agricultural drying	0.6%	0.8%
Cooling and snow melting	1.2%	0.7%
Other	0.3%	0.4%

Source: Lund, John (2007). Characteristics, Development and Utilization of Geothermal Resources. Geo-Heat Center, Oregon Institute of Technology, GHC Bulletin, p.6.

8. EMPLOYMENT AND ECONOMIC BENEFITS OF GEOTHERMAL POWER

Geothermal Industry – More than 25,000 Employed Nationwide

Several studies have examined the employment and economic benefits of geothermal energy development. Perhaps the most obvious positive byproduct is the creation of high-paying, long-term jobs. Calpine has reported that the construction of a typical 50 MW geothermal plant involves 160 people and 33 months of labor.[60] In 2008, the GEA estimates that the geothermal industry roughly accounted for 9,000 jobs in operating, construction and manufacturing and an additional 16,000 supporting positions.

These figures do not incorporate the manufacturing and installation jobs generated separately by the GHP industry. According to the EIA, direct employment in the geothermal heat pump manufacturer industry alone accounted for 1,219 person-years in 2007.[61] GHPs are a labor-intensive technology to manufacture and install. Based on estimates generated by WaterFurnace, each GHP requires 24 hours of manufacturing labor and 32 hours of installation labor, and a permanent job is created for every 18 installations.[62] GHPs require a wide range of experience, with up to 30 individuals involved with each installation.

Gross Revenue from Geothermal Royalties Increased 14% between FY 2006 and FY 2008

In addition to job creation, tax revenues from geothermal development can have a substantial impact on local economic growth. EPAct of 2005 increased these benefits such that the Federal, state and county governments will now receive 25%, 50% and 25% of geothermal revenues, respectively, from Federal leases. According to a report by the GEA, in 2008 geothermal facilities produced $9.1 million in tax revenue for 31 counties in six states—an increase of $4.3 million from the 2007 amount These counties tend to be sparsely populated rural areas where the revenue increases have noticeable positive effects; the counties overwhelmingly used the revenues to support public services, and infrastructure. In 2007 and 2008, six states received a total of $27 million in geothermal tax revenues[63], while the Federal government received $13.5 million.

9. Looking Ahead – 2009 and Beyond

Conclusion

In 2008, the United States geothermal energy industry experienced a rebirth. New research showing a dramatic increase in potential of geothermal as a major energy source, along with a volatile energy environment and climate change concerns, sparked renewed investment from government and industry. Projects currently underway may result in new breakthroughs in technology and cost efficiency for the industry, and it is poised for additional growth in 2009 despite challenging economic conditions.

At the time of publication, U.S. energy policy is rapidly evolving with significant new incentives for renewable energy development. The Recovery Act, signed by President Barack Obama on February 13, 2009, includes over $42 billion for energy programs and more than $21 billion in energy tax incentives, primarily for energy efficiency and renewable energy. The GTP received $400 million, a substantial portion of DOE's Recovery Act funds devoted to efficient and renewable energy technologies. GTP will now have more capacity to implement the major provisions of the 2007 Energy Independence and Security Act. The GTP will distribute the Recovery Act funding to partners in industry and academia through competitive awards broadly focused on EGS development, geothermal component R&D, low-temperature geothermal resources, innovative exploration techniques and geothermal heat pumps. Industry cost share will further increase investment, multiplying benefits to technology development.

The Recovery Act also includes enhanced tax provisions that provide assistance to geothermal power developers. The law extends the Renewable Energy PTCs for geothermal facilities put in place before January 1, 2013, which had been allowed to lapse, slowing industry investment. It also provides the opportunity for geothermal developers to take advantage of the ITC in lieu of the PTC when desirable, and allows the Department of the Treasury to offer grants in lieu of the tax credits. These revisions provide additional flexibility for geothermal developers. The grants are likely to be a more effective means of financing renewable energy projects since current economic conditions have largely eliminated tax equity financing as an option for developers.

GHPs also are likely to receive a big boost from the Recovery Act, which not only extended residential and commercial tax credits, but also removed a $2,000 cap that existed under EESA. Residential customers may claim a tax credit up to 30% of the installed cost of their GHP systems, and commercial customers may receive up to 10%. Other elements, such as accelerated depreciation, were also extended. Substantial funds from the Recovery Act have also been allocated to other offices within DOE, other Federal agencies, and channeled to state and local governments to improve building energy efficiency, and stimulate green jobs creation and economic growth. This new funding may also directly benefit GHPs and geothermal energy development. Some of the more notable Recovery Act funding provisions include:

- More than $11 billion is provided in grants for state and local governments through the Department of Energy's Weatherization Assistance Program, which provides energy efficiency services to low-income households; the State Energy Program, which provides states with discretionary funding for energy efficiency and renewable energy projects and programs; and the new Energy Efficiency and Conservation Block Grant Program, which seeks to limit energy use and greenhouse gas emissions. Several jurisdictions have already devoted funding through these grant programs to a variety of renewable and efficient technologies, including geothermal heat pumps.

- Approximately $8.8 billion was allocated to the Department of Education, to renovate schools and university campuses according to green standards.
- The Recovery Act sets aside $3.7 billion for energy efficiency within the Department of Defense's substantial building stock. The Department has previously been an active supporter of geothermal heat pump technology use across its facilities.
- The Departments of the Interior and Veterans Affairs both received $1 billion in multi-purpose funds that can be dedicated to renewable energy and energy efficiency projects and upgrades to their facilities.[64]

As of May 29, 2009, major legislation is currently moving through the U.S. Congress that has the potential to significantly change the way energy is produced and consumed in the United States. The American Clean Energy and Security Act of 2009 (H.R. 2454) seeks to gradually reduce carbon emissions (by 17% below 2005 levels by 2020) and increase the proportion of energy that comes from renewable sources in the United States (20% by 2020). While specific targets will likely change as the bill undergoes revisions in various congressional committees in the coming months, if enacted, it will undoubtedly lead to increased investment and deployment of renewable energy technologies in the United States.

REFERENCES

American Council for an Energy-Efficient Economy (2007). *Emerging Technologies Report - Residential Ground-Source Heat Pumps*, p. 3.

Bertani, Ruggero (2007). *World Geothermal Generation in 2007*. pp. 8-19 GHC Bulletin September 2007.

Cambridge Energy Research Associates (2008).

Commission of the European Communities. (2008). *Proposal for a Directive of the European parliament and of the council on the promotion of the use of energy from renewable sources*. Brussels, 23.1.2008. Download at: http://ec.europa.eu/energy/climate_actions/doc/2008_res_directive_en.pdf

Council of the European Union. (2007). Presidency conclusions – Brussels, 8/9 March 2007 (7224/1/07 REV 1) Section III; An integrated climate and energy policy. p. 10-14. Download at www.consilium.europa.eu/ueDocs/cms_Data/docs/pressData/en/ec/93135.pdf

Davidson, Paul (2008). *Declining energy prices extend to electricity*. USA TODAY December 17, 2008.

Database of State Incentives for Renewable Energy (DSIRE) website: www.dsireusa.org

Energy Information Administration, U.S. Department of Energy (2009). Electric Power Monthly, March 2009. www.eia.doe.gov/cneaf/electricity/epm/epm_sum.html

Energy Information Administration, U.S. Department of Energy (2009). Geothermal Heat Pump Manufacturing Activities 2007. www.eia.doe.gov/fuelrenewable.html

European Geothermal Energy Council (2008). Research Agenda for Geothermal Energy – Strategy 2008 to 2030. www.groundreach.eu/script/tool/forg/ doc795/ EGEC%20RESEARCH%20AGENDA%202009.pdf

Feo, Ed of Milbank Tweed Hadley & McCloy LLP (2008). What's Hot in Renewable Energy Project Financing. *North American Clean Energy, Vol. 2*, Issue 1.

Fleischmann, Daniel j. (2007). *Getting Geothermal Electricity Projects Online*, Presentation by ORMAT Nevada, Inc., July 23, 2007.

Gawell, Karl, Geothermal Energy Association (GEA) (2009). Personal Communication.

Geothermal Energy Association (2005). *Geothermal Industry Employment: Survey Results and Analysis*. p. 2.

Geothermal Energy Association, U.S. Geothermal Power Production and Development (March 2009) www.geo-energy.org/publications/reports/Industry_Update_ March_Final.pdf

Geothermal Energy Association (2008). Geothermal Revenue Under the Energy Policy Act of 2005: Income Distribution at Federal, State and County Levels. p. 11-12.

Geothermal Resource Council (2009). Assessment of the Price of Geothermal Power. *North American Clean Energy Magazine*, *Vol. 3* Issue 1.

Geothermal Technologies, U.S. Department of Energy website: www1.eere.energy.gov/ geothermal/

Geothermal Technologies Program, U.S. Department of Energy. BLM Offers Geothermal Leases in Utah, Idaho and Oregon. www1.eere.energy.gov/geothermal/ news_detail.html?news_id=12113

Geothermal Technologies Program, U.S. Department of Energy DOE Funds 21 Research, Development and Demonstration Projects for up to $78 million to Promote Enhanced Geothermal Systems. www1.eere.energy.gov/geothermal/news_detail.html?news_ id12018

Google.org website: www.google.org/egs/index.html

Ground Reach website: www.groundreach.eu

Hughes, Patrick (2008). Geothermal (Ground-Source) Heat Pumps: Market Status, Barriers to Adoption, and Actions to Overcome Barriers. Oak Ridge National Laboratory (ORNL/TM-0000/00).

International Partnership for Geothermal Technologies website: http:// internationalgeothermal.org

jason Gold interview, January 9, 2009.

Johnson, Katherine and Thomas, Ed. (2008). Helping Utilities and Customers Quantify the Benefits of Geothermal Heat Pumps. Market Development Group. Paper presented at IEA Heat Pump Conference, 2008.

Khan, Ali (2008). The Geysers Geothermal Field, an Injection Success Story. Geothermal Technologies Program – 2008

Lazard, (2008). www.narucmeetings.org/Presentations/2008%20EMP%20Levelized%20Cost %20of%20Energy%20-%20Master%20June%202008%20(2).pdf

Le Feuvre, P., Kummert, M., (2008). Ground Source Heat Pumps in the UK – Market Status and Evaluation. In: Proc. 9th Annual IEA Heat Pump Conference 2008.

Lovekin, james, et al. (2006). Potential improvements to existing geothermal facilities in California. GRC Transactions, *Vol. 30* pp. 885-890

Lund, john (2007). Characteristics, Development and Utilization of Geothermal Resources. Geo-Heat Center, Oregon Institute of Technology, GHC Bulletin, p.6.

Lund, john, et al. (2005). Direct application of geothermal energy: 2005 Worldwide review. Geothermics 34 pp.691-727

Martin, Keith (2008). Geothermal Deal Structures. Chadbourne & Parke, LLP. November 2008 Geothermal Investment Conference.

Mostow, P. and Braff, A (2008). Geothermal Site Acquisition and Early Development: Key Legal Issues and Emerging Strategies. Presented at Geothermal Energy 2008

New Energy Finance (2008). Meltdown: How Iceland's fall is impacting geothermal.

New Energy Finance (2009). Direct communication with Mark Taylor.

New Energy Finance (2009). To Drill or Not to Drill: Geothermal and the Credit Crunch.

Raser Technologies website: www.rasertech.com/geothermal/projects-in-development

Shepherd, William (2006). Energy Studies.London. pp. 206; Imperial College Press, 2003

Silcoff, Sean (2008). Geothermal Energy Stocks Should Recover Steam If Government Support Lasts. The Globe and Mail, October 28, 2008.

Sissine, Fred, et al. (2009). Energy Provisions in the American Recovery and Reinvestment Act of 2009. Prepared by the Congressional Research Service.

Tester, j., et al. (2006). The Future of Geothermal Energy: Impact of Enhanced Geothermal Systems (EGS) on the United States in the 21st Century.

Tharp, Ian and Mahesh, Sumeet (2008). Geothermal Energy – An Overnight Success in 104 Years. Dundee Capital Markets March 17, 2008, p. 27.

US Renewables Group (2007). Geothermal Investment: An Equity Provider's Perspective, Geothermal Investors' Forum, October 2007.

United States Geological Survey, U.S. Department of the Interior (2008). Assessment of Moderate- and High-Temperature Geothermal Resources of the United States

WaterFurnace Investors Presentation, November, (2008), www.waterfurnance.com

Woody, Todd (2008). Credit Crunch Darkens Solar's Prospects. Fortune, October 24, 2008. Via CNN.com

On the Cover

Electricity generated from U.S. geothermal sources, such as the Desert Peak geothermal field in Nevada, reached 15 billion kilowatt-hours in 2008. Courtesy of Ormat Technologies Inc.

End Notes

[1] Shepherd, William, "Energy Studies." London: Imperial College Press, (2003), p. 206.

[2] Martin, Keith, "Geothermal Deal Structures", Chadbourne & Parke, LLP, 2008 Geothermal Investment Conference, (November 2008).

[3] U.S. DOE Geothermal Technologies Program, Press Release, (2008).

[4] Lund, John, Oregon Institute of Technology, Geo-Heat Center, (2007).

[5] Geothermal Technologies Program, U.S. Department of Energy: "DOE Funds 21 Research, Development and Demonstration Projects for up to $78 million to Promote Enhanced Geothermal Systems." http://www1.eere.energy.gov/geothermal/news_detail.html?news_id12018.

[6] Google.org, http://www.google.org/egs/index.html

[7] Geothermal Technologies Program, U.S. Department of Energy: "BLM Offers Geothermal Leases in Utah, Idaho and Oregon." http://www1.eere.energy.gov/geothermal/news_detail.html?news_id=12113

[8] New Energy Finance, "Meltdown: How Iceland's Fall is Impacting Geothermal", (November 2008).

[9] New Energy Finance, "To Drill or Not to Drill: Geothermal and the Credit Crunch", (January 6, 2009).

[10] Personal Communication with Karl Gawell, Executive Director, GEA

[11] Electric Power Monthly, EIA, (March 2009).

[12] International geothermal capacity figures for 2008 are not scheduled to be released until late 2009. Source: Bertani, R. "World Geothermal Generation in 2007", (September 2007)

[13] United States Geological Survey, "Assessment of Moderate- and High-Temperature Geothermal Resources of the United States", (2008).

[14] Geothermal Energy Association, "U.S. Geothermal Power Production and Development", (March 2009)

[15] Ibid.

[16] Tester, J., et al., Massachusetts Institute of Technology, "The Future of Geothermal Energy: Impact of Enhanced Geothermal Systems (EGS) on the United States in the 21st Century", (2006).

[17] U.S. DOE Geothermal Technologies Program, (2008)

[18] Ibid.

[19] Ibid.

[20] North American Clean Energy Magazine, "Assessment of the Price of Geothermal Power", Volume 3, Issue 1. (2009) ($4,000/kW) Lovekin, J, et al., "Potential improvements to existing geothermal facilities in California", (2006) ($3,000-$3,500/kW) New Energy Finance, Direct Communication with Mark Taylor ($3,600-$3,900/kWh).

[21] Sison-Lebrilla, Elaine and Valentino Tiangco, "Geothermal Strategic Value Analysis", California Energy Commission Report #CEC-500-2005-1 05-SD, (June 2005), pgs.12-17.

[22] Davidson, Paul "Declining energy prices extend to electricity", USA TODAY, (December 17, 2008).

[23] "To Drill or Not to Drill", New Energy Finance, (January 6, 2009), p.1.

[24] "Geothermal Deal Structures", Presentation by Keith Martin, Chadbourne & Parke, LLP, Geothermal Financing Summit 2008.

[25] "What's Hot in Renewable Energy Project Financing" by Ed Feo, Milbank Tweed Hadley & McCloy LLP, published in North American Clean Energy (www.nacleanenergy.com), Volume 2, Issue 1, (2008).

[26] "Geothermal Deal Structures", Presentation by Keith Martin, Chadbourne & Parke, LLP, Geothermal Financing Summit 2008.

[27] Ibid.

[28] "Geothermal Energy Stocks Should Recover Steam If Government Support Lasts", The Globe and Mail – Globe Investor Magazine, October 28, 2008.

[29] "Geothermal Deal Structures", Presentation by Keith Martin, Chadbourne & Parke, LLP, Geothermal Financing Summit 2008.

[30] "To Drill or Not to Drill: Geothermal and the Credit Crunch", New Energy Finance, January 6, 2009, pgs. 1-2.

[31] "Credit Crunch Darkens Solar's Prospects", Fortune, October 28, 2008. Via CNN.com.

[32] "Geothermal Deal Structures", Presentation by Keith Martin, Chadbourne & Parke, LLP, Geothermal Financing Summit 2008.

[33] "Geothermal Energy – An Overnight Success in 104 Years", Dundee Capital Markets, Ian Tharp CFA, March 17, 2008, p. 27.

[34] Jason Gold interview, January 9, 2009.

[35] New Energy Finance, "To Drill or Not to Drill: Geothermal and the Credit Crunch", January 6, 2009, p. 3.

[36] Lazard Investment Bank, (2008). Download at: http://www.narucmeetings.org/Presentations/2008%20EMP%20Levelized%20Cost%20of%20 Energy%20-%20Master%20June%202008%20(2).pdf

[37] Ibid.

[38] Database of State Incentives for Renewable Energy (DSIRE). Available at: www.dsireusa.org

[39] U.S DOE EERE Network News, 2008

[40] Mostow, P. and Braff, A., "Geothermal Site Acquisition and Early Development: Key Legal Issues and Emerging Strategies", Presented at Geothermal Energy 2008.

[41] U.S. Department of the Interior, Bureau of Land Management, "Record of Decision and Resource Management Plan Amendments for Geothermal Leasing in the Western United States", (December 2008).

[42] Lund et al, "Direct application of geothermal energy: 2005 Worldwide review", Geothermics 34 (2005) 691-727

[43] Characteristics, Development and Utilization of Geothermal Resources, John W. Lund, Geo-Heat Center, Oregon Institute of Technology, GHC Bulletin, June 2007, p.6.

[44] Lund et al, "Direct application of geothermal energy: 2005 Worldwide review", Geothermics 34 (2005) 691-727

[45] WaterFurnace Investors Presentation, (November, 2008). Available at www.waterfurnance.com.

[46] Personal communication with ClimateMaster staff, (April 2009).

[47] Holihan, Peter, "Analysis of Geothermal Heat Pump Manufacturers Survey Data", Energy Information Administration, Renewable Energy 1998: Issues and Trends, p.59

[48] The American Council for an Energy-Efficient Economy, "Emerging Technologies Report: Residential Ground-Source Heat Pumps July", (2007), p. 3.

[49] Hughes, Patrick, "Geothermal (Ground-Source) Heat Pumps: Market Status, Barriers to Adoption, and Actions to Overcome Barriers, Oak Ridge National Laboratory, Report # ORNL/TM-2008/232 (December 2008), p. 17.

[50] Ibid.

[51] Personal communication with ClimateMaster staff, (April 2009).

[52] Database of State Incentives for Renewable Energy (DSIRE). Available at: www.dsireusa.org.

[53] Johnson, Katherine and Ed Thomas, "Helping Utilities and Customers Quantify the Benefits of Geothermal Heat Pumps", Market Development Group. Paper presented at IEA Heat Pump Conference, (2008).

[54] Ibid.

[55] Le Feuvre, P., Kummert, M., "Ground Source Heat Pumps in the UK – Market Status and Evaluation." Proceedings of the 9th Annual IEA Heat Pump Conference, (2008).

[56] Council of the European Union, Presidency conclusions 7224/1/07 REV 1: III. An integrated climate and energy policy, (2007) p. 10-14. Download at: http://www.consilium.europa.eu/ueDocs/cms_Data/docs/pressData/en/ec/93135.pdf

[57] "Proposal for a Directive of the European Parliament and of the Council on the Promotion of the Use of Energy from Renewable Sources", Brussels, (January 23, 2008) Download at: http://ec.europa.eu/energy/climate_actions/doc/2008_res_directive_en.pdf

[58] www.groundreach.eu

[59] "Research Agenda for Geothermal Energy – Strategy 2008 to 2030", European Geothermal Energy Council (EGEC), (2008). Download at: http://www.groundreach.eu/script/tool/forg/doc795/EGEC%20RESEARCH%20AGENDA%202009.pdf

[60] Ibid, p.3.

[61] Form EIA-902, "Annual Geothermal Heat Pump Manufacturers Survey", Released February, 2009.

[62] "WaterFurnace Renewable Energy Is Poised to Contribute to Economic Recovery and Long-Term Energy Goals", Reuters, (Feb 16, 2009). Article's figures based on an internal WaterFurnace study.

[63] "Geothermal Revenue under the Energy Policy Act of 2005: Income Distribution at Federal, State and County Levels." GEA, (2008). pgs. 11-12.

[64] Sissine, Fred et. al., Congressional Research Service, "Energy Provisions in the American Recovery and Reinvestment Act of 2009", (2009).

In: Geothermal Energy: The Resource Under Our Feet
Editor: Charles T. Malloy
ISBN: 978-1-60741-502-2

Chapter 4

GEOTHERMAL TOMORROW - 2008*

United States Department of Energy

DOE GEOTHERMAL TECHNOLOGIES PROGRAM VISION AND MISSION: A LETTER FROM THE PROGRAM MANAGER

Dear Readers and Colleagues

I am delighted to have recently taken over the Geothermal Technologies Program (the Program) in the U.S. Department of Energy's Office of Energy Efficiency and Renewable Energy. I join the Program at a very exciting time; it has been awarded a $30 million dollar budget from the Senate and $50 million from the House for the 2009 fiscal year after a two year struggle with limited funds. The budget increase demonstrates a tremendous surge of public support for geothermal energy and speaks volumes about the fantastic work the Program has accomplished. With this year's increased funding I am certain that the Program will continue to grow exponentially.

Our team has produced some great resources for the geothermal industry. We are currently developing a Multi-Year Plan to outline specific goals to accelerate commercializa tion. We are also producing a National Geothermal Database that will catalog temperature, depth, seismicity, hydropressure, and permeability throughout the United States. This will be a great resource that will mitigate risk and facilitate investment. This is going to be an asset for the geother mal community—so feel free to let us know what you feel should be included in the National Geothermal Database.

The Program has refocused on Enhanced Geothermal Systems (EGS) —a technology we see as the future of not just the geothermal industry, but of the renewable energy industry as a whole. Natural geothermal systems depend on three factors to produce energy: heat, water, and permeability. Although heat is present virtually everywhere at depth, water and permeability are less abundant. Previously, geothermal energy sources were limited to sites

* This is an edited, reformatted and augmented version of a U. S. Department of Energy publication dated September 2008.

where all three of these factors were favorable. EGS, however, consist of engineered reservoirs created to produce energy from geothermal resources deficient in economical amounts of water and/or permeability. With EGS, we can transform geothermal from an energy source useful only in a select few areas to a viable source of base load energy for much of the United States.

An EGS is created by first digging a well into hot basement rock, then injecting water into the well at a high pressure so as to promote fracturing deep within the rock, creating a reservoir. A second well is drilled to intersect the resulting reservoir, allowing water to be circulated so this heat can be extracted. Multiple wells may be drilled into the field to increase yield. The resulting energy is clean, renewable, domestic, reliable, and may be used as a source of base load power. DOE sponsored a study completed by experts from the Massachusetts Institute of Technology which found that 100,000 megawatt electrical (MWe) of geothermal energy could be produced in the United States by 2050 with sig nificant investment in EGS technologies.

Despite great potential, geothermal energy faces several barriers to growth. These issues include limited geothermal siting opportunities, inadequate technology, and high startup costs. It is my goal to lead the Program in mitigating these barriers and in developing technology that will allow geothermal resources to be explored in regions previously, and erroneously, deemed unsuitable. It is my hope that geothermal energy will no longer be tethered to a small number of naturally occurring sites, but that we will lead in finding the potential wherever it may exist. Additionally, we are working hard to develop strategic partners in government and industry to help accelerate commercialization opportunities in EGS technologies that will lower costs and expand potential. Partnered with industry and academia, the Pro gram will continue to promote this great, clean, renewable, domestic resource: geothermal energy.

Desert Peak Geothermal Plant, 65 miles northeast of Reno, Nevada.

Currently, the United States has 2,930 MWe of installed geothermal capacity and about 2,900 MWe of new geothermal power plants under development in 74 projects. Geothermal energy generated 14,885 gigawatt-hours (GWh) of electricity in 2007, which accounted for 4% of renewable energy-based electricity consumption in the United States (including large hydropower). These statistics are a great start, but we need to think bigger. We need to move beyond incrementalism and start considering growth of hundreds of megawatt units rather than 5-10 MWe.

Policy is an important tool for getting more geothermal energy into the market. We need technology-neutral, carbon-weighted, long-term incentives that account for externalities and level the playing field for renewables in today's competing power markets. The Program is working with Western Governors to support new energy corridors that will bring clean, renewable energy resources into our energy portfolio. EERE is currently evaluating options to support the development of transmission infrastructure that will further remove geothermal investment barriers.

As high energy costs and environmental concerns force us to reevaluate our energy use, the nation looks for solutions and alternatives. There is no silver bullet that will solve our problems, but an integrated portfolio of energy alternatives will help the nation to successfully navigate the coming years. Geothermal energy, through the work and support of the Program, has the promise to play a critical role in the solution. We have entered a new era in government sup port for geothermal energy. Our new goals are bigger and braver than ever—but we're more confident than we have been in the past. The geothermal industry has reached a turning point—we're going to help restore geothermal to its rightful place in the portfolio of alternative energy sources to help enhance our security, better our environment, and stimulate our economy.

Sincerely,
Ed Wall

Program Manager
Geothermal Technologies Program
U.S. Department of Energy
Energy Efficiency and Renewable Energy

Enhanced Geothermal Systems: A New Strategy for a Renewed Program

A U.S. Department of Energy-sponsored study by a panel of independent experts led by the Massachusetts Institute of Technology (MIT), *The Future of Geothermal Energy* examined the potential of geothermal energy to meet the future energy needs of the United States. The MIT study calculated the tremendous amounts of heat present at depths of 3 to 10 km below the Earth's surface (Figure 1). The panel concluded that geothermal energy could provide 100,000 MW or more in 50 years by using Enhanced Geothermal Systems (EGS).

Heat is naturally present everywhere within the Earth and is inexhaustible for all intents and purposes. Water is not nearly as abundant within the Earth as heat, and most subterranean fluids are derived from surface waters that have seeped into the Earth along porous path-ways such as faults in rock. The permeability of rock—a measure of the ease of fluid flow—results from pores, fractures, joints, faults, and other openings that allow fluids to move. High permeability implies that fluids can flow rapidly through the rock. Permeability and, subsequently the amount of fluids, tend to decrease with depth as openings in rocks compress from the weight of the earth above.

At shallow depths, typically less than 5 kilometers (km), the presence of heat, water, and porous rock can result in natural hot water reservoirs. These hydrothermal reservoirs have impermeable or low-flow boundaries that impede the movement of fluids. Often, hydrothermal reservoirs have an overlying layer that bounds the reservoir and serves as a thermal insulator, allowing greater heat retention. If hydro-thermal reservoirs contain sufficient fluids (water or steam) at high temperatures and pressures, those fluids can be extracted through wells to generate electricity and/or heat.

An alternative to dependence on naturally occurring hydro-thermal reservoirs involves engineering hydrothermal reservoirs in hot rocks for commercial use. This alternative is known as EGS.

The Promise of EGS

To achieve the goals outlined in the MIT study of large scale (100,000 MW) use of cost-competitive geothermal energy, significant advances are needed in site characterization, reservoir creation, well field development and completion, and system operation, as well as improvements in drilling and power conversion technologies. These technology improvements will also support ongoing development and expansion of the hydrothermal industry. To realize the promise of EGS as an economic national resource, researchers will have to create and sustain a reservoir over the economic life of the project.

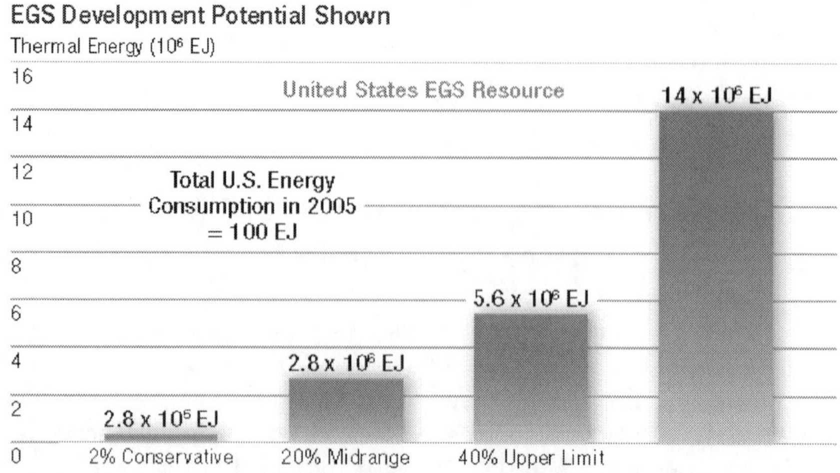

Figure 1. Potential of EGS development in the United States

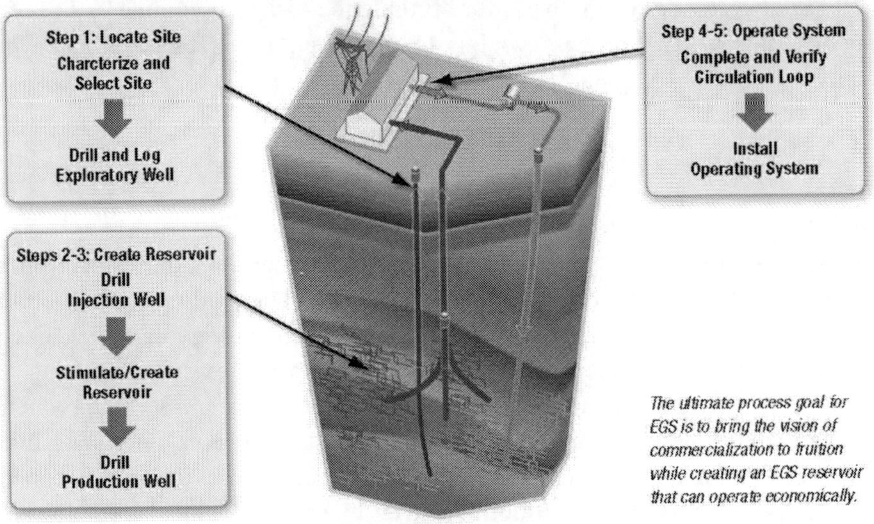

Figure 2. EGS Development Sequence

EGS reservoirs are made by drilling wells into hot rock and fracturing the rock sufficiently to enable a fluid (water) to flow between the wells. The fluid flows along permeable pathways, picking up heat from the rocks, and exiting the reservoir via production wells. At the surface,the fluid passes through power plant turbines where electricity is generated. Upon leaving the power plant, the fluid is returned to the reservoir through injection wells to complete the circulation loop (Figure 2). If the plant uses a closed-loop cycle to generate electricity, none of the fluids vent to the atmosphere. The plant will have no greenhouse gas emissions other than water vapor that may be used for cooling.

EERE Strategy

In order to achieve the maximum potential of EGS, the U.S. Department of Energy's Office of Energy Efficiency and Renewable Energy plans to advance and build from current geothermal technology to develop the sophisticated technologies required, while at the same time generating benefits in the near-, mid-, and long-term. This will require a systematic, sustained research and development effort by the federal government and a strong partnership with industry and academia to ensure full development.

A broad knowledge base about reservoir creation and operation will be essential for the eventual commercialization of EGS on a scale envisioned by the MIT study. This knowledge can be gained only by experience with field demonstrations in a variety of geologic environments reflecting a range of reservoir conditions. Immediate technology improvements are needed in reservoir predictive models, zonal isolation tools, monitoring and logging tools, and submersible pumps. These improvements and others stemming from the evaluation are essential for reaching the long-term potential of EGS.

The MIT study provides a firm basis for bringing the vision of commercialization of EGS technology to fruition. The process goal is to create an EGS reservoir that can operate economically.

EGS Future

The authors of the MIT study based their technical assump tions on results from available field tests, published reports, and well-established theory. The study's findings are cred ible, in particular the conclusion that 100,000 MW from EGS technology can be achieved within 50 years. As the study points out, significant constraints exist in creating sufficient connectivity between wells to meet economic requirements for reservoir productivity and lifetime. Over coming these constraints will require substantial reservoir testing in a number of different geothermal environments as well as research-driven improvements in technol ogy. Investments in excess of over $1 billion over 15 years will most likely be required to encourage sufficient deployment of EGS technology to produce 100,000 MW.

ACTIVITIES IN GEOTHERMAL RESEARCH AND DEVELOPMENT AT THE DOE NATIONAL LABORATORIES

The U.S. Department of Energy (DOE), through its Geothermal Technologies Program (GTP), works closely with the geothermal community to advance geothermal technologies and the U.S. geothermal industry. This includes partnering with industry, universities and colleges, research facilities, and, especially, the DOE national laboratories. Three national laboratories are particularly active in support of GTP's endeavors and have the expertise and technology in place to provide leadership where needed in research and development projects to help realize the full potential of geothermal energy. In this article, National Renewable Energy Laboratory, Lawrence Livermore National Laboratory, and Idaho National Laboratory review their activities and future plans in support of GTP.

Geothermal Technologies Program Activities at the National Renewable Energy Laboratory

The National Renewable Energy Laboratory (NREL) supports DOE in renewing and expanding its efforts to move geothermal energy forward. NREL staff has participated in and continues to lead and participate in activities related to Enhanced Geothermal Systems (EGS) technology validation and integrating outcomes into program plans.

Evaluation workshops

NREL was an active participant and leader in a series of workshops in 2007 that evaluated the assumptions, analytical methods, and conclusions presented in the 2006 report by the Massachusetts Institute of Technology, *The Future of Geothermal Energy*. This report provides a strong argument that with appropriate and reasonable investment from the public and private sectors, geothermal energy could make a substantial contribution to the nation's energy portfolio.

The workshops included discussion on reservoir creation, reservoir management and operations, and well field con struction, and led to the publication of the DOE report, *An Evaluation of Enhanced Geothermal Systems Technology.* These efforts set the stage for the renewal of the geothermal program with an emphasis on EGS development.

With this renewal, NREL has initiated 13 projects to help GTP reach its objectives. These projects support systems demonstration projects and meet program needs in systems integration, energy analysis, strategic planning, field pro gram support, and communications and outreach.

Systems demonstration support

For GTP's field demonstrations of EGS, NREL is contributing expertise in developing site selection criteria and solicitation strategies, as well as participating in the industry proposal merit reviews. NREL provides oversight and technical monitoring of field projects and management of field-related subcontracts.

Installation of a PPS-coated steam vent at Cove Fort in Utah

Strategic planning and analysis

NREL supports GTP in developing strategic plans, including the *Geothermal Technologies Program, Multi-Year Research, Development, and Demonstration Plan 2009-2015 , with Program Activities to 2025*. This plan addresses the near-term priorities for cost-shared research with indus try, as well as field projects for achieving EGS technology readiness for commercialization. NREL has participated in several internal program meetings to develop the plan and has the lead in developing the key technical section. This plan will guide GTP in fulfilling aggressive goals for making geothermal a significant contributor to the energy portfolio of the United States. NREL is also assisting the program in developing a GTP Management and Operations Plan to achieve EGS technology readiness for commercialization.

National geothermal action plan

NREL supports GTP in developing the National Geothermal Action Plan. The purpose of this document is to further the development of geothermal energy across the nation, beyond conventional geothermal fields. This Plan is intended to inform policy makers, industry executives, potential investors, and others who seek insight into the potential of generating clean, baseload energy from geothermal resources.

Program systems integration support

NREL has initiated its systems integration role in GTP's activities to help it reach the goal of developing EGS. NREL is developing an integrated baseline that will address technical scope, and will define, initiate, and man age systems-related subcontracts and integrate with in house efforts.

Analysis, evaluation, and modeling

Of the 13 projects NREL is initiating in support of GTP, seven are focused on analysis, evaluation, and modeling activities that address critical program needs for assessment of geothermal and EGS markets:

- Macro modeling of the potential geothermal energy contribution
- Techno-economic modeling of EGS
- Analyses of program risk
- Integrated energy modeling for budget support
- Analysis of geothermal CO_2 impact
- Assessment of power conversion technologies
- Assessment of data requirements for accelerating EGS commercialization.

NREL analysis is working to improve the representation of geothermal to address renewable energy technologies in evolving new energy market models. There are many energy market models that have been used to assess the potential of geothermal power and other technologies in the United States, but two shortcomings remain that are espe cially acute for geothermal power: 1) Regional aggregation doesn't allow for the consideration of more local transmis sion constraints; and 2) uncertainties in future fuel prices, technology improvements, and policies will continue to drive the energy sector.

Keith Gawlick, of the National Renewable Energy Laboratory, field testing coatings at the Mammoth Pacific Geothermal Power Plant in the Sierra Nevada Mountains

A model now exists at NREL for capacity expansion in the U.S. electric sector—the Regional Energy Deployment System—with more than 350 regions in the United States that explicitly consider transmission issues. A second model—the Stochastic Energy Deployment System—is under development and led by NREL with a team from six national labs that explicitly addresses future uncertainties in technology performance, cost, fuel prices, and policies.

Presently, a rudimentary representation using supply curves for hydrothermal power is included in each of these models. This task will improve that representation and conduct analysis of geothermal power market potential within the models.

Geothermal market, policy, and technology analysis

NREL developed the initial concept for gaining a firm understanding of the technical, economic, and market potential of all geothermal technologies (hydrothermal, EGS, heat pumps). Such an understanding is required to inform decision-makers in the identification of the most efficient use of resources. Both historical and projected metrics will be gathered to determine technology improvements and commercialization opportunities. The project will conduct analyses of market, policy, and technology status by evaluating of the impacts of research and testing options. Additionally, results of the analyses will provide information to researchers, policy-makers, and investors on areas to target for greater cost-reduction and market transformation.

Geothermal CO_2 impact analysis

This NREL project is assessing the CO_2 impact of deploying geothermal energy, specifically for EGS. The project objective is to assess the projected CO_2 impact of geothermal generation in general, and the component estimated to be the result of planned program activities, based on published studies and modeling of the geothermal representation used for the fiscal year 2010 benefits estimation process. CO_2 emission reduc tion is a key element of DOE's strategic environmental goal and an increasingly important metric for assessing the value of program activities in the budget formulation process.

Power conversion technology evaluation

NREL is also supporting GTP's EGS field experiment program by evaluating the current state of power conversion technology and assessing R&D requirements in this area. NREL is performing a detailed assessment of the needs for EGS power conversion and evaluating the ability of current technologies to meet those needs with the primary purpose of identifying gaps in technology that must be addressed for long-term EGS viability.

Growth to Assist and Represent GTP

To meet the needs of the renewed geothermal program, NREL has transitioned to a new geothermal technology manager with a strong geoscience and energy background who is adding positions to the laboratory's capabilities in support of GTP. NREL has added two energy analysts and a systems integration engineer. Staff have provided support to GTP by making or participating in presentations to key stakeholders in the geothermal community including Al Gore and the Alliance for Climate Protection, the Google Foundation, the X-Prize Foundation, and the leaders of the Hawaii Energy Initiative.

Geothermal Technologies Program Activities at Lawrence Livermore National Laboratory

Lawrence Livermore National Laboratory (LLNL) has a long history of R&D work in support of geother mal power. Key areas of research include advances in scaling and brine chemistry, economic and resource assessment, direct use, exploration, geophysics, and geo-chemistry. For example, a high-temperature, multi-spacing, multi-frequency downhole electromagnetic (EM) induction logging tool (GeoBILT) was developed jointly by LLNL and EMI to enable the detection and orientation of fractures and conductive zones within a reservoir. LLNL researchers also conducted studies on the use of geothermal energy for desalination to stave off increased salinity in the Salton Sea, an important aquatic ecosystem in California.

Since 1995, funding for LLNL's geothermal research has decreased, but the program continues to make important contributions to sustain the nation's energy future. Current efforts, as well as future research, focus on developing Enhanced Geothermal Systems (EGS) and improving tech nologies for exploration, monitoring, characterization, and geochemistry.

Techniques to assess geothermal resources

Most known geothermal resources in the Basin and Range geological province of the western United States are associated with active fault systems. Studies show that hydrother mal fluids in active fault systems circulate from deep under ground through high permeability fractures to relatively shallow levels where they can be accessed for production. For example, at the Dixie Valley field, hydraulically conductive fractures within the Stillwater fault zone are oriented so that fractures are critically stressed for normal shear failure under the regional tectonic stress field.[1] In general, the expectation is that geothermal resources occur in areas where seismic strain across faults is extremely high and where faults are favorably oriented with respect to the regional strain tensor. In the Basin and Range, these faults would strike perpendicular to the direction of maximum extension. Geothermal resources may also occur in areas where fault-normal extension associated with shear strain is the greatest.[2]

Until recently, mapping ground displacements of less than 1 centimeter (cm) was extremely difficult. LLNL is apply ing a new technique called repeat-pass Interferometric Synthetic Aperture Radar (InSAR), and refining it for geothermal applications. InSAR uses radar imaging of Earth's surface to identify potential geothermal resources. Satellite- borne synthetic aperture radar images the Earth's surface during two orbits, recording data at the surface position and using the same viewing geometry during both orbits. The maximum separation (spatial baseline) between the two orbital positions is generally 1 kilometer (km) or less, depending on the radar frequency. The difference between the phases of the two radar returns is proportional to any change in the range from the ground to the radar caused by a subsurface displacement that occurs between orbits. The topographic contribution is subtracted using a digital elevation model or additional orbits. The displacement contributions are then mapped over the entire radar scene to produce a phase difference map, or interferogram, that can be converted to a range-change map (Figure 1). Under favorable conditions, InSAR can measure displacements as small as a few millimeters. Displacement maps of geothermal regions using InSAR detect changes in elevation that can be used to manage geothermal systems and locate regions of high strain that are favorable for drilling.

GeoBILT EM Induction logging tool being deployed at Dixie Valley

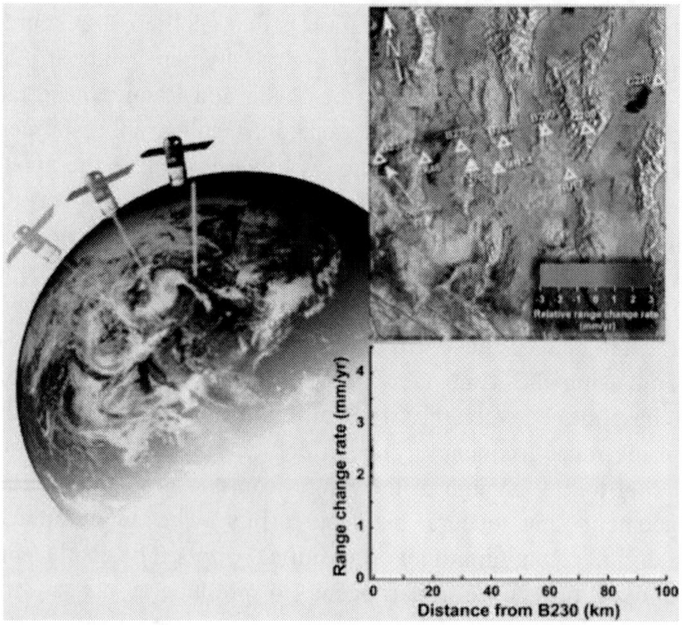

Figure 1. InSAR is used to investigate the role strain concentration plays in localizing geothermal resources in the western Basin and Range

The stochastic engine

Geophysical data are difficult to acquire and, once obtained, are often hard to interpret. Computer models can be made more meaningful when they take the uncertainty in those measurements into account and combine multiple measurements into one analysis. Stochastic models often require large numbers of calculations to evaluate many different descriptions of a problem. These evaluations allow us to understand how a small amount of data can result in a range of interpretations. Using high-performance super-computers such as Thunder and Atlas, LLNL scientists explore groundbreaking ideas in statistical theory to develop quantitative stochastic descriptions that provide a more complete picture of the subsurface.

This technology, called a stochastic engine, links predictive models, advanced statistical methods, and refined search methods. Using this technology, scientists can incorporate a proposed subsurface configuration into a computer model and produce a geophysical simulation. The simulated result is compared to actual data. If the result is consistent with observed data, it becomes part of the final analysis, leading to a clear understanding of which outcomes are very likely, less likely, and where more information could best be used.

The stochastic engine concept uses techniques developed at LLNL and has been applied to a number of research areas, including environmental remediation, CO_2 sequestration, and geothermal exploration. The power of the sto chastic engine stems from its ability to refine a model by successively narrowing possible configurations of a hypothetical model with the refinement done over progressive layers of data.

For example, suppose an area of interest is known to be composed of seven distinct rock layers that could be either highly fractured or intact. Geophysical measure ment such as EM induction, seismic velocities (Vp/Vs), or gravity of that volume, gives an observed value of 11. The stochastic approach calculates which configurations of rock layers, and in which positions, give values close to 11. Each case with a value near 11 is passed to the next stage of analysis. There, the model will continue to restrict possible configurations but base its decisions on other data types such as water, temperature, or pressure. For the simple case cited here, scientists can easily compile and compare all possible configurations. For a large area, however, the possibilities are far too numerous and we rely on computational techniques. The stochastic engine helps narrow the solutions by performing an efficient intelligent search through the collection of possible reservoir configurations, rapidly identifying the configurations that most closely match all the data. The stochastic engine is designed to choose system configurations that are consistent with observed data, allowing much more tightly constrained answers than conventional methods. The goal is to find not a single answer, but many answers. The objective is to adapt the stochastic engine to jointly invert multiple geothermal exploration data sets for better defined drilling targets to improve the success rate in finding economic geothermal resources.

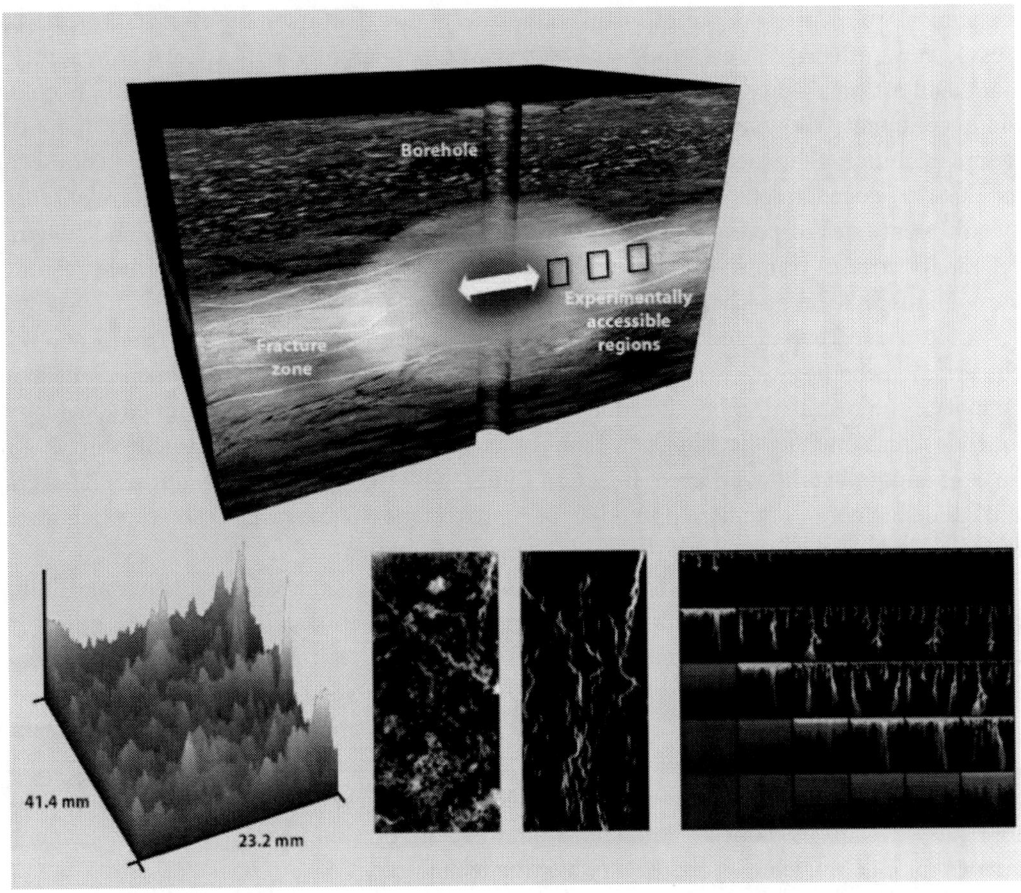

Figure 2. Top: Schematic showing an enhanced geothermal reservoir. Permeability is increased in the hot region, and fluids are pumped into the reservoir. Bottom: Integrated laboratory-scale experimental/computational investigations (fracture aperture, flow streamlines, model showing development of channeling, left-to-right) lead to better models of mechanisms that alter transmissivity in EGS and provide insights into the scaling of important coupled hydraulic/mechanical/chemical/thermal processes that aid in creating and maintaining fracture permeability

EGS—Evolution in Controlling Fracture Permeability

EGS is a technology that can be used to improve the energy recovery from a reservoir that has insufficient permeability or fluid. The use of EGS has the potential to increase geothermal electrical generation to more than 100,000 MWe in the United States by 2050. [3] One technical challenge limiting our ability to utilize enhanced geothermal energy recovery is the changing nature of fracture permeability. Mechanisms such as mineral precipitation and dissolution, flow rate, and stress can affect the underground environment, causing subsurface flow to slow over time or stop completely. EGS research performed at LLNL is aimed at understanding the mechanisms and rate of change to predict the evolution of fracture permeability and to evaluate strategies to enhance and maintain permeability in a given location.

To develop EGS, geophysicists and geochemists in LLNL's Geothermal Program combined laboratory experiments and computer modeling to characterize the hydraulic and geochemical properties of various soil samples (Figure 2). As part of this project, they assessed how effective stress, fluid chemistry, and temperature will affect permeability in natural and artificial fractures. They also used current technologies to analyze data from past field experiments, allowing them to separate the physical and chemical processes that affect fracture evolution. Statistical analysis of fracture apertures for two core samples demonstrated that EGS produced fractures with similar aperture distribution and spatial correlation will have different rates of permeability evolution depend ing on fluid composition and flow rate. Preliminary results from hydraulic mod eling indicate that variations in particle residence times will affect local geochemical reaction rates.

LLNL's expertise in geochemical modeling is criti cal to the success of EGS and other geothermal technologies. These models help researchers interpret experimental data and extrapolate the results to a broader range of expected conditions. For example, geochemical modeling can simulate the physical changes occurring in a fractured system during fluid transport and predict how different injection fluids will affect permeability during the average fluid residence period. Numerical models of representative geothermal reservoirs can also be used to optimize production and maximize reservoir lifetime.

A common problem in geothermal recovery is that minerals such as silica and lithium precipitate through flow channels, reducing fracture permeability. Removing the minerals is an expensive, time-consuming process that limits the use fulness of enhanced geothermal recovery. To improve the long-term effectiveness of an EGS reservoir, researchers need to reduce the costs for maintaining fracture permeability. One approach, developed by a team of LLNL geochemists and an industrial partner, is to extract commodity metals from the reservoir for use in other applications. This work recently led to technology licensing for a proprietary process to convert extracted lithium to lithium carbonate, a key component in batteries for electric vehicles and energy storage technology.

Geothermal research to improve energy security

Energy security is a pressing challenge for the United States—one that offers tremendous opportunity for scientific innovation. Energy recovery through EGS could help reduce the nation's dependence on imported oil. However, more work is needed before EGS can be successfully deployed on a nation-wide scale. In particular, the lifetime of EGS reservoirs is not long enough to make it a cost-effective approach for geothermal recovery. In addition, researchers need improved tools to locate the optimum sites for geothermal production and new technologies to access the energy trapped deep underground.

LLNL offers a unique combination of computational, theoretical, modeling, and experimental capabilities that directly address many of the nation's energy problems, including geothermal energy. LLNL's Geothermal Recovery Program, together with other national laboratories, industry and industrial partners, is building on its past successes in exploration technologies, geochemical analysis, and EGS processes to develop integrated geophysical approaches for geothermal energy production. Future research activities will focus on enabling technologies for better site selection, reservoir management, and EGS.

Geothermal Technologies Program Activities at Idaho National Laboratory

Idaho National Laboratory (INL) currently provides tech nical support to the U.S. Department of Energy's (DOE) Geothermal Technologies Program (GTP) to develop analytical tools that provide insight into how DOE's geothermal research can impact the cost of generating electrical power. This work is being done in support of the Strategic Planning and Analysis Project area that is managed by Arlene Anderson at DOE Headquarters.

Geothermal evaluation model

In 2005, a spreadsheet model referred to as the Geothermal Electricity Technologies Evaluation Model (GETEM) was developed to provide DOE with insight into how its research could affect the cost of producing geothermal energy. Based upon user input, model estimates are developed for costs associated with exploration, well field development, and power plant construction that are used with GETEM's esti mate of operating costs to predict a levelized cost of electricity (LCOE). The model then allows its user to evaluate how technology improvements could impact those projected power generation costs. Results help DOE prioritize research areas and identify where research is needed. The model also aids GTP in conforming to Government Progress and Results Act (GPRA) requirements for annual assessment and report ing of improvements in geothermal electric systems.

GETEM was developed by a team consisting of personnel from DOE, national laboratories, and industry, with the lead role in the development shared by Dan Entingh from Princeton Energy Resources International and Gerry Nix from NREL. A requirement for GETEM's development was that there is a referenced basis of the LCOE projections. Ideally these projections would be based upon actual cost data; unfortunately little actual data is in the public domain and, when available, frequently lacks the detail necessary to adequately characterize both cost and performance. In lieu of actual data, published engineering studies were used to develop the cost and performance correlations used in GETEM. The correlations used to characterize the energy conversion systems were based largely upon the informa tion reported in Electric Power Research Institute's (EPRI) 1995 Next Generation Geothermal Power Plant (NGGPP) study. The correlations for well costs were based on Sandia National Laboratory's analysis of historical geothermal drilling costs. These costs were reported by Mansure at the 2005 Geothermal Resources Council Annual Meeting.

Recent Development

During Fiscal Year 2008 (FY08), DOE has supported modifications to GETEM to address limitations in the model that have been identified since its original development. These changes to the model are largely being made by personnel at INL.

One of the premises in the original development of GETEM's correlations was that the costs reported in the 1995 EPRI NGGPP study were representative of the conversion system costs in 2004. Given the dramatic increases in steel costs that have since occurred, estimates using GETEM's original correlations are no longer representative of commercial plant costs. The model's LCOE projections were further limited by its use of correlations to predict plant performance and cost as functions of the resource temperature only; they did not account for the functional relationship between the plant cost and performance. Plants that are designed to more effi ciently convert geothermal fluid energy into electrical power are more expensive. Generally, more efficient (and expensive) plants are used when a resource has higher well field development costs, while less efficient plants are used with resources that are less expensive to develop.

To address these shortcomings, in 2008, GETEM was modified to improve the conversion system correlations, with a focus on air-cooled binary plants. Based on prior work done at INL, correlations were developed that predict the cost of the plant as a function of the plant performance for a given resource temperature. Prior work at INL also provided an indication of relative contributions of labor and material to the cost of major equipment items. These relative cost contributions provided a means to make adjustments to the predicted equipment costs to account for the changes in costs of fabrication materials and labor with time. This was accomplished by using the Producer Price Index (PPI) and Consumer Price Index (CPI) reported by the U.S. Department of Labor for the different materials and equipment found in the plants. The model can go either forward or backward in time from the reference year costs (2002) using these PPIs to predict equipment and plant capital costs for the desired year.

With the inclusion of the relationship between plant cost and performance, a macro was incorporated into the spread sheet model that varies the plant performance (and cost) until the LCOE is minimized. The model now trades off the additional cost of a more efficient plant with either the additional power that can be produced from a given well field or the reduced well field size (and cost) for a fixed power output. An example of this trade-off is shown in Figure 3. The lower line in this figure shows the plant contribution to the LCOE as a function of its performance. At lower performance levels, the cost of the well field necessary to support the 15 MW plant output increases due to the cost for added wells and increased geothermal pumping requirements (which also affects the plant cost contribution at lower levels of performance). The results in this figure illustrate that the minimal LCOE does not necessarily occur at conditions that produce the minimal plant contribution to generation costs.

These changes to GETEM have been completed along with some initial changes that were made to facilitate the evaluation of EGS resources. A beta version of the model has been distributed to solicit comments and feedback on the reasonableness of the estimates produced.

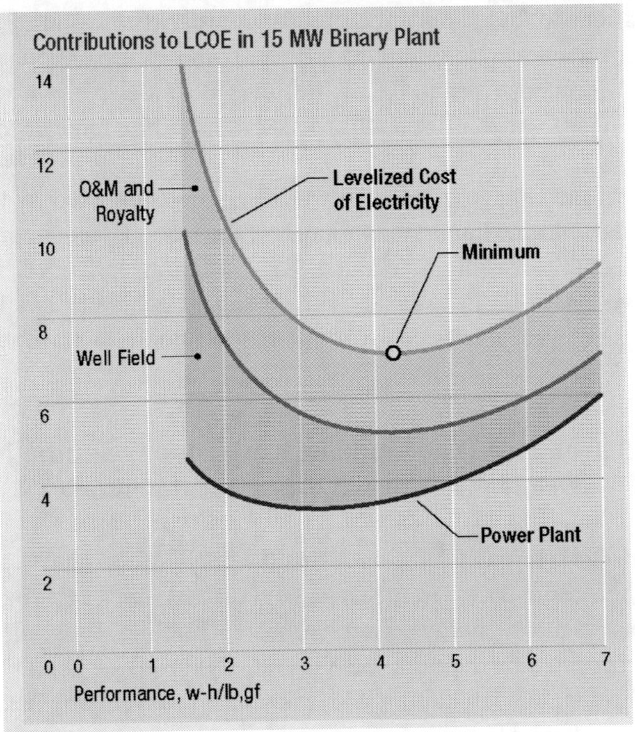

Figure 3. Effect of plant performance on LCOE for a 15 MW binary plant

GETEM's Future

Ongoing efforts are focused on modifying the GETEM model so it better represents the generation of power from EGS resources. Initially, work has focused on developing a model that characterizes the performance of the total system, including a subsurface EGS heat exchange system and production and injection wells. The production well model being incorporated predicts the necessary setting depth for the production pump based on the user's postulated hydraulic performance of the subsurface heat exchange system.

In addition, a simple model is being included for the subsurface heat exchange process. Based on a postulated fracture system, production fluid temperatures are predicted as a function of time and flow. These fluid temperatures are used to estimate the degradation in plant output with time and to establish when it is necessary to replace the EGS reservoir and/or drill additional wells. Options are also being included that allow the user to establish a stimulation cost as a function of the size of the reservoir. This model does not realistically depict a subsurface heat exchange system; however, the trends it predicts are expected to be representative of how different parameters defining the subsurface system will affect produced fluid temperature.

Estimates for pump settings, well flows, production fluid temperatures, and stimulation cost all affect the project cost and the LCOE. Incorporating these parameters into the recent version of the model allows users to vary the different postulated scenarios for the subsurface heat exchange system (including reservoir depth) and assess how these changes affect the power generation costs from EGS resources, as well as hydrothermal resources.

These modifications to the model are currently in progress; it is anticipated that the initial changes (for the binary conversion system) will be completed by the end of FY08. At that time there will be a limited distribution of the revised model for beta testing. Once those changes have been incorporated, efforts will focus on providing a means of updating the well costs.

Although most of the changes described will likely be completed by the end of FY08, as more insight is gained, and the important parameters for EGS development are better defined, further modifications will probably be needed. At some point in 2009 it is expected that the model with the revisions now in progress will be made generally available to the public. Those parties interested in receiving a copy of the model should forward their requests to:

Geothermal Research and Development Activities at Sandia National Laboratory and Lawrence Berkeley National Laboratory

The Lawrence Berkeley (LBNL) and Sandia (SNL) National Laboratories provide interdisciplinary core capabilities in geoscience (LBNL) and geo-engineering (SNL) in support of the Department of Energy's Geothermal Technology Program (GTP). In addition to technical and programmatic support, LBNL and SNL provide the GTP with intellectual continuity and a bridge for facilitating the transfer of developing technologies between academia, industry and the National Laboratories.

The Sandia and Berkeley Laboratories have a long success ful history of working with industry addressing short-term industry needs and long-term R&D efforts. Two very successful examples of short-term R&D efforts include the Geothermal Drilling (GDO) and Geothermal Technology Organizations (GTO) formed to facilitate laboratory and industry collaboration on short-term R&D projects, e.g. the use of well re-drilling technologies to minimize the cost of mitigation and the deployment and interpretation of micro-earthquake (MEQ) sensors to monitor the impact of re-injection into declining resource reservoirs.

Long-term R&D efforts include, but are not limited to: advanced methods to reduce drilling flat-time; new geophysical approaches for imaging the movement of fluids in the subsurface; development of predictive modeling capa bilities for geothermal reservoir management; investigation of alternative heat mining fluids, and development of new geochemical techniques for geothermal exploration. In addressing geothermal program challenges, both Laboratories also leverage research conducted for other sponsors, e.g. the DOE Office of Science, NNSA, other governmental agencies such as DARPA, and private concerns.

LBNL and SNL respective strengths in geoscience and geo-engineering bring a natural synergy to the geother mal program. Their combined strengths assist the DOE Geothermal Technologies Program in understanding and developing solutions to address critical issues and technical barriers associated with access, development, and evolution of geothermal resources; a synergy that will be particularly important in moving EGS technology forward. Fundamental questions in geoscience require the support of similarly motivated engineering activities.

The development of robust engineering tools and methods require a strong scientific basis. Increased fidelity in the ability to monitor seismicity, for example is clearly desirable.

Direct and collaborative interactions between scientists and engineers ensures that efforts to develop the appropriate tools compatible with the EGS environment and technical needs for development will meet the data requirements for current and future development efforts. The synergy between the two laboratories is also evident in areas where capabilities overlap (e.g., numerical modeling and material science); capabilities developed in support of various (DOE and non-DOE) customers can be exploited to provide improved solutions to the DOE Geothermal Technologies Program.

Sampling a warm spring in northern Nevada for gas and isotope analysis as part of a geochemical survey of the northern Basin and Range. Inset shows a geothermal well under test at the Dixie Valley geothermal field.

Geo-engineering core capabilities

Sandia national laboratory

Since the inception of a U.S. geothermal research program, SNL has maintained a core group of researchers and provided continuous support to promote the development of geothermal energy. Research efforts supporting the GTP activities have focused on engineering issues and cost reductions associated with the drilling, completion, and maintenance of geothermal well fields.

Drilling research (well construction)

SNL has maintained a strong program in drilling research which has followed a two-pronged approach: (1) develop technologies to realize incremental reductions in drilling cost, and (2) pursue higher-risk, longer-term R&D on advanced concepts that may ultimately lead to tre mendous reductions in cost. The focus of this work is to reduce drilling costs resulting in lower-cost power-on-line and a more competitive geothermal industry.

Geomechanics

In support of their national security mission SNL has developed and maintained broad based capabilities in geomechanics. These include materials testing in the form of a comprehensive geome chanics laboratory, theoretical modeling and regional-scale numerical simulations of geologic systems using SNL finite element computational software packages and constitutive model development (most recently, the generalized SNL GeoModel for geotechnical applications).

Geoscience core capabilities

Lawrence berkeley national laboratory

The geothermal program, which began at Berkeley Lab in 1973, has made major contributions to geothermal exploration technologies, reservoir characterization and perfor mance evaluation technologies. LBNL's core capabilities reside in three areas: subsurface geophysical imaging, hydrogeology and geochemistry.

Geophysical Imaging—advances the development of new methodologies for extracting subsurface properties, including fluid properties, saturation, porosity, pore pressure, permeability, and in situ stress. These new methodologies incorporate and couple a variety of data types, including geophysical (seismic, electromagnetic, electrical, seismo-electric, gravity, ground-penetrating radar), geomechanical (tilt, deformation), and fluid flow (pressure).

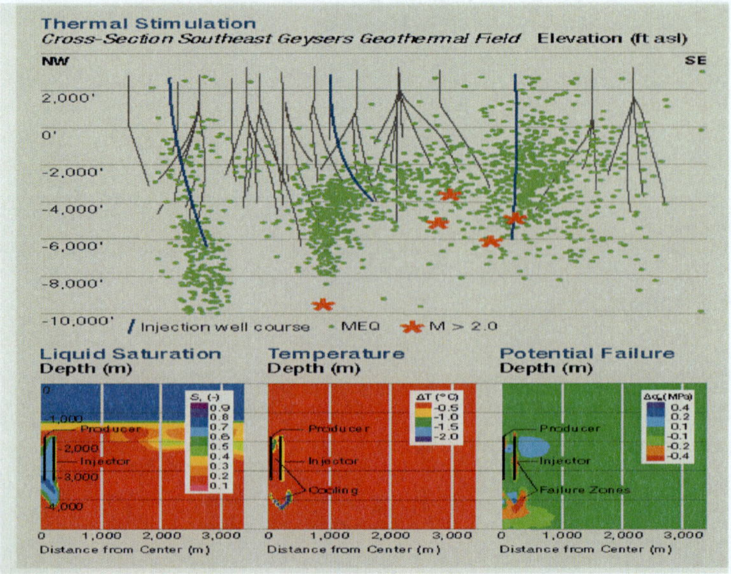

Computational capabilities developed a sophisticated coupled multi-phase, fluid flow, heat transport and mechanical deformation model (TOUGH-FLAC)

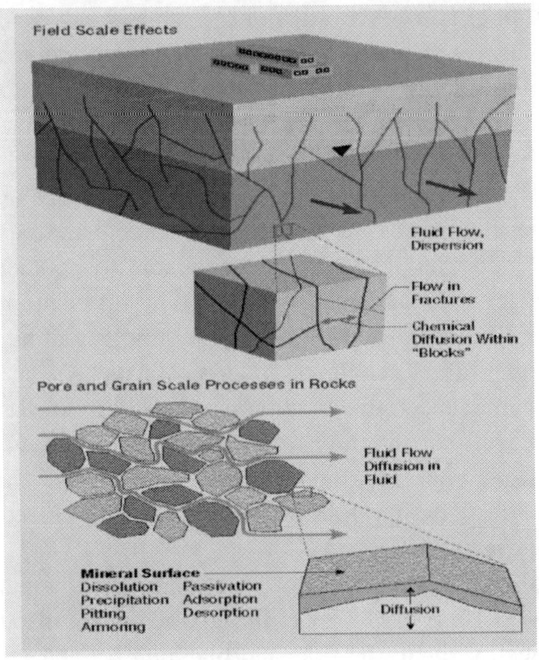

The predictive capabilities of reactive transport models for fractured media encountered in typical geothermal systems will require a thorough understanding of chemical and physical processes occurring over a broad spectrum of length scales.

Hydrogeology— conducts research in fundamental and applied hydrology with expertise in theoretical, experimental, field, and modeling approaches in a variety of research areas. The primary focus for geothermal research and development is on coupled non-isothermal, geochemi cal, and geomechanical processes. LBNL has developed a number of numerical modeling tools to study the response of EGS systems to fluid extraction and injection and predict changes in pressure, temperature, stress and fluid chemistry under different development scenarios.

Geochemistry—develops geochemical and isotopic tracers that provide constraints and information on fluid flow paths, flow rates and the rate of heat and mass transfer between fluids and reservoir rocks along fluid flow paths; advances reactive transport modeling enabling reliable prediction and quan tification of the impact reservoir stimulation will have on short and long-term permeability and the rheology of the reservoir; improves understanding of geochemical processes facilitating reservoir creation and mitigation; and, develops new tracers that reliably define reservoir attributes, such as the heat extraction efficiency along a stimulated fluid flow path.

Current LBNL and SNL Projects

Areas in which LBNL and SNL support the DOE program include a range of activities from broad planning and analysis support to specific technology focused project tasks. A number of these activities are performed in collaboration with academia, industry and other National Laboratories.

Provide technical and programmatic support

The Laboratories recently provided support to DOE management in planning activities associated with formulating the long-term GTP goals for EGS. Over the past year the LBNL and SNL technical expertise were vital in the preparation of summary and evaluation reports, white papers, and a study on the impact of induced seismicity on EGS development.

SNL and LBNL participate in professional meetings both to make presentations related to geothermal resource development and to gain an understanding of new technological advances in associated industries that may be applicable to geothermal development. The laboratories also participate with DOE in several ongoing international activities. SNL is the leader of the International Energy Agency Geothermal Implementing Agreement (IEA_GIA) Annex VII on Advanced Drilling Techniques and also assists DOE in prep aration and review of various IEA-GIA documents and the geothermal annual report. Recently, SNL provided a country presentation on well construction and completion needs during the first workshop associated with the International Partnership for Geothermal Technologies and LBNL provided a keynote address during a SEG working group meeting on unconventional resources in Vancouver, Canada.

SNL and LBNL also provide support to ongoing DOE funded field demonstration projects such as the Desert Peak EGS project through technical discussions with researchers associated with the project, visits to the sites and planning for and attending project reviews. LBNL and SNL act as the primary technical advisors to DOE for the project.

Development of improved fracture imaging of EGS sites using active and passive imaging technology

LBNL is working to extend the use of single and multi-component, 2-D, 3-D and 4D surface and borehole based seismic imaging methods and seismicity to "tomographically" image the subsurface. Time lapse imaging will be especially relevant for EGS due to the ability of geophysical methods to detect changes in subsurface properties related to massive fluid injection and subsequent fracture creation. The seismicity created from the massive injections can not only be used to map the hydrofractures (individual event analysis) but the waveforms from the events can be used to map fractures and lithology.

Seismic structures of sheared fractures during fluid injection

LBNL researchers examined seismic wave scattering by dry, sheared fractures. The experiments indicated that shearing can result in both increases and decreases in the scattering of seismic waves, depending on the micro-geometry of the fracture surface and the relative magnitude of the shear stress to the normal stress. LBNL discovered a new phenomenon of seismic wave mode conversions (normal-incidence compressional mode generates shear waves in the same and opposite propagation directions) which were induced by shear stresses. Although these experiments were done in the laboratory using high-frequency seismic waves (ultrasonic waves), the availability of high-frequency seismic waves during high-resolution fracture imaging at EGS sites can make the laboratory-observed results very applicable to the field.

Using geodetic data to improve our understanding of flow in EGS

LBNL has embarked on a project to use surface deformation patterns deduced from PSInSAR in collaboration with tilt meter data to map and model the existence and geometry of faults and fractures controlling flow following EGS reservoir stimulation. Special PSInSAR processing tech niques of remotely gathered geodetic data can resolve a few millimeters of surface deformation per year and has been shown to be very useful in mapping and characterizing geologic structures controlling fluid flow during CO_2 sequestra tion in deep aquifers.

Impact of fluid injection on natural isotopic systems at EGS sites

LBNL has coupled isotopic systematics with chemical data, to provide essential quantitative constraints upon reactive transport models, such as Toughreact, that are used to evaluate the geochemical impact on the behavior of a stimu lated reservoir and guide reservoir management. The lab is also evaluating the use of spatial and temporal shifts in the isotopic compositions of different solutes (e.g. Ca, Sr, Pb, noble gases, etc.) induced by water-rock exchange reactions to estimate the surface area of mass (heat) exchange.

Reactive geochemical transport modeling for improving injectivity of EGS reservoirs

A major concern in the development of EGS reservoirs is achieving and maintaining adequate injectivity, while avoiding the development of preferential short-circuiting flow paths. In collaboration with scientists from EGI at the University of Utah, LNBL is pursuing the feasibility of modifying the chemical composition of reinjected waters to improve reservoir performance by maintaining or even enhancing injectivity.

Development of a high temperature seismic monitoring tool

A HT downhole seismic tool is being developed by SNL to support future EGS projects. Current limits for such down-hole tools are on the order of 125 °C for long-term applications. The aim of the program is to develop high-reliability seismic tools with measurement capabilities similar to today's devices with the capability of withstanding forma tion temperatures in excess of 225 °C.

Development of a high temperature (300 °C) pressure/ temperature tool

Sandia has designed, fabricated and tested a High-Temperature (HT) pressure/temperature tool using SOI technology that can operate continuously at 300 °C. The primary pur pose of the tool is to demonstrate the ability of designing and fabricating tools for well monitoring applications that extend the limits of conventional PC board materials and to prove out the capabilities of Silicon-On-Insulator (SOI) electronics at extreme temperatures in field conditions.

EGS well construction technology evaluation

Hydrothermal geothermal wells today rarely exceed depths of 3 km. Tapping the vast thermal resource in the 3 to 10 km range using EGS represents a significant departure from current geothermal well construction practice. Sandia is in the final stages of completing a well construction technology evaluation report to address these concerns.

Critical technologies and operational practices that drive well construction economics, and can be mitigated through R&D, were ranked and targeted for future investigation. This effort included a well construction technology work shop that was convened in Houston to discuss the prelimi nary findings of the report. One of the critical findings of the well construction study, supported by the workshop group, is that there are currently many unknowns related to other aspects of EGS which can significantly influence the optimal well field specification. By anticipating these fac tors and more fully defining well construction needs, SNL hopes to help DOE increase prospects for EGS proof of concept success and commercial viability.

Creation of an advanced drilling dynamics simulator

Drillstring dynamic dysfunctions currently limit the use of advanced technology drill bits and related tools for drilling hard rock formations. Efforts have been initiated to develop an Advanced Drilling Dynamics Simulator (ADDS). The intent of this simulator is to integrate advanced technology in computational modeling and electronic controls with high-force, fast acting, servo-hydraulic actuators to represent the properties of a virtual drillstring in the laboratory. While development efforts associated with this project are not currently being pursued, industrial partners have made use of these advanced laboratory capabilities to enhance their understanding of the response of drilling tools to downhole environments.

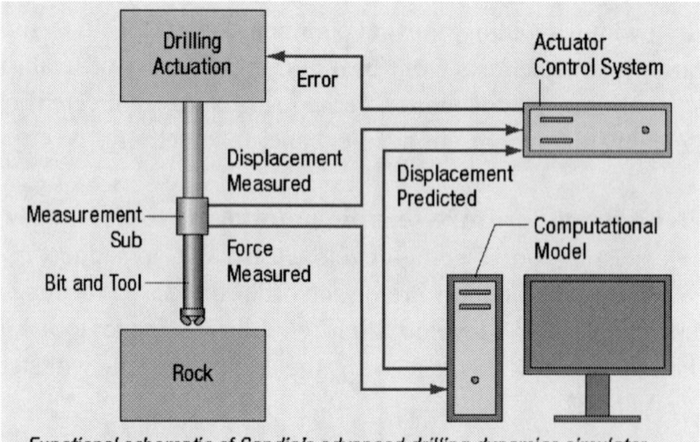

Functional schematic of Sandia's advanced drilling dynamics simulator.

Functional schematic of Sandia's advanced drilling dynamics simulator

Magneto-rheological damper

To more effectively deal with the real world evolution and variations of drill string dynamic response, a drilling tool with the ability to automatically provide the appropriate drill sting dampening is desired. To support this need, SNL participated in an effort to develop a damper with continu ously variable dampening levels. Similar to the Advanced Drilling Dynamics Simulator, this project is not currently active but through a recent license to industry, this technol ogy is soon to be commercialized.

The Price of Geothermal Power

Geothermal power plant developers are seeking ways to supply attractively priced renewable energy to utilities. The price that a geothermal power plant developer can offer to a utility in a power purchase agree ment (PPA) largely depends on a number of factors, including the power conversion technology used to generate the electricity, power plant size, and four additional factors:

1. Development costs
2. Financing charges
3. Operating and maintenance (O&M) costs
4. Resource credits.

Development costs—costs to get the plant sited, constructed, and put online—are significantly higher than those of fossil-fueled power plants. The development cost to build a natural gas power plant is about one third of the total costs. Development costs of a geothermal facility, in contrast, represent two thirds or more of total costs.

Financing charges—Customer owned utilities such as municipalities and electric cooperatives have interest rates of about 5%-6%. Independent power producers may have interest rates of 15% and higher.

O&M costs—cover charges for running the power plant and servicing and replacing equipment.

Resource credits—provide incentives for renewable energy projects.

Development Costs

Development costs include all expenditures associated with exploration, drilling, permitting, construction, and ancillary investments such as transmission costs. The development costs for a typical 20 MW power plant are shown in Table 1. These costs are rules of thumb. Actual costs can vary based on factors such as time delays, geology, environmental restrictions, project size, and transmission access.

The cost of time delays is significant, sometimes adding $10 to $20 or more per MWh or more to the cost of power. The time delays typically occur in the first two stages of devel opment where the risks are higher and the cost of capital is greater than the last two stages.

Table 1. Typical Geothermal Power Plant Development Costs

Development Stage	Cost ($/kW)
Exploration and resource assessment	$ 400
Well field drilling and development	$ 1,000
Power plant, surface facilities, and transmission	$ 2,000
Other development costs (fees, working capital, and contingency)	$ 600
Total development cost	$ 4,000

Exploration and resource assessment

Successful exploration results in the discovery of a geothermal resource capable of providing geothermal fluid to run a power plant and produce electricity. Exploration activities include regional reconnaissance and district exploration, and encompass prospecting, acquisition of rights, and field analysis. The activities are not linear, that is, developers may start acquiring land and mineral rights even before significant prospecting has begun.

Regional reconnaissance screens a large area of hundreds of square miles. The reconnaissance narrows prospecting efforts and involves geologic studies, analysis of available geophysical data, and geochemical surveys to identify more limited areas for detailed exploration.

District exploration uses geophysical surveys and temperature gradient measurements and focuses on smaller areas of 5,000 acres or more to site the first production well. Activities could include gravity surveys, ground magnetic surveys, magnetotelluric surveys, electrical resistivity surveys, and seismic surveys. The final step in district exploration is its most expensive activity—drilling the first deep exploration well.

Well field

Well field drilling and development includes siting and drilling exploration, and production injection wells, testing well flow rates and reservoir engineering. Ideally, exploration wells can be also used as production or injection wells. A successful set of wells results in high fluid temperatures and flow rates. According to the Geothermal Energy Association, exploration wells have an average success rate of 20%–25%, while production and injection wells have an average success rate of 60%–90%. Reservoir engineering determines the best location for injection wells and the flow rates that result in the most stable production.

Power plant, surface facilities, and transmission

These expenditures embrace the cost of the power plant and the geothermal fluid piping system, grid connection, ancillary infrastructure, and pollution abatement systems and environmental compliance work that include engineering, regulatory, documentation, and reporting activities.

The power plant consists of a series of unit operations and equipment such as pumps and motors, turbines, cooling towers, and transformers. The piping system connects the power plant with all production and injection wells. Grid connection includes the substation and transmission lines needed to move the output to the market and, if needed, to the well pumps. Ancillary infrastructure includes office buildings, roads, utilities, and other structures.

Pollution abatement systems and environmental compliance work is a large umbrella—geothermal power projects have to comply with federal, state, and local regulatory require ments, which vary according to location. For example, National Environmental Policy Act requirements apply if the project is on federal land.

State and local agencies, such as air and water boards, may require air and water discharge permits, and each agency may also have reporting requirements. Beginning in the exploration phase, there may be a need to consult with the state historic preservation office or Native American tribes if cultural resources are affected, and the U.S. Fish and Wildlife Service if plant or animal species of concern may be affected. Other federal regulatory agencies involved in the process include the Bureau of Land Management, the U.S. Forest Service, and the Federal Energy Regulatory Commission.

Other development costs

During geothermal development, legal services are needed to ensure quality in contract, performance, and reporting documents. A contingency reserve is necessary to provide working capital and to cover unexpected costs due to delays and unforeseen requirements. It also covers any margins required by the developer or owner.

Financing Charges

Financing charges are affected mainly by the amount of upfront capital needed to cover the first two development stages, the time line of these two stages, and the loan terms that allow the developer to repay the upfront costs and finance the second two development stages. The costs of the first two stages, exploration and well drilling and development, generally are not financed by utilities, banks, or other lending institutions. Venture and equity capital or in house reserves are sources for funding the first two devel opment stages.

For the purpose of this article, an 18% return is assumed for the investment in the first two development stages over a four-year period. Investors correctly interpret geothermal investments as high risk due to historical delays in having a project sited, developed, and online. For example, exploration at the Glass Mountain Known Geothermal Resource Area in Northern California began over 20 years ago. The exploration resulted in discovering a resource that could be commercially viable, but there is still no power plant.

Different types of utilities and lending institutions have varying interest rates and terms. For the purpose of this article a developer can compare two scenarios:

- 10% and 15 years (bank financing)
- 6% and 30 years (utility financing).

Geothermal well drilling at the Long Valley Exploratory Well near Mammoth Lakes, California

Each scenario assumes that the plant life is the same as the loan term and there is no salvage value to the plant afterwards. Zero salvage value may be a harsh assumption. However, in 15 years, the technology may have advanced to where the existing plant may need upgrading to meet new requirements.

If these two scenarios cover the boundaries of reality, the developer would expect to pay \$48 to \$87 per MWh for the project financing, assuming a 92% plant factor. Plant factor is defined as the rated capacity and the percentage of the year that the power plant is producing electricity.

O&M Costs

Fuel costs for geothermal power plants are insignificant. Expenses are costs for steam field management and geothermal fluid impacts on equipment and are covered in the power plant design costs and operations and management (O&M) costs.

O&M costs include those charges for employee salary and benefits, equipment replacement reserves, utilities, and administration. Most new geothermal power plants com ing online are going to be closed loop, air cooled, binary cycle plants. The O&M costs are assumed to be at a level of about $15 per MWh—lower than the typical plant.

Resource Credits

Geothermal power and other renewable resources may use rapid depreciation or other incentives such as the Production Tax Credit (PTC) to reduce the final costs to supply output to the market. The PTC currently is $9 per MWh. Other incentives such as pollution credits may apply in the future.

The Bottom Line

The cost of geothermal power, based on the above discussion and assumptions, is likely to be in the range of $63 to $102 per MWh, excluding any reduction due to resource credits. One caveat, a major impact on geothermal power cost is the local, regional, national, and global competition for commodities such as steel, cement, and construction equipment. Geothermal power is competing against other renewable and non-renewable power development, build ing construction, road and infrastructure improvements, and all other projects that use the same commodities and services. Until equipment and plant inventories rise to meet the increase in demand for these commodities and services, project developers can expect the costs to rise well above the background inflation level.

Jets shown prior to being attached to a geothermal system at The Geysers Geothermal Power Plant in California.

FINANCING GEOTHERMAL PROJECTS

In principle, the financing of geothermal power projects is not much different from the financing of other energy projects. It is about the allocation and management of risks among the various parties, with a few notable exceptions. The need to secure a fuel supply is essential to any energy project; whether the source of fuel is geothermal brine or conventional fuels. A distinct difference between geothermal and other energy technologies, though, is that the developer "pre-pays" the fuel cost in exploration and drilling expenses. Developers new to geothermal are sometimes surprised by the cost, lead time, and complexity involved in permitting and developing geothermal resources. Ensuring that the technology selected is compatible with the available resource is essential to long-term success. Finally, the geothermal industry has attracted a long list of developers and entrepreneurs who usually come in from other industries, often with the idea that developing a geothermal project is relatively simple compared to other, more traditional, energy projects. Some of these new developers lack the experience, skills, and capital required to deal with the considerable challenges they will undoubtedly face in the process. By employing the services of a team with geothermal experience, these "newcomers" may avoid the pitfalls that can derail potentially successful projects.

Project Financing

Classic energy-project financing involves a developer raising funds (usually through a special purpose subsidiary created and incorporated for the specific project) through various mechanisms from financing institutions, usually in the form of equity, debt, or a combination of the two.

The participation of the financial entities may be attained through various business structures such as partnerships, leases, corporate investments, and combinations of the three. Issues of tax, cost accounting, and of course econom ics are the drivers in the process, but the underlying fundamentals are clear: risk versus reward. The more confident the investors feel that there is "real" credit support behind the project, the better the financial terms they will agree to provide with their investment.

The elements that are analyzed in the financing process are those that may affect the financial performance of the project over time. Typically, the main elements are:

- Power Purchase Agreements (PPA) are a major financing support element. But for specific terms and conditions such as pricing, indexing, termination clauses, and term, the reality of the last years requires that the power purchaser have a solid credit position.
- Fuel supply availability (brine or steam), cost, and pro jected longevity of the supply. Developers should be prepared to pay a significant risk premium when trying to finance projects with unproven or undeveloped geothermal resources. The up-front cost of developing the geothermal resource means that the overall efficiency of the power plant strongly influences the return on investment.
- Site control for the entire resource, including the ability to preclude competing interests. Competition for leases and geothermal rights is often intense.
- Field performance experience of the technology which will produce the power. Few manufacturers have a long-term track record for utility-scale geothermal systems.
- Experience and credibility of the developer.
- Other elements such as structure of the deal, regulatory issues, and electricity transmission issues.

The Geothermal Resource

Geothermal projects are attractive because they produce base-load power and thus geothermal projects can generally qualify for firm long-term contracts with utilities. Despite these advantages, developing a geothermal power project is a complicated process that new developers must approach carefully. Geothermal projects have distinctly different challenges than other, more traditional, renewable technolo gies such as wind, solar, and biomass. Geothermal projects require subsurface exploration and well field develop ment and have greater upfront risk because the geothermal resource is not confirmed without drilling.

The geothermal resource is key to the success of a geothermal project and has a profound effect on financing terms. Resource critical parameters (temperature, permeability, fluid production, brine chemistry, etc.) include the ability to support power production on a sustainable long term basis.

Other details that need to be addressed include:

- Land ownership issues

- Permitting and other regulatory issues
- Transmission issues.

Financial institutions usually come on board once the geothermal resource has been fully explored and at least partially developed. At this stage, financial investors will have to be satisfied that the resource has the potential to deliver energy (and revenue) over a long period of time.

Nearly as important as the resource is the ability of the developer to demonstrate their team has the talent required to acquire, develop, and manage the geothermal field effectively. The major geothermal operators in the geothermal industry, such as Ormat, have an in-house talent base, which includes land teams, geologists, reservoir engineers, and other technical personnel. The ability of the developer's team to communicate freely among all these disciplines, on a continuous and near instantaneous basis, ensures that issues and problems in resource development and manage ment are addressed expeditiously and effectively.

Financing partners will seek assurances that the developer has:

- Critical mass of skilled, talented, and experienced personnel
- Sufficient funds to address any resource related contingencies that may arise over the term of the financing.

Condensers being retrofitted with direct contact condensers at The Geysers Geothermal Power Plant in California.

Power Plant Technology

There is no standard approach to geothermal technology. Rather, the developer must adapt the plant to the available resource. For decades, premier geothermal resources were developed using standard steam turbines. These standard steam turbines are compatible with high temperature geothermal resources, but not with cooler geothermal resources that produce mostly hot fluid and not much steam. Because the geothermal resource base in the United States is mostly composed of these cooler geothermal resources, power

plant technology has advanced to capture this resource base. Today's developers will find that steam technology is not applicable for most present-day projects; they will find that binary cycle plants will be the preferred solution. Binary cycle plants typically utilize an Organic Rankine Cycle (ORC). The ORC operates by exchanging heat between geothermal fluid and a secondary working fluid, which is typically an organic fluid (such as pentane or isobutene) with a low boiling point. The organic fluid is vaporized to drive the turbine. Although the need for an integrated solution drives many developers to binary plants using the ORC, one size does not fit all. The temperature, flow, and chemistry of the resource require specific technological and engineering solutions in order to maximize return on investment.

Drilling exploration crew performing initial well logging tests at the Desert Peak site in Nevada.

Financial institutions may be reluctant to provide non recourse financing for a power plant technology that was not field proven and tested over many years in different resource areas. The reality is that geothermal power plants are expected to be of "utility grade." Developers selecting non-utility scale technologies such as Kalina cycle and reversed refrigeration may find it difficult to secure financing.

Developers

The following are key issues a lender or investor considers before financing a project:

- **Committed developer.** One of the most important elements in putting together a geothermal power plant is a long term process. Our experience shows that the development of a geothermal project takes three to five years from site acquisition to commercial operation. One needs the capital and the commitment to go through the process, which many times does not yield a project at all. This commitment is required for the long term, as all projects have implementation issues that begin only after the financing closes.

- **Management team with the proper disciplines.** As the development and operation of a geothermal project are very challenging, the financial parties will look for developers who are astute enough to secure the services of the most experienced technical personnel in the field.

- **Well capitalized developer.** Geothermal power plants may present financial challenges during the operational phase. Geothermal projects may have long term issues with the well field and power plant. These unforeseen elements may require a temporary injection of capital on short notice.

Summary

The geothermal industry is unique within the energy market, and as such the financing process of geothermal projects contains certain unique key elements that determine the ability of a geothermal plant developer to reach suc cessful completion.

Based on Ormat's experience, we identified the main ele ments to support non-recourse project financing, as dis cussed in this article:

- Availability of the geothermal resource
- Proven technology used in the geothermal power plant
- Credibility and record of the developer.

When all of these elements are present, a geothermal proj ect is likely to be successfully financed.

INTERNATIONAL GEOTHERMAL EFFORTS – 2008

Introduction

As the global demand for clean, reliable, renewable energy increases, geothermal energy is becoming an attractive solution. This is true not only in the United States, where current production is approaching 3,000 MWe, but at numerous locations on six continents. An area of increased emphasis is Enhanced Geothermal Systems (EGS).

Some of the most integrated and profound policy developments have been occurring in Australia, where significant governmental commitment and financial support are advancing the country's EGS Hot Dry Rock efforts. Another notable development can be found in Europe, with the ENhanced Geothermal Innovative Network for Europe (ENGINE) project.

What follows is an incomplete and brief survey of geothermal developments outside of the United States in 2008.

The Wairakei Geothermal Power Plant in New Zealand

Table 1. Estimated Production of Geothermal Energy by Country and MWe in 2010[4]

Country	MWe	Country	MWe
United States	3,000	Russia	185
Philippines	1,991	Kenya	164
Indonesia	1,192	Nicaragua	143
Mexico	1,178	Turkey	83
Italy	910	Papua New Guinea	56
New Zealand	590	France	5
Iceland	580	Portugal	35
Japan	535	China	28
El Salvador	204	Germany	8
Costa Rica	197	Ethiopia	7

Australia and New Zealand

Australia and New Zealand have been consistent develop ers of geothermal energy and continue to advance their interests in development.

Australia's state and federal governments have been sup portive of developing EGS infrastructure and have set forth ambitious plans to expand geothermal development as part of the country's renewable energy portfolio. But, there has been a perceived minor setback—the latest federal budget presented a deferral of access to the A$50m geothermal fund until approximately July 2009. [5]

Australia has engaged in the Onshore Energy Security Program (OESP) and has committed A$58.9m over five years to Geoscience Australia (formerly the Australian Bureau of Mineral Resources). The Otway Basin along the Limestone Coast in South Australia is a region of interest. Three sites for development have been identified, possibly producing 1,600 MWe for 30 years. [6]

Mighty River Power in New Zealand plans to invest more than NZ $1 billion to develop 400 MWe of new geothermal generation by 2012 in the Taupo Volcanic Zone on the North Island. [7] The Kawerau plant is scheduled for completion by the end of 2008 and is expected to generate 90 MWe.

THE CREATION OF A NEW INTERNATIONAL GEOTHERMAL PARTNERSHIP

A delegation from the U.S. Department of Energy (DOE) traveled to Reykjavik, Iceland in August 2008 for the signing and initial workshop of the International Partnership on Geothermal Technology (IPGT), a new agreement between geothermal technology leaders Iceland, Australia, and the United States. The DOE's Acting Assistant Secretary for Policy and International Affairs, Katharine Fredriksen, Australia's Ambassador to Iceland, Sharyn Minahan, and Iceland's Minister of Industry Energy and Tourism, Ossur Skarphedinsson signed the Charter document on August 28, 2008. New Zealand attended as an observer.

The signing countries signaled their commitment to aggressively pursue advanced geothermal technologies, such as Enhanced Geothermal Systems (EGS), as part of a solution to energy security and global climate change concerns. "EGS has the potential to be the world's only ubiquitous form of baseload renewable energy," said Acting Assistant Secretary Fredriksen. "This partnership will bring together countries with expertise in geothermal energy to accelerate the development of EGS, bringing this technology to the market in the near-term to confront the serious challenges of climate change and energy security."

Europe

In 2007, the European Commission (EC) began funding the ENGINE program. ENGINE's main objective is to coordinate research and development initiatives for unconvention al geothermal resources and EGS and aspires to develop up to 20 demonstration sites. [7] ENGINE sets forth four research areas for geothermal development:

- Exploration: finding access to potential reservoirs at depth
- Geothermal wells: improving drilling and completion technologies

- Reservoir engineering: stimulating fluid flow underground
- Exploitation: improving efficiency.

In June 2008, the pilot plant of Soultz-sous-Forêts in Alsace, France, was inaugurated by the French Prime Minister Fran cois Fillon. The first installed Organic Rankine Cycle (ORC) module of 1.5 MWe is providing electricity to the grid. This European project, mainly funded by France, Germany, and the EC (with collaboration of several other countries includ ing the UK and Switzerland), started in 1987. Much of the technology was adopted from the Rosemanowes site in Cornwall, which benefited from earlier DOE experiments at Fenton Hill. Shell and Enel were also involved with this project. The first successful, commercially-funded EGS project is in Landau, Germany. The project was completed in 3.5 years and is producing up to 3.6 MWe. [8]

Asia

Exploration is taking place in several regions including India, the Kyrgyz Republic, China, Indonesia, and the Philippines.

In India, GeoSyndicate Power Private is working in collaboration with Panax Geothermal from Australia to exploit "wet" projects in the Godavari rift in Andhra Pradesh and the Himalayan Geothermal Province in Ladakh where there are high-heat flows.

High-heat flows have been detected in the Kyrgyz Republic and exploration licenses have been issued for several groups seeking high-heat producing granites in the Inylchek region of the republic.

Table 2. Preliminary List of ENGINE EGS Demonstration Projects by Country and Site Name

Country	Project
France	Roquette, Rhine Graben
Germany	Bruschal
	Groß Schönebeck Landau
	Unterhaching
Hungary	Fábiánsebestyén Zala County
Iceland	Icelandic Deep Drilling Program
Poland	Podhale
Slovakia	Košice
Turkey	Green Campus Izmir

Researchers have been assessing thermal resources in several regions of China and notable surveillance work has been conducted in Yunnan Province for the Rehai (Hot Sea) geothermal field of Tengchong County, where more than 815 thermal springs have been identified; 354 of which were measured at temperatures above 113°F (45°C). [6] An additional 105 MWe could be developed in Gu'an County, Hebei Province. [6]

At the time of this writing, 17 companies (including Chevron) are planning to bid for geothermal projects in West Java, Indonesia, possibly generating as much as 315 MWe. [9] The three power plants are Tangkuban Perahu (220 MWe), Cisolok Sukarame (45 MWe), and Tampomas (50 MWe). The Philippines is allowing the Philippine National Oil Company-Energy Exploration Corporation to drill geothermal wells in a restricted buffer zone adjacent to the Mt. Kanlaon Natural Park.

North America and the Caribbean

Geothermal development has been acute in Mexico and Central America. Notable developments have been produc ing geothermal energy in El Salvador, Nicaragua, and Costa Rica for many years. Canada has begun exploration with an eye to development with the Canadian Geothermal Energy Association regrouping in 2007.

The 120 megawatt Nesjavellir Geothermal Power Plant in Iceland.

Nicaragua developed a 35 MWe operation on the Momotombo reservoir in 1983. Over the years production decreased dramatically and, by 1999, power-plant generation stood at 9 MWe. Reinjection was initiated and currently the plant is producing 30-35 MWe. [6]

Interest has been developing in the Caribbean, and drilling began on Nevis at the Spring Hill site in 2008; steam was located later that year at 3,720 feet.

Africa

Another region of interest for geothermal development has been the East African Rift Valley. The United Nations estimates this region may be able to produce more than 400 MWe. In 2007, the World Bank had committed up to $13 million to the African Rift Geothermal fund, which will operate in six countries: Kenya, Uganda, Tanzania, Ethiopia, Eritrea, and Djibouti. Kenya was the first of these countries to develop geothermal energy and has the largest geothermal plant in Africa—near Naivasha (Olkaria), yield ing 130 MWe. [6, 10] Kenya Electricity Generating Company plans to install an additional 1,260 MWe by 2018 from four potential geothermal areas in the Kenya Rift—Olkaria, Menengai, Longonot, and Eburru— and also plans to develop at least 300 production and 60 reinjection wells over the next 10 years. [6]

South America

The International Geothermal Association shows no production of electricity in South America. A report indicates Bolivia may have the capacity to produce 280-370 MWe. Chile has become interested in exploring its geothermal-resource potential after the abrupt curtailment of natural gas shipments from Argentina.

Summary

As stated in the introduction, this is neither a comprehensive nor a complete survey of international geothermal development—it is merely a snapshot that provides a sample of efforts and initiatives across the globe. What is important to note is that there is an increased interest in geothermal development in 40 or more countries and all indications show this interest will be persistent and durable in light of international energy demands and the need to develop clean, renewable, baseload energy.

Utility Geothermal Working Group Update

The Utility Geothermal Working Group (UGWG) was formed in September 2005, at the Geothermal Resources Council's annual meeting in Reno, Nevada. It is a group of utilities and ancillary associations formed under the U.S. Department of Energy's Geothermal Technologies Program. UGWG is supported by six organizations:

- American Public Power Association (APPA)
- Bonneville Power Administration (BPA)
- Geothermal Resources Council (GRC)
- National Rural Electric Cooperative Association (NRECA)
- U.S. Department of Energy (DOE)
- Western Area Power Administration (Western).

The Working Group's mission is to accelerate the appropriate integration of three geothermal technologies into mainstream applications: power generation, direct use, and geothermal heat pumps (GHP). In addition to the six support organizations listed above, the UGWG members include:

- Arizona Public Service
- Ormat Technologies, Inc.
- Palo Alto Utilities
- Redding Electric Utility
- Salt River Project
- Sandia National Laboratories
- Seattle City Light
- South San Joaquin Irrigation District

- Springfield Utility Board
- State Working Groups
- Idaho National Laboratory.

Webcasts and Workshops

To help accomplish its mission, the UGWG conducts periodic training events in the form of Webcasts and workshops. Events focus on geothermal and other renewable applications, technologies, and issues. Since its formation, the Group worked with its members and GRC staff to shape utility training sessions at the 2006 and 2007 GRC meetings. These training sessions provided an opportunity for more utilities to attend the high quality meetings. Other workshops and Webcasts have focused on topics such as:

- Power Generation
- Direct Use
- Geothermal Heat Pumps
- Transmission Issues
- Project 25x25
- Renewable Energy
- Coal Fired Power Plants
- Public Participation
- Clean Renewable Energy Bonds
- Geothermal Heat Pump Economics.

Power Generation and Direct Use Findings

Utilities are continuing on the path of integrated resource planning (IRP) to provide energy services to their custom ers. IRP demonstrates that energy efficiency remains the first choice in a utility resource portfolio and that direct use is an application that utilities continue to avoid. On the other hand, geothermal power generation is of interest to utilities. Geothermal power plants are capital-intensive, requiring most of the funding up front before the project produces any revenue. Utilities are more confident in the plants and are willing to negotiate a financeable power purchase agreement (PPA) with a developer, if the following five condi tions are met:

- Delineated geothermal resource with a bank able report that defines probable long term performance
- Defined permitting path without pitfalls
- Credible developer with a proven project management track record
- Control of entire geothermal resource to preclude competing interests for same fluid/ steam supply
- Use of proven technologies.

The Creamery Brewpub and Grill is one of the case histories described in UGWG workshops. It uses geothermal energy from the Klamath Falls, Oregon, geothermal district heating system for all its heating purposes. Uses of geothermal energy include space heating of approximately 11,000 ft^2 (1,022 m^2) of restaurant/pub space, snow-melting of about 1,000 ft^2 (93 m^2) of sidewalks, and generation of hot water for the brewing process. In cold months the brewery saves about $1,100 in space heating expenses and saves around $300 per month in energy used to brew the beer

The UGWG conducts several geothermal technology workshops each year

Utilities are willing to enter into PPAs if the output compares favorably with the "default power plant", which is currently a gas-fired combined cycle plant. Utilities estimate purchasing power from the default choice in the range of $65 to $90 per MWh, which includes capital, O&M, and fuel costs.

Table 1. Costs For a Typical 20 MW Power Plant

Development Stage	Cost (Millions of $)
Exploration and resource assessment	$ 8
Well field drilling and development	$ 20
Power plant, surface facilities, and transmission	$ 40
Other costs (fees, operating reserves, and contingencies)	$ 12
Total cost	**$ 80**

The price that a geothermal power plant developer can offer to a utility in a PPA largely depends on 1) the exploration, drilling, and development costs of getting the project online and 2) the financing charges associated with the costs. Costs for a typical 20 MW power plant are shown in Table 1.

Using the above costs as a basis, a typical geothermal power plant has a capital cost of $4,000/kW. This capital cost is translated to a MWh cost by applying an annual factor reflecting interests rates for financing the total capital cost.

- At an annual factor of 0.2, reflecting an interest rate of 18% – 20%, capital costs are $104/MWh.
- At an annual factor of 0.15, reflecting an interest rate of 13% – 1 5%, capital costs are $ 76/MWh.

There are no fuel costs and the typical O&M cost for a plant is about $15/MWh. The O&M costs assume that the power plant uses Organic Rankine Cycle (ORC) technology for energy conversion with air to air cooling towers. ORC technology uses a moderately high molecular mass organic fluid such as butane or pentane to absorb the heat from geothermal fluid and drive the turbine. The technology has the benefits of high-cycle and turbine efficiencies, low turbine mechanical stress, reduced turbine blade erosion, and the fact that a full time operator need not be present.

If the power plant uses a different technology or water to air cooling towers, the O&M costs are likely to be higher. Using these two annual factors and adding the O&M cost to the annualized capital costs, the developer may be able to offer a utility output in the range of $91 to 119/MWh. This price could be lowered if the utility were to finance the power plant construction.

Geothermal Heat Pump Findings

GHPs represent an energy-efficient technology making strong gains as a viable alternative heating and cooling system, both in the United States and around the world. [11] Although this technology has been in existence since the 1940s, it still has not realized its full market potential, but the technology is gaining ground. The UGWG and one of its four major support organizations, Western, developed a report that describes the reasons why geothermal heat pump technology appeals to electric utilities and end users, and explains why this appeal has not been enough to sustain a national market.

Western also developed two worksheets that provide the economics of GHP versus other heating, ventilation, and air conditioning (HVAC) options from the customer and utility perspective. This report and the spreadsheets help readers to:

- Understand the benefits geothermal heat pumps offer cus tomers and electric utility providers
- Describe market potential and appeal of geothermal heat pumps
- Document tactics and strategies that some electric utili ties have used to develop sustainable and effective geothermal heat pump programs.

Twelve utility programs with successful geothermal heat pump installations were selected to be included in this report. These are not all the utilities currently offering geothermal heat pump programs. Nor are they some of the geothermal "pioneers" that first established utility programs. Rather, these are the utilities still committed to selling and promoting this technology. The selected utilities featured in this report have found the right alchemy of program elements to create innovative and successful geothermal programs.

The report identifies one major barrier to expanding GHP applications –costs that the customer must incur without utility financing. The GHP typically has a 20% premium when compared to traditional air-source heat pump system installations. [12] Cost premiums are associated with designing and installing ground loop systems that operate year-round without auxiliary back-up units. According to one Energy Information Administration (EIA) report, these systems have a payback period of two to 10 years when energy and maintenance costs are accounted for. [13] Other reports have indicated simple payback periods of five to eight years. The large variance in payback discourages implementing these systems. Typically, businesses and individuals look for a return on an investment within a two to three year payback, and a longer payback is highly unattractive for consumers and businesses alike.

If the utility were to step in and finance all or part of the GHP system for customers, the customers may likely enjoy a positive cash flow from the start of the system operation. The utility could place a lien on the customers' properties and charge an interest rate, in the form of a loop lease, which is digestible for the customer and financially prudent for the utility.

To illustrate a typical residential application, the following assumptions are used and compare to a GHP system with a conventional HVAC system that uses a natural furnace for heating and electrically served air conditioning for cooling.

Sources for assumptions are DOE and EIA. If the conven tional source is propane, oil, or electric resistance for heat ing, GHP economics are better.

- Electric Rate = 10′ per kWh
- Electric AC Use = 1,660 kWh per year
- Gas Rate = $1.50 per therm
- Gas Heating Use = 900 therms per year
- GHP System Cost = $10,000

Using the above assumptions, conventional HVAC costs $1,516/yr. The GHP costs are $1,390/yr, assuming that a loop lease is available to finance the GHP system costs. Loop leases vary due to loan terms. If the utility offers 6% financing and 30 year terms, the loop lease is \$330/yr.

Does it make sense for a utility to offer a GHP program that includes a loop lease to the customer? Utility economics are less straight forward than customer economics. The utility needs to assess how the program affects its peak period (summer vs. winter), including the impact of the default heating option (electric resistance vs. other fuel sources such as natural gas or propane.

If the GHP system is replacing electric resistance heating, the utility saves about 40% in peak demand in the summer and winter, and loses about 70% of revenues from kWh sales. GHP makes sense if the peak demand savings and interest revenues from the loop lease more than offset the revenue losses and any other losses resulting from imple menting the program. Other revenue losses include actions such as rebates, rate reductions, or lower interest rates.

Conclusions

The UGWG finds utility members are interested in two of the three geothermal technologies—power generation and geothermal heat pumps. Direct use appears to be too far afield from their core business to pursue at this time. Based on the results of training and interaction with the members over the past year, the UGWG plans to continue promoting the two geothermal technologies of interest to its members. The focus will be on workshops, training programs, and field assessments that cause more geothermal power plants to be developed and more geothermal heat pumps to be put into the ground.

State Policies Provide Critical Support for Renewable Electricity

Growth in renewable energy in the United States over the past decade has been propelled by a number of forces, including rising fossil fuel prices, environmental concerns, and policy support at the state and federal levels. Arguably, the two most-important types of state policies for supporting electricity generation from geothermal and other forms of renewable energy are renewables portfolio standards (RPS) and utility integrated resource planning (IRP) requirements. Within the western United States, where the vast majority of the nation's readily-accessible geothermal resource potential resides, these two types of state policies have been critical to the growth of renewable energy and promise to continue to play a fundamental role for the foreseeable future.

Renewables Portfolio Standard

A renewables portfolio standard (RPS) requires utilities and other retail electricity suppliers to produce or purchase a minimum quantity or percentage of their generation supply from renewable resources. RPS purchase obligations generally increase over time, and retail suppliers typically must demonstrate compliance on an annual basis. Mandatory RPS policies are backed by various types of compliance enforce ment mechanisms, although most states have incorporated some type of cost-containment provision, such as a cost cap or a cap on retail rate impacts, which could conceivably allow utilities to avoid (full) compliance with an RPS target.

Currently, 27 states and the District of Columbia have man datory RPS requirements. Within the 11 states of the contiguous western United States all but three (Idaho, Utah, and Wyoming) now have a mandatory RPS legislation (Utah has a voluntary renewable energy goal), covering almost 80% of retail electricity sales in the region. Although many of these state policies have only recently been established, the impact is already evident: almost 1,800 MW of new renew able capacity has been installed in Western states following the implementation of RPS policies. To date, wind energy has been the primary beneficiary of state RPS policies, representing approximately 83% of RPS-driven renewable capacity growth in the West through 2007. Geothermal energy occupies a distant second place, providing 7% of RPS-driven new renewable capacity in the West since the late 1990s, though geothermal's contribution on an energy (MWh) basis is higher because the nameplate capacity of a generator (essentially its maximum instantaneous output) is expressed in units of megawatts, while the amount of energy produced by a generator over some period of time is expressed in megawatt-hours.

Looking to the future, a sizable quantity of renewable capacity beyond pre-RPS levels will be needed to meet state RPS mandates: about 25,000 MW by 2025 within the Western United States (Figure 1). Geothermal energy is beginning to provide an increasingly significant contribution, as evi denced by the spate of new projects recently announced to meet state RPS requirements. Most of this activity has been driven by RPS policies in California and Nevada, where the Geothermal Energy Association has identified 47 new geothermal projects, in various stages of development, totaling more than 2,100 MW. [14] Additional geothermal projects in Arizona, New Mexico, Oregon, and Washington are also under development to meet those states' RPS requirements.

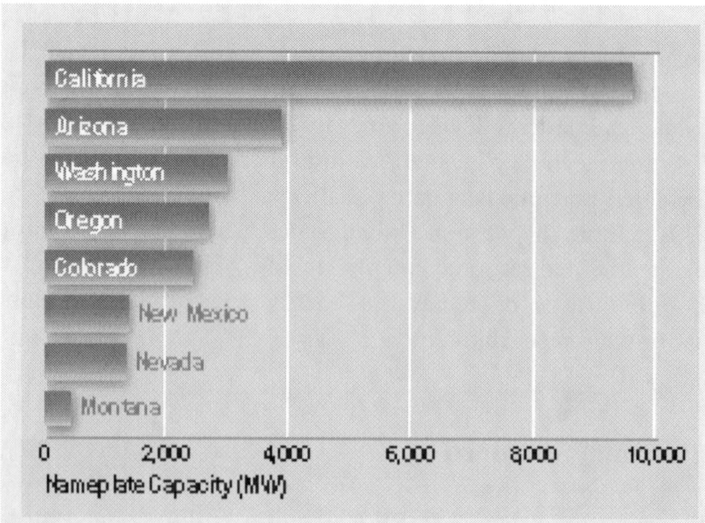

Figure 1. New renewable capacity needed by 2025 to meet western RPS requirements (beyond pre-RPS levels).

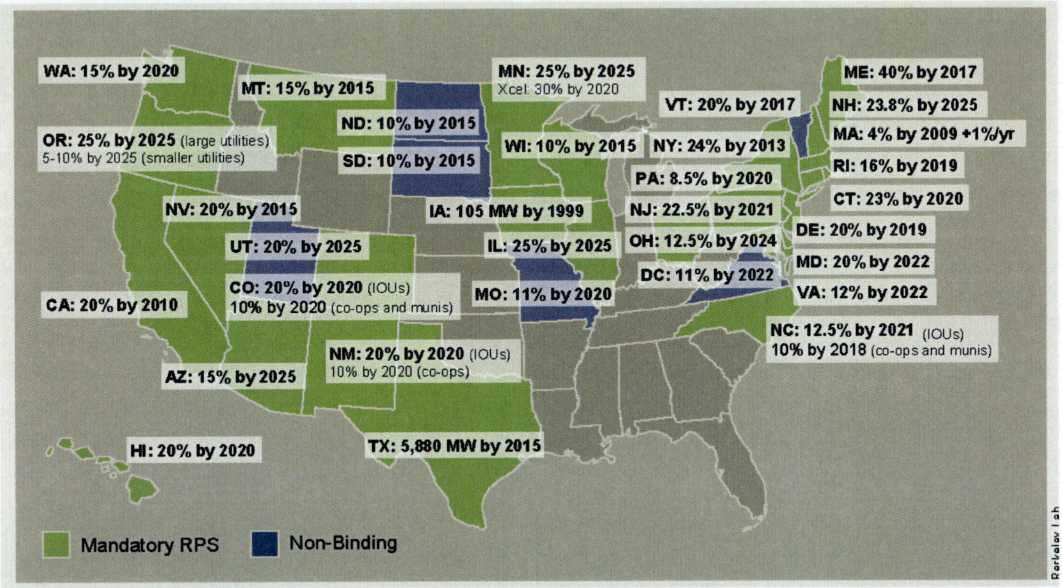

Renewable Portfolio Standards by state

Integrated Resource Planning

The other major state policy driver for renewable electricity growth, particularly in the West, is IRP. (IRP is also alternatively referred to as least-cost planning, long-term procurement planning, and default supply resource procurement planning). IRP was first formalized as a practice in the 1980s, but the practice was suspended in some states as electricity restructuring efforts began. A renewed inter est in IRP has emerged in the past several years with three western states (California, Montana, and New Mexico) re-establishing IRP. Other states are developing new rules to strengthen their existing processes.

In its barest form, IRP simply requires that utilities periodically submit long-term resource procurement plans in which they evaluate alternative strategies for meeting their resource needs over the following 10 to 20 years. However, many states have developed specific requirements for the IRP process that directly or indirectly support renewable energy. The most general of these is an explicit requirement that utilities evaluate renewables, and that they do so on an equivalent or comparable basis to conventional supply-side generation options. Many states also require that utilities include various types of risk analyses within their IRP. For example, utilities are often required to evaluate fuel price risk within their resource plan, which can reveal the value of renewables as a hedge against rising fuel prices.

Of particular importance for supporting renewable energy is the increasingly common requirement that utilities evaluate the potential costs and risks associated with future greenhouse gas regulations. Virtually all of the major western utilities that prepare IRPs incorporated future carbon dioxide regulations in their analyses of alternative resource strategies in their most recent resource plans. Some state public utility commissions (California, New Mexico, and Oregon) have even specified particular carbon dioxide emission allowance prices utilities are required to include in their analyses, or have established other requirements related to how utilities undertake analysis of carbon regulation risk.

The impact of IRP on renewable energy development is most apparent in states without an RPS, where the IRP process has often led directly to procurement or con struction of new renewables. For example, in its 2004 IRP, Idaho Power selected a preferred resource portfolio containing new geothermal resources, and subsequently issued a Request for Proposals for 100 MW of geothermal energy that has since culminated in the signing of at least one power purchase agreement (for the output from a new geothermal unit at the Raft River Project in Idaho). Similarly, many of the Washington and Oregon utilities were actively procuring new renewable resources prior to enactment of those states' recent RPS laws, in part as a result of IRP. Even in states with an RPS, IRP has played an important role in supporting renewables development, in some cases leading utili ties to pursue greater levels of renewables than is strictly required for compliance with the RPS. For example, in its most recent IRP, Public Service Company of Colorado opted for a resource portfolio—including 20 MW of new geothermal power—that far exceeded the quantity of renewables needed to meet the state's RPS requirements.

Conclusion

Together, state RPS policies and IRP requirements are creating strong demand for new renewable electricity generation capacity, which is driving the development of new geothermal resources in the Western United States. Both types of policies are relatively stable and are therefore likely to continue to support new renewable electricity generation for the foreseeable future. The extent to which geothermal energy ultimately benefits from these policies will depend largely on how well it can compete against other renewable resource options.

Contributors

Roger D. Aines H
Lawrence Livermore National Laboratory

Cynthia E. Atkins-Duffin
Lawrence Livermore National Laboratory

Galen Barbose
Lawrence Berkeley National Laboratory

Doug Blankenship
Sandia National Laboratory

Mark Bolinger
Lawrence Berkeley National Laboratory

Bill Foxall
Lawrence Livermore National Laboratory

David Hicks
National Renewable Energy Laboratory

Nalu Kaahaaina
Lawrence Livermore National Laboratory

B. Mack Kennedy
Lawrence Berkeley National Laboratory

Greg Mines
Idaho National Laboratory

Guy Nelson
Utility Geothermal Working Group

Jihan Quail
Sentech, Inc.

Jeffery J. Roberts
Lawrence Livermore National Laboratory

Curt Robinson, Ph.D.
Geothermal Resources Council

Charles Visser
National Renewable Energy Laboratory

Ryan Wiser
Lawrence Berkeley National Laboratory

John J. Zucca
Lawrence Livermore National Laboratory

A Strong Energy Portfolio for a Strong America

Energy efficiency and clean, renewable energy will mean a stronger economy, a cleaner environment, and greater energy independence for America. Working with a wide array of state, community, industry, and university partners, the U.S. Department of Energy's Office of Energy Efficiency and Renewable Energy invests in a diverse portfolio of energy technologies.

REFERENCES

[1] Barton, C. A.; Hickman, S.; Morin, R.; Zoback, M. D.; Finkbeiner, T.; Sass, J.; and Benoit, D (1997). "*In-Situ Stress and Fracture Permeability Along the Stillwater Fault Zone*, Dixie Valley, Nevada." Presented at the Twenty-Second Workshop on Geothermal Reservoir Engineering. Stanford University. Stanford, California. January 27–29.

[2] Blewitt, G.; Coolbaugh, M.; Holt, W.; Kreemer, C.; Davis, J.; and Bennett, R. (2002) "Targeting of Potential Geothermal Resources in the Great Basin from Regional Relationships Between Geodetic Strain and Geological Structures." *GRC Transactions;* Vol. 26, pp. 523-526.

[3] Tester, J. W. et al. (2006) *"The Future of Geothermal Energy: Impact of Enhanced Geothermal Systems (EGS) on the United States in the 21st Century."* Presented in an assessment by an MIT-led interdisciplinary panel. MIT. Cambridge, Massachusetts. 2006.

The work discussed in this section was performed with the support of the U.S. Department of Energy Office of Energy Efficiency and Renewable Energy (EERE), under National Renewable Energy Laboratory Contract DE AC36-99GO10337, Lawrence Livermore National Laboratory Contract DE AC52-07NA27344, Idaho National Laboratory Contract DE AC07-05ID14517, Lawrence Berkeley National Laboratory Contract DE AC02-05CH11231, and Sandia National Laboratory Contract DE AC04-94AL85000.

[4] Bertani, R. "Geothermal Power Plants Commissioned in the Third Millenium." *IGA News;* Vol. 72, April-June 2008; pp. 5-10.

[5] *HDRPL Geothermal Newsletter.* Issue 20, June 2008.

[6] *GRC Transactions;* Vol. 32 (preprint), 2008.

[7] Robertson-Tait, A.; Ledru, P.; Goldstein, B.; and Nathwani, J. "Current EGS Funding Initiatives in Europe, Australia, and the United States." *GRC Bulletin;* Vol. 37/3, May-June 2008; pp. 18-24.

[8] Baria, R. Personal communication by e-mail. 18 June 2008.

[9] "Chevron, Medco to Tap RI Geothermal Potential." *The Jakarta Post* online http://old.thejakartapost.com/headlines.asp. Accessed June 7, 2008.

[10] "UN to Ensure Geothermal Explorers in East Africa." EcoEarth.Info online http://www.ecoearth.info/. Accessed March 22, 2007.

[11] Johnson, K. *Geothermal Heat Pump Guidebook.* 3rd Edition. 2007; p. 3

[12] "Geothermal Heat Pumps." Sacramento Municipal Utility District. January 2007.

[13] Kavanaugh, S. *Ground-Coupling with Water Source Heat Pumps*. 2004; p.10.

[14] "Developing Plants in the U.S." *The Geothermal Energy Association* online http:www.geo-energy.org/information/developing.asp.

In: Geothermal Energy: The Resource Under Our Feet
Editor: Charles T. Malloy
ISBN: 978-1-60741-502-2

Chapter 5

AN EVALUATION OF ENHANCED GEOTHERMAL SYSTEMS TECHNOLOGY*

United States Department of Energy

FOREWORD

This document presents the results of an eight-month study by the Department of Energy (DOE) and its support staff at the national laboratories concerning the technological requirements to commercialize a new geothermal technology, Enhanced Geothermal Systems (EGS). EGS have been proposed as a viable means of extracting the earth's vast geothermal resources. Those who contributed to the study and authored portions of the report include: Allan Jelacic, Raymond Fortuna, Raymond LaSala and Jay Nathwani (DOE); Gerald Nix (retired), Charles Visser, and Bruce Green (National Renewable Energy Laboratory); Joel Renner (Idaho National Laboratory); Douglas Blankenship (Sandia National Laboratories); Mack Kennedy (Lawrence Berkeley National Laboratory); and Carol Bruton (Lawrence Livermore National Laboratory). Richard Price (TMS Inc.) and Clifton Carwile (consultant) also made substantial contributions. Michael Reed, Jim McVeigh, Jihan Quail, and Christina Van Vleck (SENTECH, Inc.) and Raymond David (National Renewable Energy Laboratory) contributed to the design and production of this report.

EXECUTIVE SUMMARY

A DOE-sponsored study, ***The Future of Geothermal Energy***[1], by a panel of independent experts led by the Massachusetts Institute of Technology (MIT), examined the potential of geothermal energy to meet the future energy needs of the United States. The panel concluded that geothermal energy could provide 100,000 MWe or more in 50 years by using advanced technology known as Enhanced Geothermal Systems (EGS). EGS are fractured, hot-rock

* This is an edited, reformatted and augmented version of a U. S. Department of Energy publication dated 2008.

reservoirs that have been engineered to extract heat by the circulation of water between injection and production wells.

This report briefly reviews the assumptions and conclusions of the MIT study. On the whole, those assumptions were reasonable and within the bounds of a balanced systems analysis. Conclusions about the amounts of investment needed to achieve competitiveness and produce 100,000 MWe were not supported.

The report's primary purpose is to evaluate relevant technology from today's commercial geothermal industry and other related industries. Much of the information covered here was developed through workshops attended by experts from the geothermal industry and related industries.

The steps in EGS reservoir development involve identifying a suitable site, creating the reservoir, and operating and maintaining the reservoir. Each step requires implementation of technologies specialized for the uniquely challenging geothermal environment. Currently available technologies are identified and assessed relative to their ability to satisfy the needs of EGS reservoir development. The adequacy of technology has been determined for both near- and long-term applications.

To achieve the goals outlined in the MIT study of large scale (100,000 MWe) use of cost-competitive geothermal energy, significant advances are needed in site characterization, reservoir creation, wellfield development and completion, and system operation, as well as improvements in drilling and power conversion technologies. These technology improvements will also support ongoing development and expansion of the hydrothermal industry. To realize the promise of EGS as an economic national resource, we will have to create and sustain a reservoir over the economic life of the project. The DOE strategy is to leverage and build from current geothermal technologies and resources to develop the advanced technologies required for EGS, while at the same time generating benefits in the near-, mid-, and long-term. This will require a systematic, sustained research and development effort by the Federal government in strong partnership with industry and academia to ensure full development of EGS.

A broad knowledge base about reservoir creation and operation will be essential for the eventual commercialization of EGS on a scale envisioned by the MIT study. This knowledge can only be gained by experience from field demonstrations in a variety of geologic environments reflecting a range of reservoir conditions. Immediate technology improvements are needed in reservoir predictive models, zonal isolation tools, monitoring and logging tools, and submersible pumps. These improvements and others stemming from the evaluation are essential for reaching the long-term potential[2] of EGS.

1. PURPOSE

In January 2007, a comprehensive study, ***The Future of Geothermal Energy***, was released by the Massachusetts Institute of Technology (MIT). The DOE-sponsored study, performed by a panel of independent experts, examined the potential of using the Earth's heat to help meet the future energy needs of the United States. The panel concluded that geothermal energy is capable of providing at least 100,000 MWe within 50 years by using advanced technology known as Enhanced Geothermal Systems (EGS). The MIT panel also concluded that: "Most of the key technical requirements to make EGS work economically over a wide area of the country are in effect, with remaining goals easily within reach."[3]

This report examines the conclusions of the MIT study and the critical assumptions that led to those conclusions, and determines whether those conclusions are well-founded. The report also evaluates the state of technology available in today's commercial geothermal industry and other EGS-related industries. Technologies from those industries will be essential to meeting the 100,000 MWe goal envisioned by the study.

Technology is constantly evolving; improvements and new approaches are introduced as market conditions warrant or dictate. This report only considers the adequacy of technology available today to bring EGS projects to market. With the success of the first commercial projects, technology improvements beyond those envisioned here can be expected to meet future needs.

In developing information for this report, DOE sponsored four workshops at which experts from today's geothermal industry and representatives from allied industries, such as the oil and gas industry, were asked to give their candid opinion of EGS technology. These workshops provided the basis for many of the conclusions presented here. Summaries of the workshops are available at the DOE geothermal web site: http://www.eere.energy.gov/geothermal/development_workshops.html. The summaries contain additional discussion of relevant issues.

2. Geothermal Energy and the EGS Concept

Heat is naturally present everywhere in the earth. The MIT study calculated the prodigious amounts of heat at depths from 3 to 10 km (Figure 1). Given the current state of knowledge about the earth's thermal properties, this analysis is sound and was endorsed at the EGS workshops. For all intents and purposes, heat from the earth is inexhaustible.

Water is not nearly as ubiquitous in the earth as heat. Most aqueous fluids are derived from surface waters that have percolated into the earth along permeable pathways such as faults. Permeability is a measure of the ease of fluid flow through rock. The permeability of rock results from pores, fractures, joints, faults, and other openings which allow fluids to move. High permeability implies that fluids can flow rapidly through the rock. Permeability and, subsequently, the amount of fluids tend to decrease with depth as openings in the rocks compress from the weight of the overburden.

At shallow depths, typically less than 5 km, the coincidence of heat, water (usually with dissolved minerals and gases), and permeable rock can result in natural hot water reservoirs. These hydrothermal reservoirs have impermeable or low-flow boundaries such as structural discontinuities or other geological features that impede the migration of fluids. Often, hydrothermal reservoirs have an overlying layer or caprock that bounds the reservoir and also serves as a thermal insulator, allowing greater heat retention. If hydrothermal reservoirs contain sufficient fluids (water or steam) at high temperatures and pressures, those fluids can be produced through wells to generate electricity or, for process heat.

Today, the geothermal industry is a thriving commercial enterprise in the United States and throughout the world. The installed domestic capacity of geothermal power plants exceeds 2,800[4] MWe total in five western states, and the Geothermal Energy Association

(GEA) predicts that generating capacity in the United States will double over the next five years, driven in large part by incentives such as the Production Tax Credit. These power plants use hot water and steam from hydrothermal reservoirs as their energy source.

Beginning with national resource assessments by the U.S. Geological Survey (1975 and 1978[5]), various studies have estimated the developmental potential from identified hydrothermal resources to be in the range of tens of thousands of megawatts. After some 30 years of exploration, the estimated total potential has not increased significantly, leading some analysts to conclude that the occurrence of natural hydrothermal reservoirs is limited. The natural hydrothermal resource is ultimately dependent on the coincidence of substantial amounts of heat, fluids, and permeability in reservoirs, and the present state of knowledge suggests that this coincidence is not commonplace in the earth. An alternative to dependence on naturally occurring hydrothermal reservoirs involves human intervention to engineer hydrothermal reservoirs in hot rocks for commercial use. This alternative is known as "Enhanced Geothermal Systems," or EGS. EGS reservoirs are made by drilling wells into hot rock and fracturing the rock sufficiently to enable a fluid (water) to flow between the wells. The fluid flows along permeable pathways, picking up *in situ* heat, and exits the reservoir via production wells. At the surface, the fluid passes through a power plant where electricity is generated. Upon leaving the power plant, the fluid is returned to the reservoir through injection wells to complete the circulation loop (Figure 2). If the plant uses a closed-loop binary cycle to generate electricity, none of the fluids vent to the atmosphere. The plant will have no greenhouse gas emissions[6] other than vapor from water that may be used for cooling.

A complete geothermal system includes both surface and underground components, and the MIT study analyzed elements of both components. DOE has focused this technology evaluation on the underground component (i.e., the EGS reservoir), rather than the energy conversion surface component. The surface component represents a significant fraction of the overall cost of a commercial EGS and will be a major factor in ultimately determining economic viability. However, by far the greater knowledge gaps and technology uncertainties involve the reservoir.

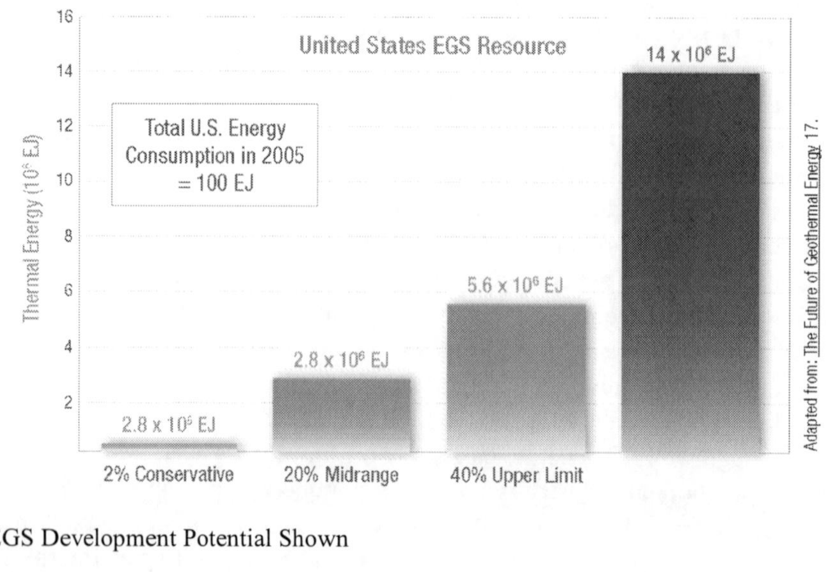

Figure 1. EGS Development Potential Shown

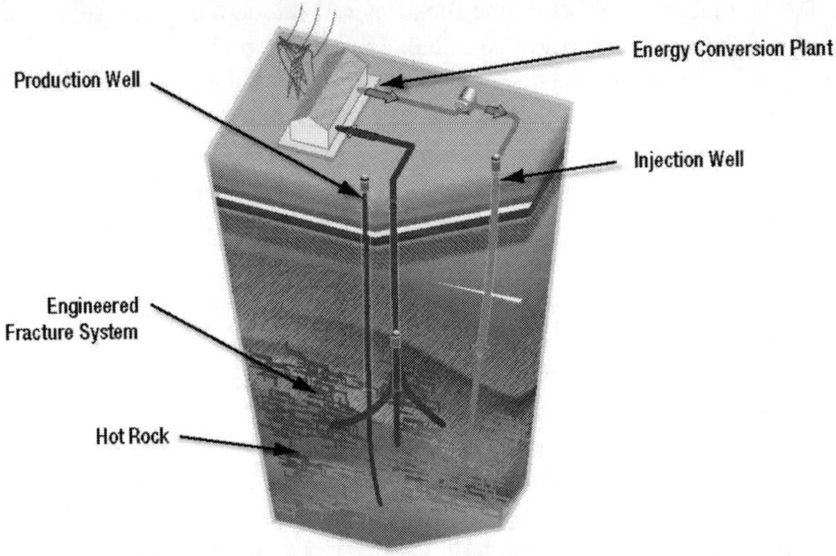

Figure 2. EGS Cutaway Diagram

3. Assessment of Assumptions in the MIT Study

The DOE Geothermal Technologies Program (GTP) hosted four workshops, starting with a workshop on Enhanced Geothermal Systems on June 7-8, 2007, in Washington, DC, with invited experts from geothermal and related industries. Key individuals from the MIT panel described the study's assumptions and uncertainties, methods, and results to the participants. Summaries of the workshops are available at the DOE geothermal web site: http://www1.eere.energy.gov/geothermal/egs_technology.html.

The MIT study made a number of assumptions regarding geothermal resources, EGS technology, and the economics of EGS. The more critical of those assumptions (italicized) are examined below. Further discussion about the assumptions can be found in the workshop summaries.

3.1. Geothermal Resource

The study used the most current data available on subsurface temperatures across the United States to estimate heat in place at depths of 3 to 10 km. The analytic technique combined heat flow data, a general representation of geology, and thermal conductivities for the rock underlying the contiguous United States in a geographic information system (GIS) model to calculate the temperature at depth[7]. Oil, gas, and water well temperatures were used to validate the model's predictions, but those data are limited in depth and geographic extent.

Estimation of the resource required assumptions regarding *thermal gradient and conductivity of rock and the ability to extrapolate limited data to ensure complete coverage*. Because rock types at depth are uncertain for many locations, conductivity was estimated for most sites. The conductivities of common rock types at the depths of the study vary within a narrow range, and the impact on resource calculations is minimal.

Although few direct measurements of temperature exist at depths greater than 5 km in the United States, the estimates reported by the MIT study are sound and based on verifiable theory. The energy calculations are conservative in that the MIT panel discounted the considerable heat known to be present in the 0 to 3 km depth interval as well as the heat beneath certain dedicated lands such as national parks and wilderness areas.

3.2. Recoverable Resource

Literature estimates[8] of recoverability range as high as 90% of the heat in place (at uneconomic low flow rates) to a more credible 40%, given an optimal production strategy. The MIT study examined the recovery factor parametrically[9]. *The recovery factor of 2% is a conservative assumption* in the analysis. No geothermal reservoir has operated long enough to demonstrate an actual long-term recovery factor. Even at 2% recovery, the amount of resource in place is prodigious, well in excess of that needed to produce 100,000 MWe.

While seemingly conservative for overall resource calculations, the choice of recovery factor is critical for EGS reservoir creation and operation and ultimately for economic viability. At the reservoir scale, the MIT panel apparently assumed *a local depletion factor for their economic analysis* based on modeled results by Sanyal and Butler[10]. Without long-term reservoir performance data, the choice of recovery factor remains somewhat arbitrary. Some reservoir engineers have calculated that at The Geysers geothermal field in northern California, the world's largest geothermal power complex, only about 10% of the heat in place has been recovered after nearly 50 years of production. Given the field's continued operation, albeit at a reduced production level, 10% might be taken as a conservative estimate.

3.3. EGS Well Drilling

Drilling is an essential operation in creating and sustaining EGS reservoirs. Today's oil and gas drilling technology can routinely reach depths of 4 to 5 km. The MIT study concludes that drilling costs rise exponentially for oil and gas wells while costs for geothermal wells remain linear. Estimates of drilling costs with depth were calculated by a parametric cost model using a rather limited data base from shallow wells[11]. The assumption of *linear well cost with depth* is not realistic beyond 5 km, given the rigors of the geothermal environment (temperature, pressure, hard crystalline rock, reactive fluids) and current state of technology. Hence, the projected well costs with depth are considered optimistic.

The MIT study modeled improvements in drilling based on an analysis of experience gained through case studies. This concept of experience driving improvements is termed the "learning effect." While there are uncertainties in the impact of learning, available studies tend to validate the assumption that learning reduces well costs, especially within a given field. Although a number of technological improvements are examined which would reduce cost, those improvements are correctly not included in the economic analysis. Ultimately, drilling costs will have to be reduced to produce 100,000 MWe from EGS.

3.4. Reservoir Creation

The MIT study assumed that *the principal means of EGS reservoir creation will be hydraulic stimulation* or the pumping of large volumes of fluids into the reservoir rock thereby fracturing the rock or opening pre-existing fractures. Hydraulic stimulation is a standard, mature technology[12], used in oil and gas fields to enhance production. This technology has been applied at all the EGS field projects to date with varied success. The MIT study contains an excellent summary of those stimulation experiments.

The key assumption associated with reservoir creation is that sufficient volumes of rock can be stimulated with enough fracture surface area and permeability to enable the extraction of large quantities of heat. This assumption is partially corroborated by EGS field experiments around the world, notably at Soultz-sous-Forêts, France. Rock volumes on the order of cubic kilometers can be stimulated, assuming that observed microseismic events are indicators of shear fractures and hence, correlate with reservoir volume. However, the assumption that the reservoir volume will have adequate interconnectivity or permeability at commercial scales has not yet been proven. The results to date are based on pilot-scale experiments. An attempt to expand the reservoir at Soultz by connecting a third well to two others was largely unsuccessful, probably due to a previously unknown permeability barrier within the reservoir[13].

Assumptions about the ability to create EGS reservoirs of sufficient volume, surface area, permeability, and inter-well connectivity for commercial applications are reasonable, but optimistic, given the current state of knowledge. These assumptions have not been corroborated by large, well-documented field projects in a number of different geologic settings.

3.5. Reservoir Operation and Maintenance

The MIT study of reservoir performance under production conditions contains significant uncertainties that derive from reservoir geometry and permeability. The flow rate of circulating fluid in an EGS reservoir and the thermal drawdown associated with this flow rate are major unknowns. The analysis assumed *a flow rate of 80 kg/sec at 200°C from each production well*, equivalent to a commercial hydrothermal reservoir. This is a reasonable target, given that EGS reservoirs are intended to serve as enhanced or augmented hydrothermal reservoirs. At present there is no experimental evidence to verify that this level of productivity can be achieved. As pointed out in the analysis, the EGS project at Soultz, which is the best-performing project to date, has had a maximum well productivity of about 25 kg/s. Well productivity remains the greatest technological challenge for the commercialization of EGS.

Besides productivity, the analysis assumed that the fracture system will provide sufficient thermal stability for long-term production. This derives from the total effective surface area of the reservoir. The MIT study assumed *a conservative reservoir lifetime of six years, where lifetime is defined as a 10°C decline in fluid production temperature*, after which the reservoir would have to be re-drilled and re-stimulated. This temperature decrement is conservative because greater amounts of cooling have been observed in commercially operating reservoirs. The reservoir lifetime and other parameter values for the base case EGS economic models are unknown in that no commercial-scale EGS plant has operated with sufficient thermal drawdown to establish reliable lifetime performance data.

The analysis assumed that *the system loses up to 2% of total injectate during reservoir operation*. For some systems, the cost of water could dominate stimulation costs. Water losses during operation are also a potentially important cost. Lacking knowledge about water consumption in various EGS environments, the assumption to limit water losses is optimistic if water must be accounted for in project costs.

For energy conversion, the assumption was that *the engineering systems would be the same as those used for liquid-dominated hydrothermal resources at similar temperatures (flash steam and binary power cycles)*. This is a reasonable assumption since differences between fluids produced from hydrothermal and EGS reservoirs should be minimal once chemical stability is attained during circulation. The thermodynamic analyses are based on well-understood and well-founded theory and data. Because energy conversion efficiencies have a linear influence on the calculated recoverable resource, errors in assumed energy conversion efficiencies may represent a minor source of error in resource calculations. The overall approach and the cost and performance results obtained are sound.

3.6. EGS Economics

Every element of the economic analysis has a different level of risk and different calculation requirements. Assumptions were made regarding many parameters for each EGS system element, including reservoir productivity, drilling, plant cost, resource depth, interest rates, and so forth. Interconnection with the power grid was assumed not to be an issue. Although stimulation and reservoir connectivity remain as major issues, reservoir creation efforts are assumed to be consistently effective. The technical parameters with the highest uncertainty and risk are flow rate per production well and thermal drawdown rate (i.e. reservoir lifetime.)

The analysis includes different learning curves for each technology element. *The learning curve for achieving 80 kg/sec flow rates was assumed to be a one-time*[14] *effect that is achieved fairly quickly.* The opinion of the experts in the workshops was that learning curves based on oil and gas experience may be optimistic for EGS well drilling[15]. The study uses an equity rate of return of 17%, which corresponds to a fairly risky venture. The drilling cost model uses a conservative contingency factor of 20% for trouble costs. For this type of economic modeling, the surface plant design is not specified in detail, so correlations must be used. Taken in total, the learning curve for plant costs is somewhat optimistic, and the long-term cost was based on the judgment of the MIT panel. Sensitivity analyses were performed to identify the variables most responsible for uncertainty and risk.

Some important assumptions were made regarding future baseload supply and demand. Of the 90 GWe of nuclear power in the existing power plant fleet, about half are assumed to be retired in the time frame of the study, and about 50 GWe of coal generation is also projected to be retired. This turnover in existing plant inventory provides an opportunity for replacement with EGS, concomitant with the development goal of 100 GWe.

The study assumed that *a carbon tax of $10/ton equivalent (equivalent to seven to eight mills/ kWh) would be introduced beginning in year 10*. This tax was included on the expectation that policies would be adopted that would make thermal generation pay some externality cost (either through carbon capture and sequestration, or through a tax on CO_2 emissions). Without the carbon tax assumption, the predicted economic advantage of EGS over conventional electric generation is reduced and the rate of EGS penetration would be lessened.

3.7. Summation

The authors of the MIT study based their technical assumptions on results from available field tests, published reports, and well-established theory. On the whole, those assumptions were reasonable and within the bounds of a balanced systems analysis. The study's findings, in particular that 100,000 MWe from EGS technology can be achieved within 50 years, are credible. But as the study points out, significant constraints exist in creating sufficient connectivity between wells to meet economic requirements for reservoir productivity and lifetime. Overcoming these constraints will require substantial reservoir testing in a number of different geothermal environments as well as research-driven improvements in technology. Consequently, the conclusion that a $300 million to $400 million investment over 15 years will be needed to make early-generation EGS power plants installations competitive is overly optimistic. This level of investment and the combined public/private investment of $800 million to $1 billion over 15 years to encourage sufficient deployment to produce 100,000 MWe are not supported by the analysis. Investments in excess of these amounts will probably be required.

4. Technology for EGS Reservoir Development

The MIT study provides a firm basis on which to consider how to bring the vision of commercialization of EGS technology to fruition. The remainder of this report evaluates available technologies from various fields and estimates their utility for EGS applications.

In conducting this technology evaluation, EGS reservoir development has been represented as a multi-step decision process. The process goal is to create an EGS reservoir that can operate economically. The logical steps that must be taken to complete an EGS economic reservoir project are: (1) finding a site; (2) creating the reservoir; and (3) operating the reservoir. The steps are illustrated in Figure 3, showing the tasks that must be performed. The decision process that should be followed to complete each step and task is illustrated in Appendix A.

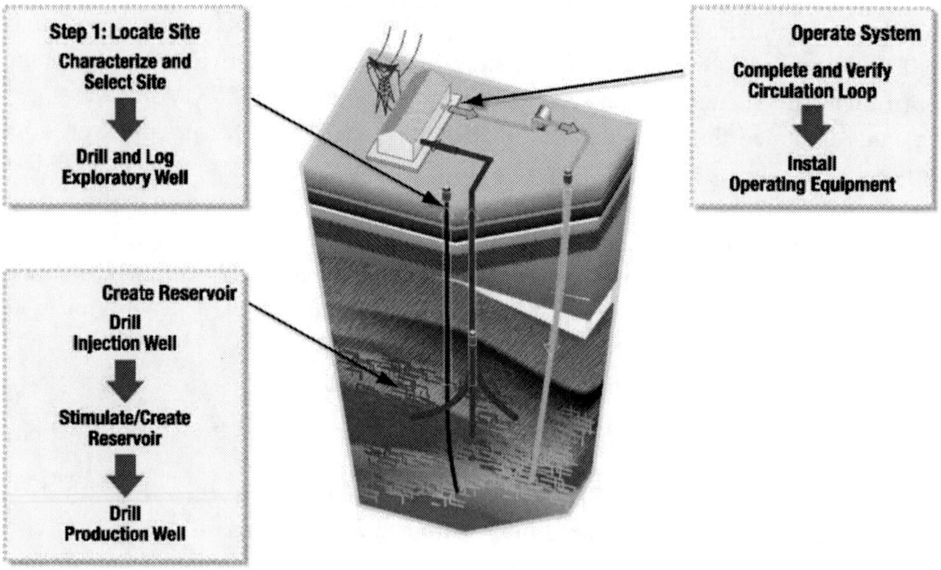

Figure 3. EGS Development Sequence

At each step, a certain measure of performance must be achieved to allow the project to proceed to the next step. Those performance measures will usually depend on conditions at the site and the desired operational properties of the EGS reservoir. Ideally, they would be determined a priori by the model used to simulate the reservoir, but no such model yet exists at this level of detail. The MIT study used two independent models to consider EGS economics, but those models did not specify performance measures for each step of the development process. While lack of well-defined measures and a suitable reservoir model are drawbacks, the outcome of the technology evaluation should not be significantly affected at this level of analysis.

Results of the technology evaluation are presented here in Tables 1-5 which list the *Required Tasks* that must be performed to develop a commercial-scale reservoir, the *Available Technologies* to do the task, the current *Status* of those technologies, and the *Adequacy* of the technologies to complete the task satisfactorily. In the tables, *Near-Term* adequacy refers to the ability of available technology to complete the required task independent of cost at a few selected demonstration sites. *Long-Term* indicates the ability to commercialize EGS on a scale commensurate with the MIT study. (Note: Color coding in the table indicates the relative degree of adequacy /readiness for each technology, with green denoting most adequate and red denoting least adequate.) While some technologies are adequate for the job at hand in the near-term, they may not be suitable for large-scale EGS deployment because of performance limitations or cost. Where applicable, such deficiencies are identified.

4.1. Finding the Site – Site Characterization

The first step in creating an EGS reservoir is to find an appropriate site. At this time, lack of experience with EGS development presents a problem in defining what is "appropriate."

Site characterization will draw on existing knowledge of the site and its surroundings to the extent data are available. Depending on the quality of the databases, substantial pre-development information may be obtained about a number of technical and non-technical properties. Adequate heat obviously must be present for the desired application, but the depth to the target temperature is important for economic reasons. Various site properties that should be known for successful creation of the reservoir include:

- Temperature gradient and heat flow
- Stress field
- Lithology and stratigraphy
- Structure and faulting
- *In situ* fluids and geochemistry
- Geologic history
- Seismic activity
- Proximity to transmission
- Land availability
- Demographics

The list is illustrative of the range of properties, rather than exhaustive. Some or all of these properties may already be known if wells have previously been drilled at the site and suitable data collected. Lacking such wells or data, prospective reservoir properties must be inferred from the surface. Additional information can be gathered by site analyses and surveys, such as seismic reflection and geologic mapping. Surface-based technologies are available that can provide information about many site characteristics, but that information becomes more problematic with depth. Remote technologies have proven successful in finding new hydrocarbon resources for companies involved with the oil and gas industry. However, these technologies have not yet been successfully applied to EGS, especially regarding expected reservoir properties. For example, the MIT study states that: "Exploration methods that can effectively tell us the stress field at depth from the surface are not currently available"[16]. There currently appears to be no known technological solution to remotely characterize important EGS reservoir properties with confidence.

Table 1 summarizes the information needs for site characterization and the technology available to meet those needs. The table and following commentary suggest potential areas of improvement in technology.

As Table 1 notes, current technology can be used to characterize potential EGS sites. As EGS commercialization grows, new technology will be needed that will enable site characterization to be done in a less costly manner and with greater confidence. The table indicates target areas of needed technological improvement:

- Better models of the appropriate geologic settings for EGS.
- Improved geophysical methods for finding fluid-filled fractures.

Geologic models are only as good as the quality and adequacy of the information used to construct them. Existing models are based on oil and gas field properties that cannot predict the potential for EGS. Models for most appropriate EGS geologic settings must be developed.

Table 1. Finding the Site – Site Characterization

■ = Most Adequate ■ = Some Degree of Adequacy ■ = Least Adequate

REQUIRED TASK	AVAILABLE TECHNOLOGIES	TECHNOLOGY STATUS	ADEQUACY Near-Term	ADEQUACY Long-Term
Determine temperature gradient and predict temperature at depth	Various temperature measurement tools in shallow boreholes	Commonly used throughout industry. Gaps are primarily data, not technology.	YES	YES
	Geothermometry (chemical and isotopic measurements)	Some interpretation of geothermometry requires sophisticated understanding of numerous interacting factors, such as shallow equilibration.	YES	NO
Determine stress field using surface-based technology	InSAR	The strength of Interferometric Synthetic Aperture Radar (InSAR) is its ability to provide observations of ground displacements with a precision of a few millimeters in images with 20-meter spatial resolution covering 100-km distances.	YES	YES
	Global Positioning System (GPS)	The GPS provides only regional coverage unless many instruments are used with close spacing. GPS may provide regional indication of stress.	YES	YES
Determine geologic characteristics and history (lithology and structure)	Geophysical surveys	Surveys are routinely used in mineral and petroleum exploration. Seismic interpretation is inadequate for many geothermal environments. Magnetotelluric surveys can be improved. Electrical resistivity surveys are technically acceptable, but equipment can be improved. Magnetic surveys are technically acceptable.	YES	NO
	Lithologic analysis	The science, methods, and equipment used for this analysis are mature.	YES	YES
	Geologic mapping	The science, methods, and equipment used for mapping are mature.	YES	YES
Detect fluid-filled fractures	Self-potential; streaming potential	Self-potential is commonly used for shallow hydrothermal systems.	NO	NO
Evaluate background seismicity	Seismometers located in shallow surface holes	Surface seismometry is a mature technology.	YES	YES
Predict potential for stimulation	Geologic models from the oil and gas industry	Data and experience are inadequate for modeling of most projected EGS environments due to lack of sufficient measurements under geothermal conditions.	NO	NO

Table 2. Finding the Site – Exploratory Well and Reservoir Characterization

■ = Most Adequate ■ = Some Degree of Adequacy ■ = Least Adequate

REQUIRED TASK	AVAILABLE TECHNOLOGIES	TECHNOLOGY STATUS	ADEQUACY	
			Near-Term	Long-Term
Characterize subsurface conditions	High-temperature logging and imaging tools Log interpretation methods	Some tools for one-time measurements of wellbore and system parameters are available, but they may be deficient at high temperatures. Geothermal log interpretation methods are derived from oil and gas experience.	YES	NO
	Stress measurement inferred from natural breaking of rocks in the wall of the wellbore and "mini-frac"	Principal stress direction and magnitude are estimated from limited testing capabilities. Imaging tools for breakouts currently require a heat shield.	YES	NO
	Core sampling and evaluation	Routinely used in mineral exploration. Interpretive techniques for geothermal applications are still evolving.	YES	NO
Isolate zones within the well for mini-fracs and flow testing	High-temperature packers	A prototype high-temperature packer for low-pressure applications is available, but little field work has been performed.	NO	NO
Conduct flow tests	Pressure, temperature, and fluid-flow measurement tools	Suitable downhole instrumentation for standard flow tests available up to 200°C.	YES	NO
Perform stress measurements	Micro-fracs and borehole breakouts, core-based measurements	Suitable technology available for lower-temperature applications. Technology lacks zonal isolation capability.	YES	NO
Interpret data to plan stimulation	Models predicting the effect of stimulation on fracture formation and growth	Current models have not been effective in a geothermal environment.	YES	NO

Significant research is required to develop and demonstrate surface-based technology with adequate resolution at reservoir depth. At present, remote determination of key EGS reservoir rock characteristics appears out of reach in the near-term.

4.2. Finding the Site – Exploratory Well and Reservoir Characterization

While remote techniques for determining in-situ reservoir properties give some first-order knowledge about the site, they do not confirm the site's suitability for development at the projected reservoir depth. This must be done with an exploratory well, which may be either a slim hole or a full-scale injection well. This well is intended to measure and/or confirm reservoir properties; it is not necessarily part of the final EGS reservoir. A slim hole has the advantage of lower cost, while a large diameter well may eventually be used for the final reservoir. The difference reflects the degree of confidence (and financing) the developer has in completing the project. The situation would be improved by the presence of a pre-existing well that can be reworked (i.e., deepened, diverted, or perforated) as an exploratory well.

Taking rock core and a suite of well logs is a vital element of drilling an exploratory well to fully characterize the reservoir rock. Because stress field azimuths guide the drilling program for the remainder of the wells in the field, a small hydraulically created fracture (a "mini frac") is induced in the reservoir rock to determine the in-situ stress field. Flow tests of fluid between the rock and the wellbore should also be conducted as a means of determining reservoir productivity prior to stimulation.

The aforementioned techniques will probably be used in the first wells drilled at any project. Table 2 shows technology availability and adequacy for gaining knowledge of the subsurface that will be useful in planning the stimulation. The table indicates that adequate technology exists to characterize early EGS sites, but broad-scale improvements in technology will be required in the long-term.

The MIT study and Table 2 note that high-temperature instrumentation for borehole imaging and other purposes is a key technology deficiency. Though tools exist that can perform satisfactorily for short periods, instruments capable of collecting data in place for protracted periods (i.e., days to years) for well stimulation and, more importantly, for reservoir operation and management remain elusive. Until methods for reliable zonal isolation are available for high-temperature applications at high differential pressures, all stimulation attempts, including mini-fracs, will be limited to open-hole or low- temperature applications.

The procedures for drilling the exploratory well should not differ greatly from those used in drilling any moderate to deep well. If the well has an intended use as a production or injection well, precautionary steps should be taken in well completion (casing strings and cementing, open-hole section). The technologies, materials, and services needed to construct the well are available from commercial suppliers.

4.3. Creating the Reservoir – Injection Well

Once preliminary characterization activities have been completed, reservoir development can proceed with drilling of the initial injection well. Information about the reservoir rock (e.g., temperature, stress field, lithology, and structure) is valuable in planning the drilling campaign. As with the exploratory well, drilling technology is fully commercial, though long-term improvements can be made, especially by adapting equipment and tools to the high-temperature geothermal environment. Table 3 addresses the adequacy of current drilling technology for EGS applications.

EGS well construction activities resemble those employed in the oil and gas industry, but there are substantive differences. Geothermal wells are typically drilled at higher temperatures with larger diameters in harder rock. These differences and the small size of the geothermal industry have retarded geothermal drilling technology relative to oil and gas technology. The geothermal industry has moved forward despite the disadvantages. For purposes of this evaluation, technologies associated with drilling and completing injection and production wells are taken to be the same, and this report does not differentiate between them.

Table 3. Creating the Reservoir – Injection Well

■ = Most Adequate ■ = Some Degree of Adequacy ■ = Least Adequate

REQUIRED TASK	AVAILABLE TECHNOLOGIES	TECHNOLOGY STATUS	ADEQUACY	
			Near-Term	Long-Term
Reduce rock	Drilling bits	Roller bits are used in hard rock. Advanced bits (e.g., PDC-based drag bits) are used in oil and gas, and they drill 60% of footage worldwide. Alternatives to mechanical methods (flame jet, etc.) are in experimental stages.	YES	NO
Steer the direction of wells	Advanced steering tools	Wireline based systems are used in geothermal. Commercial advanced steering tools allow control over well trajectories. Tools providing limited steering control are in use by one geothermal firm. Commercial tools are limited to ~150°C.	YES	NO
	Logging while drilling/diagnostics while drilling	The technology is commonly used in the oil and gas industry. Commercial tools are limited to ~150°C.	YES	NO
Complete wells	Metal casing in various diameters and production tubing (e.g., slotted liner)	Fully commercial systems to complete wells are available. Advanced technology, such as expandable tubulars and casing-while-drilling and low-clearance casing systems, is emerging in oil and gas applications. Underreamers work only in "soft" rock. Elastomers used in these systems fail at high temperatures.	YES	NO
	Design methods for selective cementing	Various high-temperature cement formulations are available from the drilling service industry. Design methods for selective cementing of casing exist for wells with small temperature fluctuations.	YES	NO
Isolate zones during drilling	Open and cased hole packers and expandable tubulars and screens	Elastomer and cement packers commonly used for low-temperature applications. Experimental versions of low-pressure packers developed for geothermal applications are not generally available. Retrievable packers for high pressure operations in high temperature (>150°C for extended operation) wells are not available.	NO	NO
Log well	Logging tools (e.g., tools to measure downhole pressure, flow, temperature, image fractures)	Logging tools and sensors are available for operation up to about 150°C. Higher temperature versions of some tools are available but have limited lifetimes or require heat shielding. Unshielded prototypes for pressure and temperature are experimental.	YES	NO
Monitor well parameters	Micro-fracs and borehole breakouts, core-based measurements	Suitable technology available for lower-temperature applications. Technology lacks zonal isolation capability.	YES	NO
Interpret data to plan stimulation	Monitoring tools, sensors (e.g., tools to measure downhole pressure, flow, temperature, and seismicity)	Monitoring tools and sensors are available commercially for sustained operation up to about 150°C. Tools capable of operation at >200°C are still experimental.	YES	NO

The following commentary elaborates on points made in Table 3:

- PDC bits dominate drilling because of increased rate of penetration and longevity, but these bits have yet to be proven in geothermal environments. Roller cone bits are used in geothermal hard rock environments, a century-old technology that is robust but slow. Advancements in rock reduction technologies will probably be needed for EGS commercialization.
- High temperatures have hampered the introduction of oil and gas related technologies into geothermal well construction. The target operating temperatures of EGS wells (≥ 200°C) are greater than those of almost all oil and gas wells. Steering tools used at The Geysers geothermal field are primitive (i.e., first generation steering tools that use a cumbersome detachable wireline for power and communication), and attempts to use more advanced tools have failed. Operators have been able to achieve adequate results from old technologies, but better steering and logging while drilling (LWD) tools are desirable.
- Casing and cementing costs are responsible for roughly 30% of the cost of constructing a geothermal well. Reducing the amount of steel used is a goal for all drilling operations. "Lean" casing designs such as expandable tubulars, casing-while drilling, and low-clearance casing systems (i.e., with a minimal annulus between casing strings) now emerging in the oil and gas industry offer additional advantages. These technologies increase options for dealing with difficult well conditions. Transferring this technology to the geothermal environment to bridge rock strength and temperature issues will require hard-rock underreamers. Some casing schemes, such as expandable tubulars, employ elastomer sealing elements that are not suitable at geothermal temperatures.
- Geothermal wells are cemented to the surface to constrain casing and stabilize wellheads. Reducing the amount of cement used would simplify cementing and provide greater predictability in casing-rock interactions. Robust designs proposed by industry experts will let the casing float or expand and contract freely due to thermal fluctuations. Similar strategies would also benefit injection well designs.
- Zonal isolation is essential for many EGS reservoir development activities. Well control problems often require direct measures (e.g., cementing for lost circulation control) at specific zones of the wellbore. Open-hole zonal isolation tools include packers and alternative casing systems. Experimental packer systems developed for geothermal environments only operate at low pressure, or they are not retrievable and are not commercially available. Current best practices address wellbore problems during drilling, but alternatives that employ retrievable open-hole packers will be needed in the future.
- Logging tools for measuring temperature, pressure, flow, fracture imaging, and other formation characteristics require heat shielding and can only be used for brief periods. While the drilling industry works within these limitations, more robust tools capable of operating in >200°C environments are needed. Monitoring tools emplaced in the wellbore for long-term operations measure many of the same parameters recorded during transient logging activity, but can also include other reservoir monitoring sensors such as those for monitoring induced seismicity. Advances in components, battery technology, materials, and fabrication methods are desirable.

- Economic viability will require design and construction of wells and well fields that efficiently exploit the geothermal resource. The design space for EGS well construction should include options for highly deviated directional wells, multilateral completions, multiple completion zones, and so forth.

Current technology gaps will hamper, but not prevent, implementation of EGS demonstrations and supporting experimental projects. Economic development of EGS will require advances in a number of technologies and well construction schemes that maximize the effectiveness of stimulation, injection, and production of EGS reservoirs.

4.4. Creating the Reservoir – Stimulation

Once the first injection well has been drilled and completed, the reservoir rock can be stimulated. Stimulation usually requires an open-hole section through the targeted fracture zone, which has been determined from logs, core, and other information gathered during site characterization.

1. Drill the first deep well (injection) with the casing set at appropriate depth to give required mean reservoir temperature.
2. Obtain basic fundamental properties of the underground such as stress field, joint characteristics, *in situ* fluid characteristics, mechanical properties of the rock mass, and the identification of flowing/open zones where appropriate.
3. Having established the best positions for the sensors of the microsensor array, install an appropriate instrumentation system to yield the best possible quality of microseismic sensor array, install an appropriate instrumentation system to yield the best possible quality of microseismic event locations, not only during the first stimulations, but for all events likely during the reservoir's lifetime.
4. Conduct stepped flow rate injections until the pressure for each injection's step becomes steady. The maximum injection pressure should exceed the minimum formation stress at the point of injection.
5. Maintain high flow rate injection until the seismicity migrates to the distance necessary for targeting the second well.
6. Depending on the relationship of the *in situ* stress an the density of *in situ* fluid, it may be possible to influence the vertical direction of reservoir growth by selecting an appropriate stimulation fluid density.
7. Perform a shut-in test to assess the size of the reservoir.
8. Carry out flow logs in the injection well to identify the main flowing zones.
9. Let the reservoir deflate and then make injection tests at lower flow rates to assess the permanent residual enhancement of permeability, i.e., flow against injection pressure.
10. Target the second well (production) into the periphery of the seismically activated structure, with well separation appropriate to suit economic targets. At the same time, ensure that the well has a downhole pumping chamber incorporated in its completion plan.
11. Stimulate the second well in a stepped manner as described about to improve access to the previous stimulated zone and eventually permit the recovery of the mobile *in situ* fluid (carry out diagnostic technique as in steps 3 and 8).
12. Conduct short circulation tests to assess the connectivity between the injector and the producer.
13. Perform tracer tests to evaluate reservoir flow-through volume, to characterize the residence time distribution, and to identify any short circuit paths.
14. Repeat steps 10 to 14 for the third well, i.e., the second production well, and for a fourth and even fifth, if the system warrants this.

Adapted from: The Future of Geothermal Energy 12-14.

Figure 4. MIT Study Reservoir Creation Process Tasks

The tasks required to stimulate the reservoir as determined by the MIT study are shown in Figure 4 (see Tasks 1-9). Carrying out these tasks should create a large fractured volume of rock. The fracturing fluid must be pumped at high pressures and flow rates, and a high-fidelity seismic monitoring network is critical to the stimulation (Task 5).

Table 4. Creating the Reservoir – Stimulation

■ = Most Adequate ■ = Some Degree of Adequacy ■ = Least Adequate

REQUIRED TASK	AVAILABLE TECHNOLOGIES	TECHNOLOGY STATUS	ADEQUACY	
			Near-Term	Long-Term
Plan and design stimulation (e.g., zones, pressures, volumes, fluids, proppants)	Stimulation models for oil and gas; basic numerical models for geothermal applications	Stimulation modeling techniques for geothermal systems are not a mature technology.	**YES**	**NO**
Imaging and mapping of fractures	Microseismicity, gravimetry, self-potential, tiltmeter arrays	Surface microseismic and gravity tools are adequate for most purposes, but the resolution may be insufficient for EGS. Self-potential is not proven for this purpose. Tiltmeter results are difficult to interpret in zones of multiple fractures.	**YES**	**NO**
Identification of flow paths during and post-stimulation	Microseismicity, gravimetry, SP, tiltmeter arrays	The utility of existing techniques for tracking fluid flow has not been demonstrated. Microseismic techniques are not hardened for downhole use.	**NO**	**NO**
Effective real-time decision-making capability for stimulation	Oil and gas industry stimulation modeling and control technology	The oil industry has modeling and control capability for petroleum environments, but experience in geothermal systems is lacking.	**YES**	**NO**
Zonal isolation for stimulation	Stimulation packers, slotted liners	Packers that can operate at stimulation pressures and temperatures are not available. Slotted liners and related technologies may not perform adequately for EGS	**NO**	**NO**
Create/enhance flow paths	Hydraulic stimulation; chemical stimulation; and rate controlled explosives	Geothermal stimulations for EGS use water or water weighted with dense chemicals such as barium sulfate salts. Chemical and other stimulation methods have been used in hydrothermal systems.	**YES**	**NO**
Keep flow paths open	Proppants for both near well bore and far field use scaling, dissolution, and permeability control	Proppants are typically used in oil and gas stimulations. Temperature-hardened proppants have not been evaluated in geothermal environments. Scaling and dissolution control technologies are available, but may not be adequate for EGS conditions.	**YES**	**NO**

The knowledge base for stimulation of geothermal systems remains limited. Some experts believe that successful stimulation will require favorably oriented pre-existing fractures or zones of weakness. Furthermore, stimulation would be dominated by shearing rather than tensile fracturing. They advocate applying only enough hydraulic pressure to shear existing zones of weakness, rather than using higher pressures that would induce tensile fractures. Hydraulic stimulations in oil and gas fields, which are typically done at pressures well in excess of the rock strength, appear to include a combination of both shearing and tensile fracturing. For EGS to become a universal technology, stimulation must succeed in a variety of stress environments.

The technology is available to stimulate both petroleum and geothermal reservoirs, though stimulation is not commonly practiced by the geothermal industry. The ability to create a circulation system with both high productivity and thermal stability over time has not been demonstrated.

Table 4 identifies the adequacy of technologies needed to complete the stimulation successfully.

The purpose of reservoir stimulation is to provide abundant fluid flow paths between the injection well(s) and the production well(s). These flow paths should have minimal impedance to reduce pumping power needs, but adequate residence time and surface area to sustain the production of hot fluids.

The following commentary elaborates on points made in Table 4:

- Stimulation design involves selecting zones of the wellbore to be targeted for stimulation as well as the rate, pressures, and volumes of injectate. The type(s) of stimulation fluids and use of proppants must also be considered. The oil and gas industry routinely employs application-specific design tools for its stimulations[17]. While design codes exist, robust tools that couple hydrological-thermal-mechanical chemical phenomena are not available. Revised and/or new design tools will be required for commercial EGS development. Field experience and data are vital to developing such tools.
- Mapping the evolution or growth of fractures during stimulation is very important. Surface (or near-surface) seismic monitoring, gravimetry, and tilt measurements are generally considered adequate for fracture imaging. The utility of these measurements depends on the fracture population; for example, tilt measurements are most useful in tracking single, rather than multiple, fractures. Increased resolution and accuracy of these mapping techniques requires downhole tools that can withstand the temperatures associated with EGS.
- While remote sensing of fracture growth via microseismic analysis indicates possible fluid flow paths, the ability to directly map the flow through the created reservoir does not currently exist. Methods to track fluid flow should be investigated[18].
- The oil and gas industry has demonstrated that real-time control and adjustment of the stimulation process is vital to success. Development of intelligent control systems requires both theoretical understanding and practical knowledge obtained from multiple stimulations in a variety of situations.
- Zonal isolation is required for selective stimulation of target wellbores. For open-hole applications, zonal isolation tools include packers and alternative casing designs. While experimental packer systems have been developed for geothermal

environments, they were either designed for low pressure applications (e.g., for lost circulation control) or were not retrievable. General purpose open hole packers do not exist for geothermal environments, with the primary barrier being the poor stability of elastomeric seals at high temperatures. Cased hole isolation tools suitable for high-temperature environments are emerging, and these tools have the advantage of metal-to-metal seals.

- The purpose of stimulation is to create well-to-well flow paths that minimize impedance but meet operational needs. As previously suggested, the stimulation method may be important in the development of an optimum fracture network. Variable fluid weights and viscosities for hydraulic stimulation may be needed. Chemical stimulation techniques may be needed to increase permeability near the wellbore and in far-field fractures. More novel techniques using deflagration methods could also be used to enhance the effective radius of the wellbore.
- The MIT study assumes that EGS should target geologic settings where fractures are critically stressed. By stimulating such zones, reductions in the normal stress across fractures will induce shear failure. If the rock's matrix strength is sufficient, this shearing will result in dilation and increased fracture aperture; the fracture effectively "self-props." In settings where the stress state and fracture characteristics are not optimum, proppants may have to be employed to prevent fractures from closing. In addition, it may be advantageous to create secondary fractures that link to the existing network, again requiring proppants. While proppants are used extensively in the oil and gas industry, they are not chemically stable for high-temperature applications. Temperature hardened proppants may be required.

Creation of the reservoir through stimulation is considered to be a critical aspect of EGS development. Technology needs, such as real-time control of the stimulation, are extensions of current petroleum industry capabilities. EGS-specific needs, those with general applicability, include understanding fluid flow paths through use of imaging techniques not currently available.

4.5. Completing the Well Field

Once the initial volume has been stimulated by either opening existing fractures, creating new fractures, or both, circulation can be established by drilling a production well. Care must be taken in drilling this well; directional drilling may be required to intersect the fractures created during stimulation of the initial well. Measurements of acoustic signals generated by the stimulation are used to delineate the zone of fracturing. At present, insufficient knowledge about fracture behavior is available to pinpoint the target zone. The current strategy seeks to penetrate the fringe of the acoustic signals or microseismic cloud, rather than the zone of highest event density, to maximize the inter-well distance.

The remaining tasks to complete the reservoir are described in the MIT study (see Figure 4, Tasks 10-14). These tasks are necessary to establish sufficient connectivity between wells. As suggested by Task 14, the number of wells drilled (both injection and production) will depend on the size of the reservoir, the productivity of the wells, and the development plan. In early stages, especially at a new site, the number will likely be in the range of three to six wells until working experience is gained.

Production wells will follow much the same drilling procedures as injection wells (see Table 3). Differences are likely in the extent of directional drilling, completion, and cementing to accommodate stresses due to thermal cycling. Temperature-hardened proppants may be required to maintain fluid flow between wells (see Table 4).

Technology issues related to completing the EGS reservoir with additional wells appear at this point to be minimal. The greatest concern is positioning additional wells to optimize energy production while minimizing opportunities for short circuiting (premature breakthrough of injected fluids in production wells.) The best tool to avoid such unwelcome consequences is a reliable reservoir model which can predict flow between wells.

4.6. Operating and Maintaining the Reservoir

The economics of EGS hinge on the ability to produce energy for extended periods without resorting to expensive remedial actions, such as unscheduled drilling of additional wells. Experience in operating and maintaining EGS reservoirs is limited, and currently there is no knowledge base from which to make decisions. Critical issues at this step of development include short circuiting, dissolution or precipitation of minerals altering the reservoir's plumbing, buildup of dissolved solids and gases in the circulating fluid, inefficient recovery of heat, induced seismicity, and fluid losses from the reservoir. The petroleum industry has dealt with some of these issues, albeit in non-geothermal environments. In the absence of working experience with EGS reservoirs, responses to these issues must largely be trial and error.

Operationally, EGS wells will have to function at high pressures and flow rates in both injection and production modes for years. Pumping of fluids is expected to be necessary to maintain adequate flow rates. Reliable high-temperature submersible pumps have been consistently mentioned by experts as a technology gap throughout this analysis. The adequacy of technology to meet the operational requirements of EGS reservoirs is indicated in Table 5.

The management goal for reservoir operation is to sustain rated output for the design lifetime. To meet this goal it is necessary to optimize the extraction of heat with respect to temperature drop, maintain the rate of fluid production, prevent subsurface fluid loss, and minimize parasitic power losses. As heat extraction proceeds, temperatures in the reservoir will decline. Successful EGS reservoir management will require careful monitoring and mitigation to control the impact on heat recovery and power plant efficiency.

At this stage, every barrier to successful management and operation of EGS facilities has probably not been identified. However, as shown in Table 5, several technological advancements will clearly be needed to ensure economic success:

- Submersible electrical pumps will control fluid loss and minimize parasitic losses associated with high injection pressures. To meet EGS needs associated with long-term, high-temperature, deep-well operation, technical advancements in pump connections, materials, seals and controllers are required. Submersible electric pumps with 1000-3000 horsepower motors must survive ~3 years at ≥ 200°C.

Table 5. Operating the Reservoir

■ = Most Adequate ■ = Some Degree of Adequacy ■ = Least Adequate

REQUIRED TASK	AVAILABLE TECHNOLOGIES	TECHNOLOGY STATUS	ADEQUACY	
			Near-Term	Long-Term
Maintain acceptable flow rates and reduce or eliminate fluid loss	Submersible Electrical Pumps	SEPs have operating temperatures of 175°C with power ratings of 1500 kW. Pump technology is not adequate for long-term high-temperature deep operation.	NO	NO
	Packers for fracture isolation	High temperature zonal isolation tools are not available to allow control of flow from multiple zones.	NO	NO
	Lineshaft pumps	Lineshaft pumps are a fully commercial technology, but depth-limited.	YES	NO
Maintain reservoir and track reservoir evolution	Monitoring tools, sensors (e.g., tools to measure pressure, flow, temperature, seismicity)	Monitoring tools and sensors are available for sustained operation up to about 150°C. Downhole monitoring tools capable of sustained operation at >200°C do not exist.	NO	NO
Monitor rock/fluid interactions	Geochemical analytical techniques; geochemical models	Geochemistry is understood for a large subset of relevant chemicals, but real-time detection technology has limited scope and poor accuracy. Geochemical models lack confirmatory field data.	YES	NO
Mitigate reservoir and surface problems (e.g., short circuiting, pressure drop)	Coupled modeling tools and simulators; cements; zonal isolation tools	Cements are routinely used in the hydrothermal industry for lost circulation. Zonal isolation tools that operate at high temperatures are not available.	YES	NO
Manage induced seismicity	Pressure control; rock mechanics models	Operation protocols that limit injection/ production pressures are considered a useful management tool. Rock mechanics models are available, but cannot predict seismicity.	YES	NO
Control scaling	Scale control technology	Chemical management technologies (e.g., additives for pH control) are used in the hydrothermal industry to mitigate well bore and near-field scaling. Technologies for hydrothermal systems may not be as effective for EGS which will operate in chemical disequilibrium.	YES	NO
Validate reservoir model using field data	Monitoring tools and sensors (e.g., tools to measure pressure, flow, temperature, seismicity); tracers	Few sensors can operate at high temperature for long periods. A temperature sensor is available, but it must be hardened for geothermal conditions. Tracer tests are an established method of validating reservoir models. Smart tracers have not been developed.	YES	NO
Design field expansion	Reservoir simulation models	Existing models are not fully coupled to enable planning of field expansion. Sufficient data to validate models is not available.	YES	NO
Generate electricity	Heat exchanger and power plant	Power conversion technology is relatively mature in the hydrothermal industry. Currently evaluating for EGS applications.	YES	TBD

- High-temperature packers and other zonal isolation tools will reduce or eliminate fluid loss, help identify and mitigate short circuiting of flow from injectors to producers, and target individual fractures or fracture networks for testing and validating reservoir models.
- Temperature-hardened tools for real-time down-hole monitoring of temperature, pressure, and flow, along with in-stream surface monitoring of fluid chemistry, would significantly enhance the ability to track the hydrologic and thermal evolution of the reservoir, monitor rock-fluid interactions, and provide the appropriate field data for validating and updating reservoir models and simulators.
- Operation of an EGS reservoir will require injection of fluids that are not at equilibrium with the reservoir rock mass. As a result, scaling and/or dissolution will likely occur in the wellbore or the reservoir. Treatments available today may not be adequate for long-term operation.
- Although research sponsored by DOE has significantly advanced the sophistication and use of tracers for characterizing hydrothermal systems, development of new "smart" tracers is warranted. For instance, reliable tracers that can measure and/or monitor the surface area responsible for rock-fluid heat and mass exchange do not exist, limiting the ability to quantify and predict heat extraction efficiencies.
- Induced seismicity is an issue with potential to halt if not end a project, as demonstrated in Soultz, France, and Basel, Switzerland. Studies of the issue, including one released under the auspices of the International Energy Agency, conclude that damaging earthquakes as a result of EGS reservoir operation are unlikely. Nevertheless, the issue could have strong negative consequences on the acceptability of EGS projects near population centers. Initial impact in the U.S. is believed to be low since many candidate sites for early development are in unpopulated areas. The current state of knowledge does not point to technological solutions. Protocols for operation of EGS facilities have been proposed, but they have not yet been generally adopted or proven to be effective.
- Reservoir management and operation relies heavily on models and simulators that can accurately predict reservoir behavior. For optimum EGS operation, fully coupled Hydrologic-Thermal-Mechanical-Chemical (H-T-M-C) models and simulators will be necessary to predict fluid flow, heat extraction, temperature drawdown, rock-mechanical processes, and chemical processes that will have either beneficial or deleterious impacts on reservoir performance and longevity.
- The technical advances for submersible pumps and high-temperature isolation tools are likely to be unique to EGS. Development of temperature hardened tools for real-time, down-hole monitoring will be readily transferable to conventional geothermal systems. Smart tracer technology and fully coupled H-T-M-C models and simulators are multi-use technologies that will have value beyond the EGS domain.

Specific technology requirements at this stage of EGS reservoir development remain uncertain due to limited operating experience. Operational experience is measured in months rather than years. The longest period of continuous performance was at Rosemanowes, U.K. Fluids were circulated at Rosemanowes for three years, during which production temperatures fell from 80°C to 55°C, suggesting to some experts a probable short circuit in the reservoir.

Technology solutions to address short circuiting, like other concerns with long-term operation, will require a much larger and broader experience base.

5. Conclusions

The MIT study was comprehensive, the assumptions and models were properly addressed and applied, and the study is suitable as a starting point for identification and prioritization of technology improvements required to commercialize EGS. While there are uncertainties in the analysis and gaps in knowledge, the study presents the present understanding of the EGS opportunity in a realistic manner. EGS can contribute substantially to meet future U.S. energy needs.

There are three critical assumptions about EGS technology that require thorough evaluation and testing before the economic viability of EGS can be confirmed:

1. **Demonstration of commercial-scale reservoir** – This requires stimulation and maintenance of a large volume of rock (equivalent to several cubic kilometers) in order to minimize temperature decline in the reservoir. Actual stimulated volumes have not been reliably quantified in previous work.

2. **Sustained reservoir production** – The MIT study concludes that 200°C fluid flowing at 80 kg/sec (equivalent to about 5 MWe) is needed for economic viability. No EGS project to date has attained flow rates in excess of ~25 kg/sec.

3. **Replication of EGS reservoir performance** – EGS technology has not been proven to work at commercial scales over a range of sites with different geologic characteristics.

These assumptions can be tested with multiple EGS reservoir demonstrations using today's technologies. However, as this evaluation shows, Research and Development should be conducted in parallel with field projects to fill some long-term technology gaps. The key technology requirements for immediate development stemming from this evaluation include:

- Temperature-hardened submersible pumps
- Zonal isolation tools
- Smart tracers
- Monitoring and logging tools
- Coupled models to predict reservoir development and performance

Experience from the conventional geothermal and petroleum industries provides a solid foundation from which to make technology improvements. In the long-term, significant reduction in drilling costs will be necessary to access deeper resources, and the cost of conversion of the energy into electricity must be reduced. These improvements will rapidly move EGS technology forward as an economically viable means of tapping the nation's geothermal resources.

APPENDIX A. LOGIC PROCESS FOR DEVELOPMENT OF AN ENHANCED GEOTHERMAL SYSTEM (EGS) FACILITY

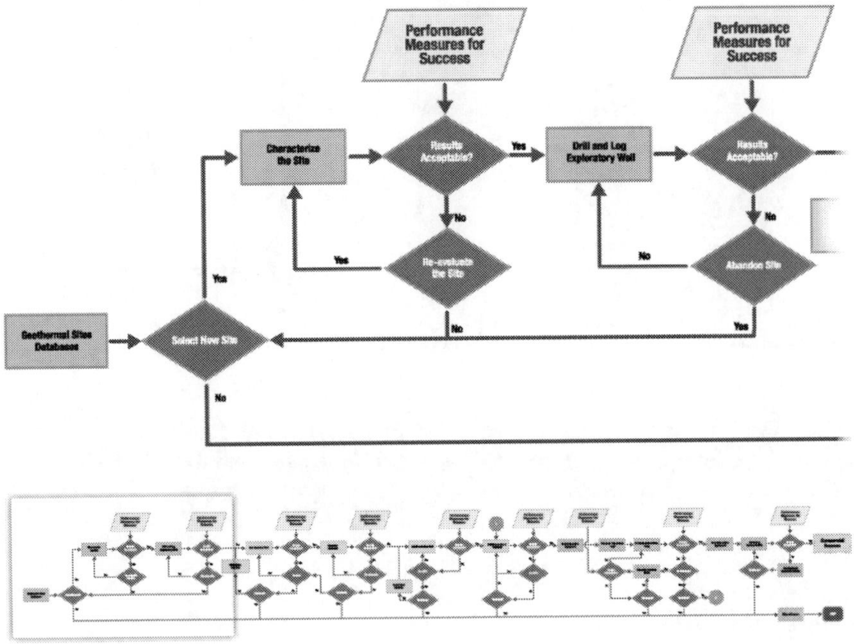

Step 1. Finding a Site

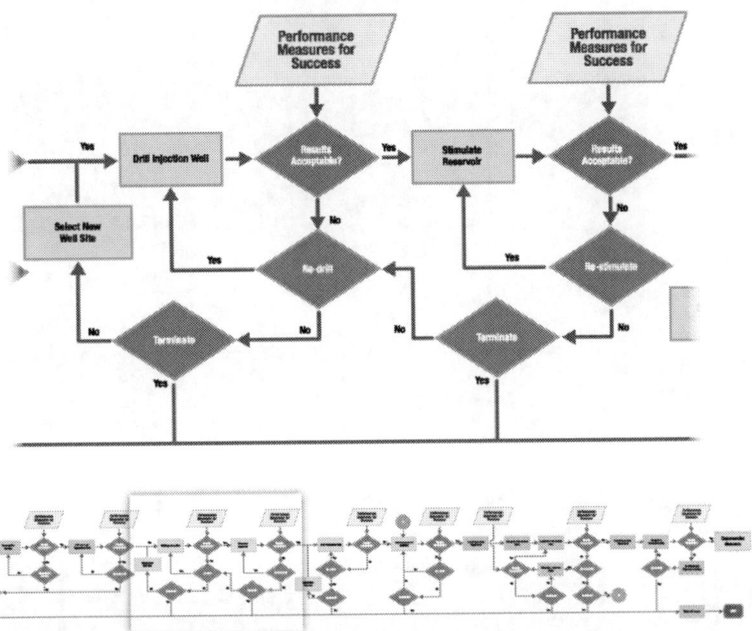

Step 2. Creating Reservoir

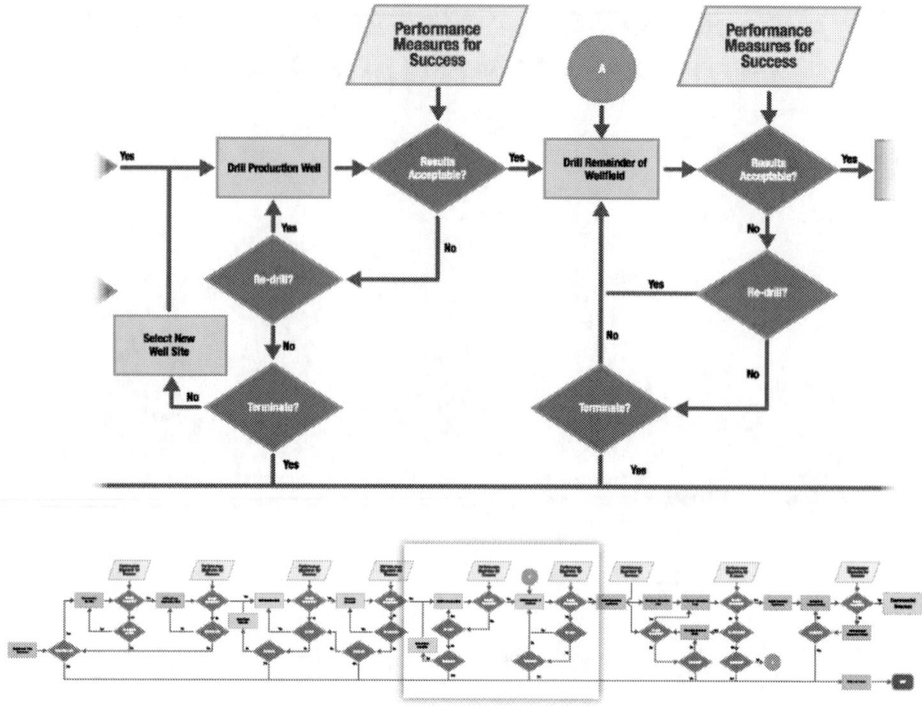

Step 3. Completing Wellfield

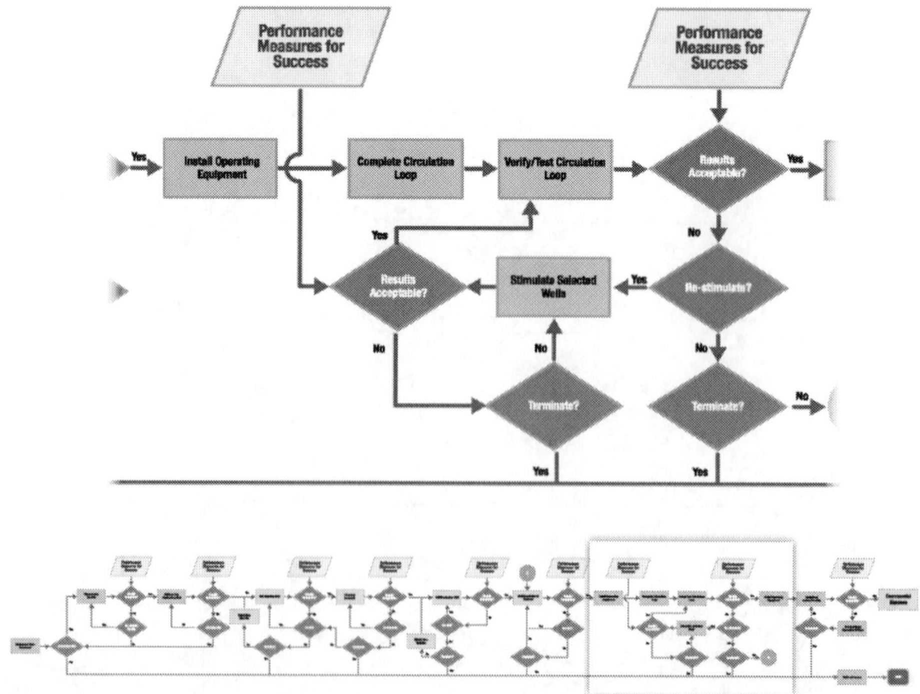

Step 4. Operating Reservoir

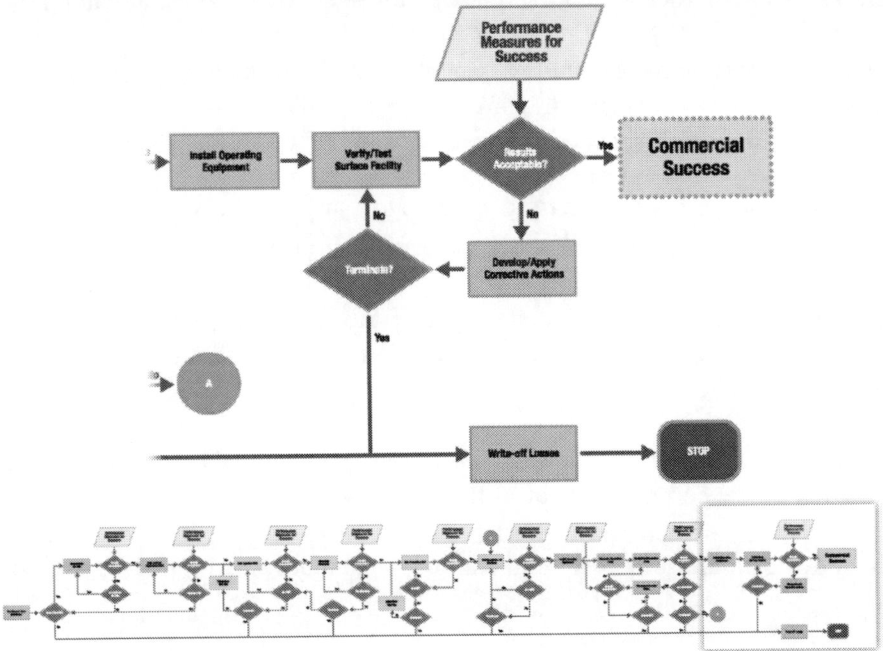

Step 5. Operating Facility

GLOSSARY

Borehole Breakouts Failure of the borehole wall which forms because of stress in the rock surrounding the borehole. The breakout is generally located symmetrically in the wellbore perpendicular to the direction of greatest horizontal stress on a vertical wellbore.

Binary Cycle Binary geothermal systems use the extracted hot water or steam to heat a secondary fluid to drive the power turbine.

Casing Pipe placed in a wellbore as a structural interface between the wellbore and the surrounding formation. It typically extends from the top of the well and is cemented in place to maintain the diameter of the wellbore and provide stability.

Core A cylinder of rock recovered from the well by a special coring drill bit.

Depletion Factor Annual percentage of the depletion of the thermal resource.

Drag Bit Drilling bit that drills by scraping or shearing the rock with fixed hard surfaces, "cutters." See also *rotary cone bits* and *polycrystalline diamond compact bits.*

Enhanced Geothermal Systems (EGS) Engineered reservoirs that can extract economic amounts of heat from geothermal resources.

Fault A fracture in rock exhibiting relative movement between the adjoining surfaces.

Fracturing Treatments Fracturing treatments are performed by pumping fluid into the subsurface at pressures above the fracture pressure of the reservoir formation to create a highly conductive flow path between the reservoir and the wellbore.

Global Positioning System (GPS) A navigational system using satellite signals to fix the location of a radio receiver on or above the earth's surface.

Geothermal Resources The natural heat of the earth that can be used for beneficial purposes when the heat is collected and transported to the surface. See also *EGS* and *hydrothermal reservoir.*

Gravimetry The use of precisely measured gravitational force to determine mass differences that can be correlated to subsurface geology.

Hydraulic Stimulation A stimulation techniques performed using fluid. See *Stimulation.*

Hydrothermal Pertaining to hot water.

Hydrothermal Reservoir An aquifer, or subsurface water that has sufficient heat, permeability, and water to be exploited without stimulation or enhancement.

Induced Seismicity Induced seismicity refers to typically minor earthquakes and tremors that are caused by human activity that alters the stresses and strains on the Earth's crust. Most induced seismicity is of an extremely low magnitude, and in many cases, human activity is merely the trigger for an earthquake that would have occurred naturally in any case.

Interferometric Synthetic Aperture Radar (InSar) A remote sensing technique that uses radar satellite images to determine movement of the surface of the earth.

Line Shaft Pump Fluid pump that has the pumping mechanism in the wellbore and that is driven by a shaft connected to a motor on the surface.

Liner A casing string that does not extend to the top of wellbore, but instead is anchored or suspended from inside the bottom of the previous casing string.

Lithology The study and description of rocks, in terms of their color, texture, and mineral composition.

Lost Circulation Zones in a well that imbibe drilling fluid from the wellbore, thus causing a reduction in the flow of fluid returning to the surface. This loss causes drilled rock particles to build up in the well and can cause problems in cementing casing in place.

Magnetic Survey Measurements of the earth's magnetic field that are then mapped and used to determine subsurface geology.

Magneto-telluric An electromagnetic method of determining structures below the earth's surface using electrical currents and the magnetic field.

Matrix Treatments Treatments performed below the reservoir fracture pressure, and generally are designed to restore the natural permeability of the reservoir following damage to the near wellbore area. Matrix treatments typically use hydrochloric or hydrofluoric acids, to remove mineral material that reduces flow into the well.

Micro-seismicity Small movements of the earth causing fracturing and movement of rocks. Such seismic activity does not release sufficient energy for the events to be recognized except with sensitive instrumentation. See also *seismicity*.

Mini-frac A small fracturing treatment performed before the main hydraulic fracturing treatment to acquire stress data and to test pre-stimulation permeability.

Packer Device that can be placed in the wellbore to block vertical fluid flow so as to isolate zones.

Permeability The ability of a rock to transmit fluid through its pores or fractures when subjected to a difference in pressure. Typically measured in darcies or millidarcies.

Polycrystalline Diamond Compact Drilling Bit (PDC) A drilling bit that uses polycrystalline diamond compact inserts on the drill bit to drill by means of rotational shear of the rock face. See *drag bits.*

Proppant Sized particles mixed with fracturing fluid to hold fractures open after a hydraulic stimulation.

Recovery Factor The fraction of total resource that can be extracted for productive uses.

Resistivity Survey The measurement of the ability of a material to resist or inhibit the flow of an electrical current, measured in ohm-meters. Resistivity is measured by the voltage between two electrodes while an electrical current is generated between two other electrodes. Resistivity surveys can be used to delineate the boundaries of geothermal fields.

Roller Cone Bit Drill bit that drills by crushing the rock with studded rotating cones attached to the bit.

Resource Base All of a given material in the Earth's crust, whether its existence is known or unknown, and regardless of cost considerations.

Seismic Pertaining to, of the nature of, or caused by an earthquake or earth vibration, natural or man-made.

Seismicity The phenomena of earth movements. Also the frequency, distribution and intensity of earthquakes. Syn. *seismic activity*.

Seismometer Electrical device that is used on the surface and within wellbores to measure the magnitude and direction of seismic events.

Self-potential Self-potential in geothermal systems measures currents induced in the subsurface because of the flow of fluids.

Spinner Survey The use of a device with a small propeller that spins when fluid passes in order to measure fluid flow in a wellbore. The device is passed up and down the well continuously measuring flow to establish where and how much fluid enters or leaves the wellbore at various depths.

Slim Hole Drill holes that have a nominal inside diameter less than about 6 inches.

Slotted Liner Liner that has slots or holes in it to let fluid pass between the wellbore and surrounding rock.

Smart Tracer Tracer that is useful in determining not only the flow path between a well injecting fluid into the subsurface and a well producing fluid from an adjacent well, but which can also be used to determine temperature along the flow path, the surface area contacted by the tracer, the volume of rock that the tracer interacts with, and the relative velocities of separate phases (gas, oil and water in petroleum fields; steam and liquid water in geothermal systems).

Stimulation A treatment performed to restore or enhance the productivity of a well. Stimulation treatments fall into two main groups, hydraulic fracturing treatments and matrix treatments.

Stress The forces acting on rock. In the subsurface the greatest force or stress is generally vertical caused by the weight of overlying rock.

Structural Discontinuity A discontinuity of the rock fabric that can be a fracture, fault, intrusion, or differing adjacent rock type.

Submersible Sump Pump with both the pumping mechanism and a driving electric motor suspended together at depth in the well.

Tiltmeter Device able to measure extremely small changes in its rotation from horizontal. The "tilt" measured by an array of tiltmeters emplaced over a stimulation allow delineation of inflation and fracturing caused by the stimulation.

Thermal Gradient The rate of increase in temperature as a function of depth into the earth's crust.

Thermal Drawdown Decline in formation temperature due to geothermal production.

Tracer A chemical injected into the flow stream of a production or injection well to determine fluid path and velocity.

Under Reamer A drilling device that can enlarge a drill hole. The device is placed about the drill bit and can be opened to drill and then closed to be brought back up through smaller diameter hole or casing.

Well Log Logging includes measurement of the diameter of the well and various electrical, mass, and nuclear properties of the rock which can be correlated with physical properties of the rock. The well log is a chart of the measurement relative to depth in the well.

Zonal Isolation Various methods to selectively partition portions of the wellbore for stimulation, testing, flow restriction, or other purposes.

Energy Units[19]

Joule (J)

This is the basic energy unit of the metric system, or in a later more comprehensive formulation, the International System of Units (SI). It is ultimately defined in terms of the meter, kilogram, and second.

$$1 \text{ Exajoule (EJ)} = 10^{18} \text{ J}$$

British Thermal Unit (Btu)

This is a basic measure of thermal (heat) energy. A Btu is defined as the amount of energy required to increase the temperature of 1 pound of water by 1 degree Fahrenheit, at normal atmospheric pressure.

BTU is the English system analog of the calorie. For specific heat capacities to be the same, whether expressed in Btu/lb-°F or in cal/gm-°C:

$$1 \text{ Btu} = 251.9958 \text{ cal.}$$
$$1 \text{ Quadrillion Btu (Quad)} = 10^{15} \text{ Btu} = 1.055 \text{ EJ}$$

Kilowatt-hour (kWh)

The kilowatt-hour is a standard unit of electricity production and consumption. By definition, noting that 1 kilowatt = 1000 watts:

$$1 \text{ kWh} = 3.6 \times 10^6 \text{ J (exact).}$$

The relationship between the kWh and the Btu depends upon which "Btu" is used. It is common, although not universal, to use the equivalence:

$$1 \text{ kWh} = 3412 \text{ Btu.}$$

This corresponds to the International Table Btu. [More precisely, 1 kWh = 3412.14 Btu (IT).]

$$1 \text{ Terawatt-year (Twyr)} = 8.76 \times 10^{12} \text{ kWh} = 31.54 \text{ EJ} = 29.89 \text{ quad.}$$

End Notes

[1] *The Future of Geothermal Energy: Impact of Enhanced Geothermal Systems (EGS) on the United States in the 21st Century* (Massachusetts Institute of Technology, 2006). http://www1.eere.energy.gov/geothermal/future_geothermal.html.

[2] Ibid. 14-19.

[3] Ibid. 1-3.

[4] Geothermal Energy Association (GEA), "All About Geothermal Energy: Current Use," 6 Apr. 2008. http://geo-energy.org/aboutGE/currentUse.asp.

[5] L.J. Patrick Muffler, ed. *Assessment of Geothermal Resources of the United States* (1978). http://pubs.er.usgs.gov/usgspubs/cir/cir790.

[6] Enex Binary Plants, "Benefits of the Enex Binary Plant," 6 Apr. 2008. http://www.enex.is/?PageID=191.

[7] Petty, Susan and Gian Porro. 2007. *Updated US Geothermal Supply Characterization*. 22-24 Jan, at Stanford University, Stanford, California.

[8] Williams, Colin F. 2007. *Updated Methods for Estimating Recovery Factors for Geothermal Resources*. 22-24 Jan, at Stanford University, Stanford, California.

[9] Thorsteinsson, Hildigunnur, et al. 2008. *The Impacts of Drilling and Reservoir Technology Advances on EGS Exploitation.* 28-30 Jan, at Stanford University, Stanford, California.

[10] Sanyal, Subir K. and Steven J. Butler. 2005. *An Analysis of Power Generation Prospects from Enhanced Geothermal Systems.* 24-29 Apr, at World Geothermal Congress 2005, Antalya, Turkey.

[11] Augustine, Chad et al. 2006. *A Comparison of Geothermal with Oil and Gas Well Drilling Costs.* 30 Jan - 1 Feb, at Thirty-first Workshop on Geothermal Reservoir engineering, at Stanford University, Stanford, California. http://conferences-engine.brgm.fr/contributionListDisplay.py?confId=3.

[12] Fokker, Peter A. 2006. *Hydraulic Fracturing in the Hydrocarbon Industry.* 29 Jun - 1 Jul, at Enhanced Geothermal Innovative Network for Europe (ENGINE) Workshop 3, Zurich, Switzerland.

[13] Takatoshi, Ito. 2006. *Detection of Flow-pathway Structure upon Pore-pressure Distribution Estimated from Hydraulically Induced Micro-seismic Events and Applications to the Soultz HDR Field.* 29 Jun – 1 Jul, at Enhanced Geothermal Innovative Network for Europe (ENGINE) Workshop 3, Zurich, Switzerland.

[14] "Hydraulic Fracturing," *Permian Basin Oil and Gas Magazine* Apr. 2007:14-17. http://www.pbpa.info/newsletter/0704.pdf.

[15] Ibid.

[16] *The Future of Geothermal Energy* 5-8.

[17] Yoshioka, Keita et al. 2008. *Optimization of Geothermal Well Stimulation Design Using a Geomechanical Reservoir Simulator.* 28 – 30 Jan, at Thirty-Third Workshop on Geothermal Reservoir Engineering, at Stanford University, Stanford, California.

[18] Karner, Stephen L. 2006. *Correlating Laboratory Observations of Fracture Mechanical Properties to Hydraulically-Induced Microsensitivity in Geothermal Reservoirs.* 30 Jan – 1 Feb, at Stanford University, Stanford, California.

[19] Units were taken from American Physical Society Web site, http://www.aps.org/policy/reports/popa-reports/energy/units.cfm.

In: Geothermal Energy: The Resource Under Our Feet
Editor: Charles T. Malloy
ISBN: 978-1-60741-502-2

Chapter 6

GROUND-SOURCE HEAT PUMPS: OVERVIEW OF MARKET STATUS, BARRIERS TO ADOPTION, AND OPTIONS FOR OVERCOMING BARRIERS*

William Goetzler, Robert Zogg, Heather Lisle and Javier Burgos

LIST OF ACRONYMS

AFUE	Annual fuel utilization efficiency
ARI	Air-Conditioning and Refrigeration Institute
ASHP	Air-source heat pump
BT	U.S. Department of Energy, Office of Energy Efficiency and Renewable Energy, Building Technologies Program
CDD	Cooling degree day
CHP	Combined Heat and Power
COP	Coefficient of Performance
DOD	U.S. Department of Defense
DOE	U.S. Department of Energy
EERE	U.S. Department of Energy, Office of Energy Efficiency and Renewable Energy
EIA	U.S. Department of Energy, Energy Information Administration
GHP	Geothermal Heat Pump
GSHP	Ground-source Heat Pump
GT	U.S. Department of Energy, Office of Energy Efficiency and Renewable Energy, Geothermal Technologies Program
HDD	Heating degree day
HPWH	Heat Pump Water Heater
HSPF	Heating Seasonal Performance Factor

* This is an edited, reformatted and augmented version of a U. S. Department of Energy publication dated February 2009.

IEA	International Energy Agency
kWh	Kilowatt-hour
MMBtu	Million British thermal units
MW_t	Megawatt (thermal)
NREL	National Renewable Energy Laboratory
ORNL	Oak Ridge National Laboratory
Quad	Quadrillion (10^{15}) British thermal units
R&D	Research and development
SEER	Seasonal energy efficiency ratio
TBtu	Trillion (10^{12}) British thermal units
UEC	Unit energy consumption
UES	Unit energy savings
WSHP	Water-source heat pump
ZEB	Zero-energy building
ZEH	Zero-energy home

EXECUTIVE SUMMARY

We conducted this investigation to:

- Summarize the status of ground-source heat pump (GSHP) technology and market penetration globally
- Estimate the energy saving potential of GSHPs in the U.S.
- Identify and describe the key market barriers that are inhibiting wider market adoption of GSHPs
- Recommend initiatives that can be implemented or facilitated by the DOE to accelerate market adoption.

Of the 15,400 MWt (4.38 x 10^6 tons) global installed base of GSHPs, about 56 percent of this capacity is installed in the U.S., corresponding to about 65 percent of the GSHP unit installations. Europe follows, with about 39 percent of the installed capacity, and Asia has about 5%. In Europe, Sweden is the dominant player in the GSHP market, with almost 2500 MWt (711,000 tons) installed— more than double any other European country.

The U.S. GSHP market is split roughly evenly between residential and commercial applications, with only a very small market for industrial applications.

GSHPs can provide significant primary unit energy savings compared to typical ASHPS or typical furnaces with air conditioners. Savings are often in the range of 30 to 60 percent of space-conditioning energy consumption, depending on GSHP efficiency, technology replaced, climate, and application.

Our energy-savings and economics analysis compares two high-efficiency technologies (GSHPs and advanced ASHPs) to two typical-efficiency baseline systems (typical ASHPs, and furnaces with air conditioners). We used general relationships between fundamental (unsubsidized) economics and market penetration to project ultimate market penetrations of GSHPs and the associated national primary energy savings. Table E-1 summarizes the technical potential energy savings (savings if all technically applicable applications are converted to GSHPs) and projected primary energy savings based on the ultimate market penetration, predicted based on economics.

In addition to high energy efficiency, GSHPs offer two key benefits:

- Can have factory-packaged refrigeration loop
- Can reduce peak electric demand.

GSHPs face three key barriers:

- High equipment costs compared to ASHPs
- Cost and difficulty of evaluating the suitability of individual installation sites
- Installation-specific design and engineering of the ground loop is generally required
- Space requirements for ground coupling can be problematic in densely built areas.

Table E-1. National Primary Energy Savings Potential of GSHPs

Sector	Technical Potential (Quad)	Market Potential (Quad)
Residential	3.1	0.1
Commercial	0.6	0.05
Total	3.7	0.15

While advanced ASHPs offer lower unit energy savings compared to GSHPs, they tend to be more economically attractive and may be able to save similar amounts of energy on a national basis. We, therefore, recommend that DOE support advanced heat pumps in general, rather than supporting only one type. Incentives such as federal tax credits or utility rebates can be based on energy efficiency achieved, rather than type of heat pump. R&D projects can be pursued based on the individual merit of each prospective project, rather than type of heat pump. This will require close coordination between the DOE Geothermal Technologies Group (which is responsible for GSHPs) and DOE Building Technologies Group (which is responsible for ASHPs). This coordination will help ensure that both types of heat pumps are developed, evaluated, and promoted based on apples-to-apples cost and performance comparisons, and that duplication of effort is avoided to the extent possible.

1. INTRODUCTION

The U.S. Department of Energy (DOE), Office of Energy Efficiency and Renewable Energy (EERE), Geothermal Technologies Program has commissioned this study of ground-source heat pumps (GSHPs), also known as geothermal heat pumps, to help assess whether new initiatives are appropriate to further the development and market adoption of this advanced, energy-efficient technology. The report that follows provides an overview of the technology, describes that status of the international GSHP market, assesses the energy savings potential of GSHPs in the U.S., explains key barriers to widespread adoption of the technology, and suggests some initiatives that might help accelerate market adoption.

1.1. Background

The Geothermal Technologies Program has in the past addressed GSHPs, but in recent years has focused primarily on conversion of high-temperature supplies of geothermal energy into electricity. However, the increasing importance of building energy efficiency generally, as well as EERE's programmatic focus on net-zero energy homes (NZEH) and net-zero energy commercial buildings (NZEBs), suggest that the topic of GSHPs should be reassessed to determine whether any new DOE initiatives are warranted to increase the relatively small market penetration of GSHPs.

Residential, commercial, and institutional buildings account for about 40% of US primary energy consumption and carbon emissions, 72 percent of electricity consumption, 55 percent of natural gas consumption, and significant oil consumption in the Northeastern U.S. (DOE 2008). Over the long term, buildings are expected to continue to be a significant component of increasing energy demand and a major source of carbon emissions, driven in large part by the continuing trends of urbanization, population and GDP growth, as well as the longevity of building stocks. However, because building equipment and many structural features are frequently upgraded, the short term potential for improving the energy integrity of the existing building stock is substantial.

Over the past several decades GSHP systems have gradually improved and have achieved a small but growing share in heating, cooling and (in some cases) water heating equipment markets, with modest policy emphasis and research to accelerate technology improvement or enhance affordability. Yet large energy savings have been demonstrated at the individual project level, suggesting that even today's proven GSHP technology may be underutilized. In areas like the Northeast where many building owners are dependent on fuel oil, high oil prices may create unprecedented demand for high efficiency heating and cooling solutions such as GSHPs. New initiatives may be needed to effectively address the barriers that continue to inhibit greater adoption of GSHPs in applications where they are cost-competitive.

1.2. Objectives

The objectives of this effort were to:

- Summarize the status of GSHP technology and market penetration globally
- Estimate the energy saving potential of GSHPs in the U.S.
- Identify and describe the key market barriers that are inhibiting wider market adoption of GSHPs
- Recommend initiatives that can be implemented or facilitated by the DOE to accelerate market adoption.

The project took a national perspective. However, the investigation paid particular attention to Northeastern markets where heating oil and propane are common fuel sources and have become very expensive in recent years, which could provide an opportunity for GSHPs to improve their market penetration.

This investigation is meant to provide an overview of current market conditions and make recommendations to DOE policymakers for improving market penetration. It is based on readily available data and information on market penetration and energy consumption but does not include extensive, detailed new modeling.

Appendix A contains the complete statement of work for this analysis.

1.3. Overall Approach

Figure 1-2 summarizes the overall approach to this investigation. To understand the benefits associated specifically with coupling a heat pump to the ground, we compare the potential impacts of GSHPs to those for advanced air source heat pumps (ASHPs). Historically, ASHPs have been used for heating and cooling primarily in moderate climates such as the Southern and Western U.S., but have not been very common in cold Northern climates. The relatively high electricity rates in the Northeast, combined with the need for expensive and inefficient resistance heating during cold weather, typically made heat pumps unattractive in the Northeast. However, this regionality has begun to change in recent years, as high natural gas prices and advanced technology which avoids the need for resistance heating during cold weather, have combined to make ASHPs much more attractive in colder climates. In fact, some manufacturers have introduced "cold climate" air source heat pumps that are suitable for virtually any climate.[1] Such technology is expected to continue to advance, and we can expect to see far more air source heat pumps used in cold climates in the future. Heat pumps have historically comprised 20-25% of U.S. unitary space-conditioning equipment sales (Figure 1-1). However, in the past five years heat pump market share has risen to approximately 35% (AHRI, Appliance Magazine).

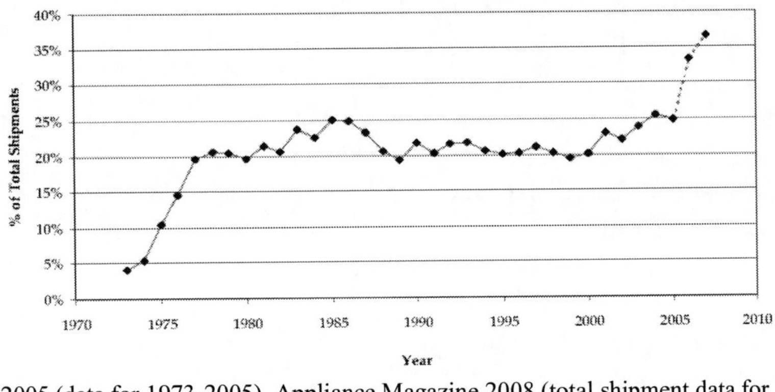

Source: AHRI 2005 (data for 1973-2005), Appliance Magazine 2008 (total shipment data for 2006- 2007).
Note: ASHP data for 2006-2007 is projected by NCI based on the average annual growth rate for 2001-2005 of 10%.

Figure 1-1. ASHP Heat Pump Market Share of Total Unitary A/C and Heat Pump Sales

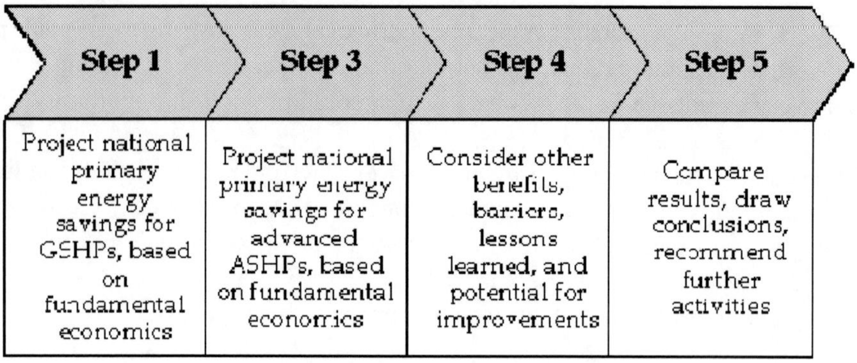

Figure 1-2. Overall Approach to GSHP Evaluation

Our process starts with a comparison of the fundamental economics of each technology in the residential and commercial markets and the likely national impacts. While GSHPs are generally more energy efficient compared to the best- available ASHPs, national impacts also depend on likely market penetrations of each alternative. We also consider other benefits and barriers that are not reflected in the economic analysis, including lessons learned from global experience with GSHPs. We then compare results, draw conclusions, and make recommendations.

In gathering information for this investigation, we:

- Conducted interviews with the following organizations and companies:
  - Advanced Hydronics, Inc. (http://advancedhydronics.com/)
  - CDH Energy Corp. (http://www.cdhenergy.com/)
  - Geothermal Heat Pump Consortium (http://www.geoexchange.org/)
  - Geo-Heat Center, Oregon Institute of Technology (http:// www.geoexchange.org/)
  - Major Geothermal (http://www.majorgeothermal.com/)
- Reviewed the proceedings from the following conferences:
  - 7th IEA Heat Pump Conference, Beijing 2008
  - 8th IEA Heat Pump Conference, Las Vegas 2005
  - 9th IEA Heat Pump Conference, Zurich 2008
  - World Geothermal Congress 2005
- Conducted internet searches and reviewed websites of several organizations involved in GSHP development or promotion, including:
  - International Energy Agency
  - Oak Ridge National Laboratory
  - National Renewable Energy Laboratory
  - Geothermal Heat Pump Consortium
  - California Energy Commission Consumer Energy Center
  - American Council for an Energy Efficient Economy
  - European Heat Pump Association
  - International Ground Source Heat Pump Association
  - Natural Resources Canada

Table 1-1. Report Organization

Section	Content/Purpose
1	Introduction—Describes work scope, objectives, and overall approach
2	Status of Global GSHP Markets—Summarize the global market situation for GSHPs and identify lessons learned that are applicable to the U.S., if any
3	National Energy-Savings Potential—Documents analysis of unit energy savings, technical potential, likely ultimate market based on economics, and likely national primary energy savings. Residential and commercial building examples used.
4	Other Benefits of GSHPs—Briefly describes benefits of GSHPs that are not captured in our economic analysis
5	Key Barriers to GSHPs in the U.S.—Briefly discusses various barriers to GSHPs
6	Applicability to Zero-Energy Homes and Buildings—Briefly discusses GSHP implications for ZEH and ZEB
7	Summary/Conclusions
8	Recommendations
References	--
Appendix A	Scope of Work
Appendix B	Residential Primary Unit Energy Consumptions
Appendix C	Residential Primary Unit Energy Savings
Appendix D	Commercial Primary Unit Energy Consumptions
Appendix E	Commercial Primary Unit Energy Savings
Appendix F	Residential Electricity Price Projections—EIA projections of residential electricity prices for three cases/scenarios
Appendix G	Residential Annual Energy Costs
Appendix H	Commercial Annual Energy Costs

1.4. Report Organization

This report is organized as shown in Table 1-1. This structure is consistent with the work statement.

2. Status of Global GSHP Markets

Global Overview

Ground-source heat pumps (GSHP) are a small but growing fraction of the global installed base of space-conditioning equipment. The global installed capacity has reached about 15,400 MWt, and annual energy use is estimated to be 87,500 TJ (Lund 2005). The global GSHP capacity has seen tremendous growth in recent years. Annual growth rates have exceeded 10% over the last 10 years (Le Feuvre 2008), mostly in North America and Europe. Figure 2-1 shows the increase from 1,900 MWt in 1995 to 5,300 MWt in 2000 and 15,400 MWt in 2005. As of 2005, 33 countries had installed at least 100 MWt of GSHP capacity.

As shown in Figure 2-2, North America represents the largest portion of installed GSHP capacity at 56%, followed by Europe at 39% and Asia at a modest 5%.

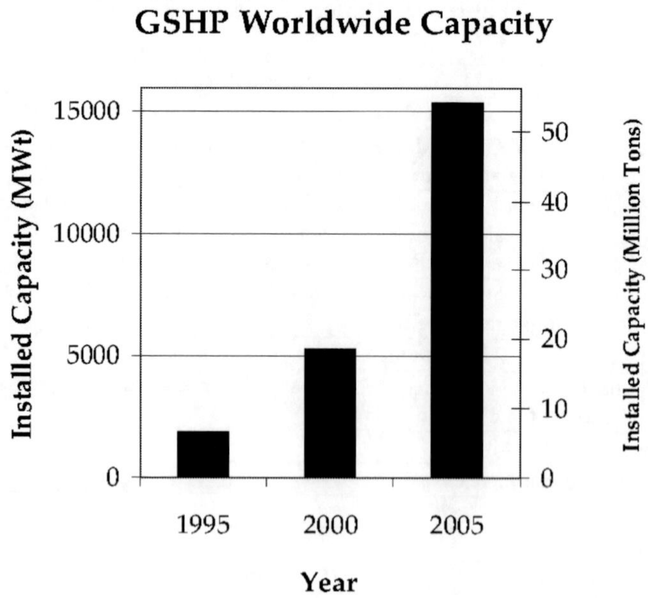

Source: Lund, et al. "Direct application of geothermal energy: 2005 Worldwide review" (2005).

Figure 2-1. GSHP World -wide Installed Capacity in MWt

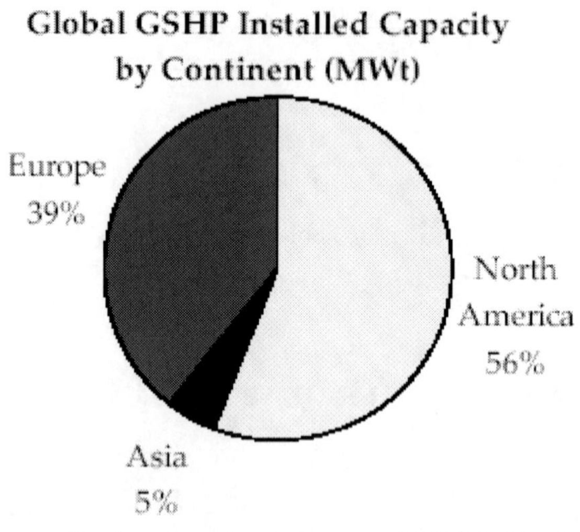

Source: Lund, et al. "Direct application of geothermal energy: 2005 Worldwide review" (2005).

Figure 2-2. Global GSHP Installed Capacity (MWt) by Continent

Figure 2-3 shows the GSHP installed base by country in terms of MWt of capacity. The United States comprises approximately two thirds of the installed base and Sweden, the leading European country, represents one fifth. In total, 900,000 individual units were estimated to be installed as of 2005.

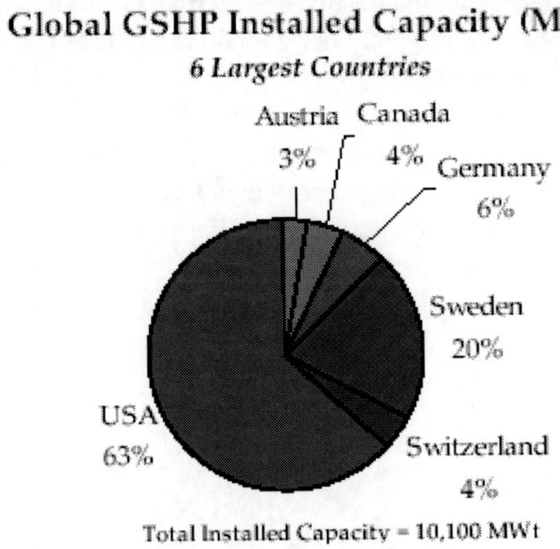

Source: Curtis, et al. World Geothermal Congress, 2005.

Figure 2-3. Global GSHP Installed Capacity by Country (MWt)

Equipment Description

Ground source heat pumps are generally classified by the type of ground loop (see Figure 2-4). Market share of each type varies by country depending on site characteristics, promotion, and applications.

Open loop systems, or "groundwater-source" heat pumps, shown in Figure 2-4(a), are the oldest and cheapest type of GSHP system, assuming the groundwater is suitable for use. Open loop systems have been in common use since the 1970's and currently represent approximately 10-20% of the U.S. market (Lund 2005). In such systems, groundwater is used as the heat carrier and is brought directly to the heat pump. The water is discharged either back into the well or into a body of surface water. These systems require an ample, shallow, and pure supply of groundwater. Because of their effect on the community groundwater, municipal regulations sometimes inhibit the installation of open loop systems.

Closed loop, or ground-coupled, systems use a loop containing water or a glycol solution through the ground loop and use a refrigerant loop to transfer the heat to the heat pump (Figure 2-4b). The ground loop can be laid vertically or horizontally in the ground, or occasionally laid in a pond or lake.

The vertical configuration involves a borehole drilled to a depth of 150 to 220 ft per ton of capacity (Rafferty 2008). The vertical loop has a smaller ground surface area requirement, typically 200-400 ft^2 (5-10 m^2/kW), which makes it more feasible for small properties, but it adds on significant drilling costs to the total installation cost of the system (ASHRAE 1995).

The horizontal loop is usually a less expensive option, because it only involves digging a 4-5 ft trench as opposed to a deep well. However, it requires much more space, and the ground temperature is subject to seasonal fluctuation at shallow depths. The horizontal trench length ranges from 125 to 300 ft per ton of capacity (Rafferty 2008). The length of pipe necessary is a function of system size, climate, soil/rock thermal characteristics and loop type. The ground surface area necessary for a typical horizontal loop ranges from 2000 ft^2 to 3500 ft^2 per ton (50- 90 m^2/kW) (ASHRAE 1995).

A variation of the horizontal loop is the spiral, or "slinky", loop configuration in which the piping is laid out in an overlapping circular fashion. This configuration requires less ground area but more pipe length and pumping energy than a basic horizontal setup.

In a pond loop, the ground loop is submerged in a lake or a pond. If a suitable body of water is available, this design is an economical option, because it involves minimal digging.

Direct exchange systems run refrigerant through the ground loop to exchange heat directly. Such systems do not have to use a pump, but require a much greater copper tube length and refrigerant charge. They are not commonly used.

Source: Water Furnace 2008

(a) Open Loop

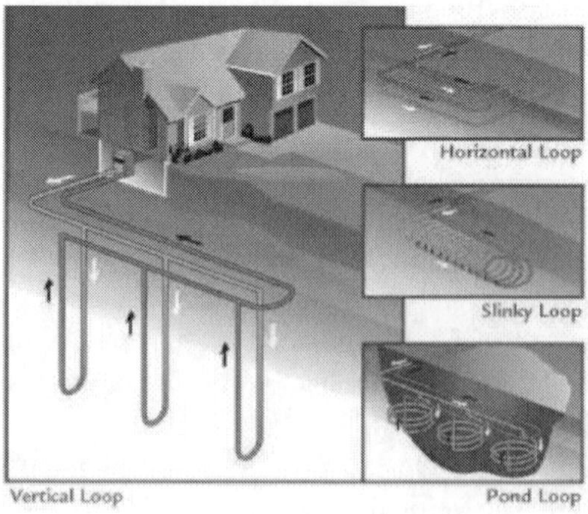

Source: KCPL 2008.

(b) Closed Loop

Figure 2-4. Residential Ground Loops

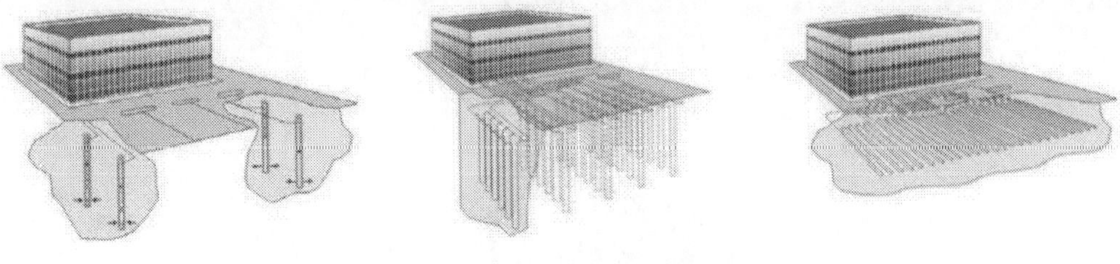

(a) Open Loop (Groundwater) (b) Vertical Closed Loop (c) Horizontal Closed Loop

Source: NRCan 2008

Figure 2-5. Commercial/Institutional Ground Loops

Ground-source heat pumps can be applied in a variety of residential, commercial, and institutional settings. In addition, a number of community - based systems have been installed in various countries around the world. The size of individual units ranges from about 1.5 tons for small residential applications to over 40 tons for commercial and institutional applications. As shown in Figure 2 -5, larger commercial applications can involve numerous rows of piping connected either in series or in parallel.

In the U.S., the capacity of most units is sized for the cooling load and is consequently oversized for the heating load, except in northern climates where the primary load is the heating load. In Europe, the capacity is usually sized for heating load, often to provide base load, with peak load provided by fossil fuel.

Europe

Europe has seen significant growth in the GSHP industry in the past 10 years. GSHPs represent 25% of all heat pumps sold in Europe (Forsén 2008). Over 690,000 units, representing 7,300 MWt of capacity, have been installed in Europe through 2006 (EUObserv'ER 2007, EHPO 2008).

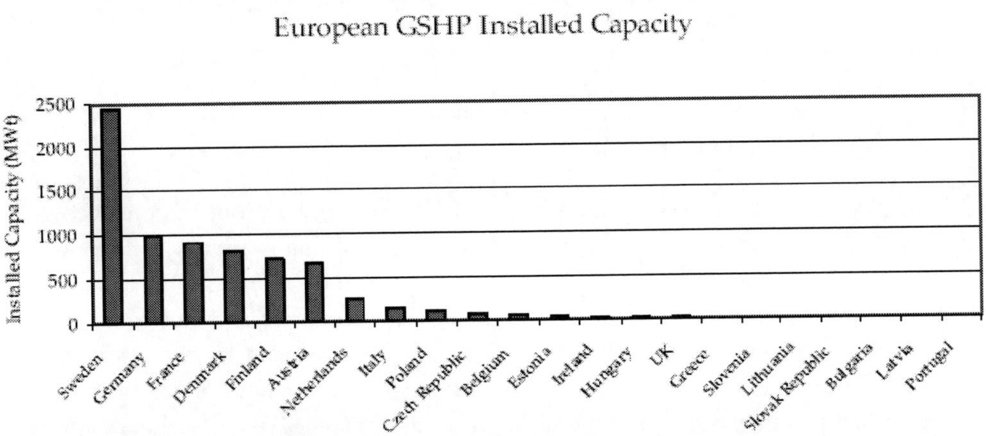

Figure 2-6. Global GSHP Installed Capacity by Country

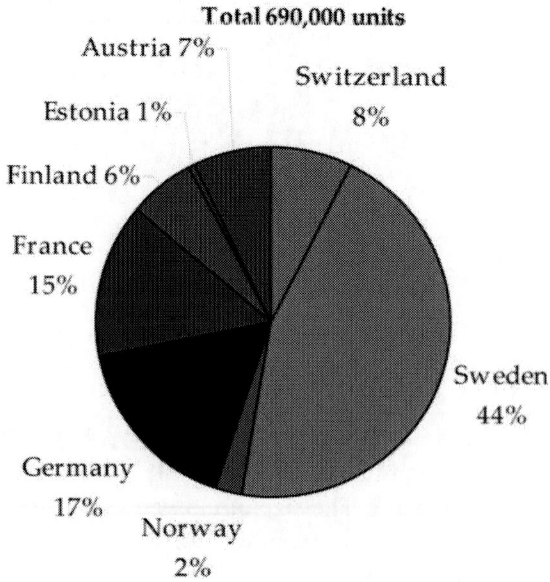

Figure 2-7. European GSHP Installed Base (units)

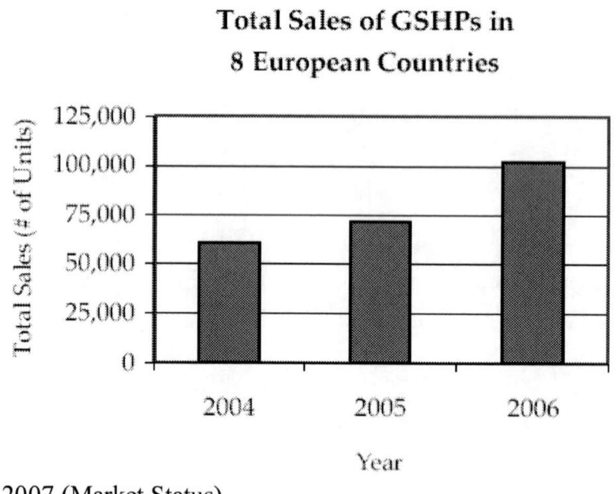

Source: GroundReach 2007 (Market Status)

Figure 2-8. Total Sales of GSHPs in 8 European Countries[2]

Figure 2-8 shows the rise in annual sales GSHPs in eight European countries between 2004 and 2006. The compound annual growth rate for this period is approximately 30%.

Figure 2-9 shows the estimated market penetration for the same eight European countries for both the new construction and retrofit markets. The countries are plotted as along both curves, although only Sweden, Switzerland, Finland, and Norway have significant retrofit markets. The figure shows how successful Sweden has been in penetrating the retrofit market with over 75% market penetration. In addition, it is clear that GSHPs have a very strong stake in the Swiss new construction market.

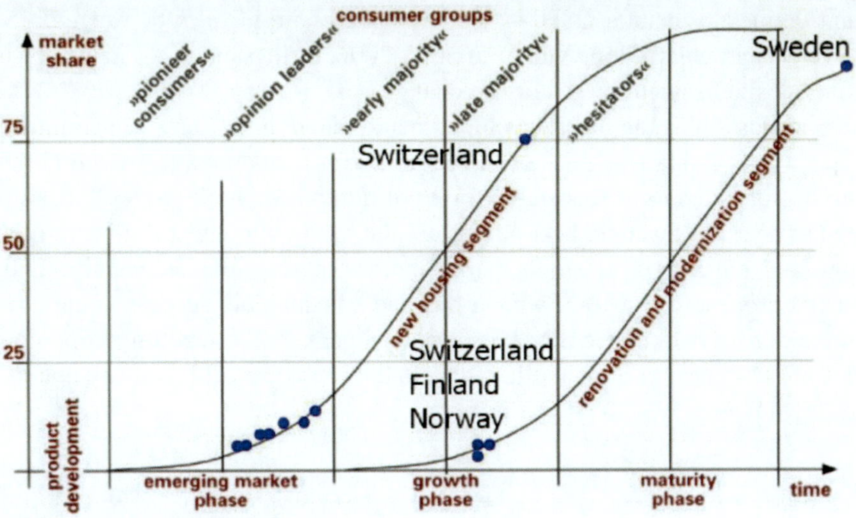

Source: GroundReach 2007 (Market Status)

Figure 2-9. Market Penetration for GSHPs in 8 European Countries, 2006

Individual European Countries

Austria has the fifth highest capacity density per land area worldwide with 23,000 heat pumps installed (Le Feuvre 2008). Approximately 95% of heat pumps used in the Austrian housing market are ground -source (Le Feuvre 2008). As in so many countries during the late 1970's and early 1980's, the heat pump market was plagued by poor performance and reliability once less -experienced companies started entering the explosive market during the oil crisis. The LGW (Leistungsgemeinschaft Wärmepumpe) trade association was formed in 1990 to promote heat pumps and develop education and training programs and has helped the market achieve renewed growth.

The French heat pump market first developed between 1975 and 1985 during the oil crisis. After this initial boom, the market essentially disappeared due to a lack of skilled installers and poor equipment quality. The market was jumpstarted in 1997 by an initiative of Electricite de France (EDF), the national French electricity company, in association with ADEME (French environment and energy management agency) and BRGM (French mining and geological research board). In 2005, public authorities including the French Electricity Board and the French Environment and Energy Management Agency have implemented a substantial subsidy scheme for heat pumps in general which will continue through 2009.

This subsidy has helped grow the retrofit GSHP market from 2% of the total GSHP market before the subsidy to 13% in 2007. In addition, France has set the objective for 2010 to equip 20% of all new single family homes with GSHPs (~40,000 units per year).

Germany has the second largest installed base in Europe. Figure 2-10 shows the growth of annual heat pump sales in Germany since 1997. Electric utilities have been an ally to the industry through promotion of heat pump benefits. Several utilities offer special heat pump tariffs that benefit the consumer (EHPO 2008). In 2008, the German government instituted a new market incentive program to support renewable energy systems, which the German government defines to include GSHPs. The dramatic boom in sales between 2005 and 2006 was caused by a particularly long winter in 2005-2006, further increases in energy prices, and the considerable media attention to climate change.

Sweden stands out as the most developed market among all European countries, with the highest capacity per capita worldwide (Le Feuvre 2008). The dramatic growth in the domestic market can be attributed to the escalating price of oil and electricity as well as an increase in energy related taxes. In Sweden, heat pumps are the most common space-heating in both new construction and retrofitting of single family homes, at approximately 34% (EHPO 2008). Unlike other European countries, Sweden has had considerable success at capturing a large portion of the retrofit market. GSHP sales reached a peak in 2006, before dropping by 30% in 2007. Sales are estimated to drop another 20% in 2008 (Figure 2-11), mostly due to the global economic slowdown.

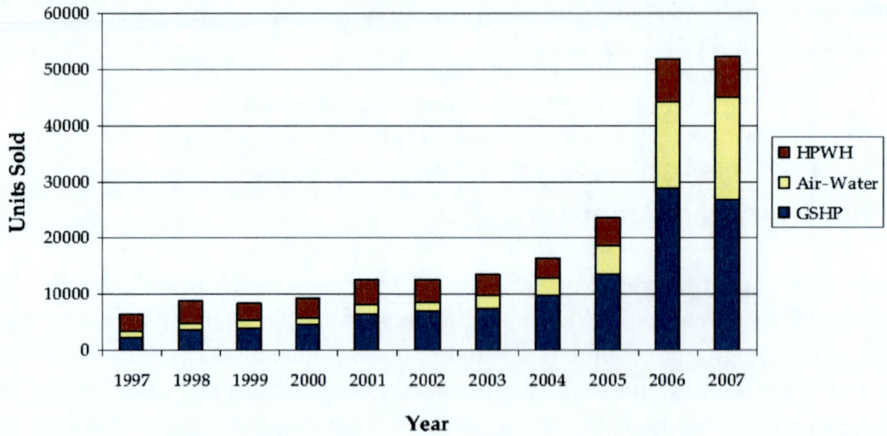

Source: European Heat Pump Outlook 2008

Note: Heat pump water heaters (HPWH) are used for water-heating only, as opposed to space- heating. Air-water heat pumps heat a hydronic circuit.

Figure 2-10. German Heat Pump Market Development 1997-2007

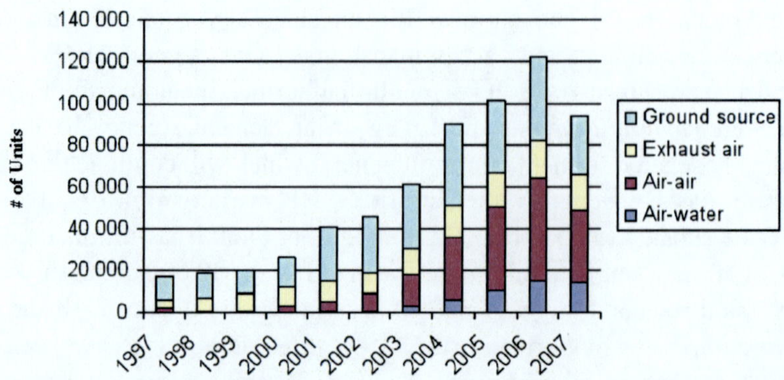

Source: European Heat Pump Outlook 2008

Figure 2-11. Swedish Heat Pump Market Development 1997-2007[3]

Switzerland, with over 25,000 GSHP systems in operation, is estimated to have the highest installed density in world, with an average of more than one unit per 2 km^2 (Curtis 2005). Figure 2-12 shows the how the installed capacity has grown since 1980. Swiss public utilities have used a system called "energy contracting" to effectively provide an incentive for the adoption of GSHPs, which involves planning, installing, operating, and maintaining GSHP systems at their own cost and selling the heat (or cold) to the property owner at a contracted price in cents per kilowatt -hour (Curtis 2005). In general GSHPs are installed primarily in a decentralized manner to meet individual needs, which avoids the cost of heat distribution associated with district heating.

Asia

The Asian heat pump market is currently much less established than that of Europe and North America, but there has been some recent growth in China and Japan, as well as active research and development.

China has approximately 630 MWt of installed GSHP capacity.

Applications include residences, office buildings, schools, hotels, commercial buildings, hospitals, and banks. In Beijing, China, over 38,000 ft^2 of the Olympic Village was air-conditioned by GSHPs (Zheng, 2008).

The central government maintains a GSHP policy with ambitious expectations in all provinces. For public buildings, such as schools, hospitals, and administrative buildings, the government will cover the initial investment for a GSHP. For other buildings, the government will subsidize the cost by $4/$ft^2$ of building floorspace for a surface or groundwater heat pump and $6/$ft^2$ for a ground-coupled heat pump. In 2005, the city of Ningbo in the Zhejiang province included a 20% subsidy of installed cost for GSHPs as part of their "Measures to Administrate the Particular Fund to Develop Energy-Saving and Clean Production".

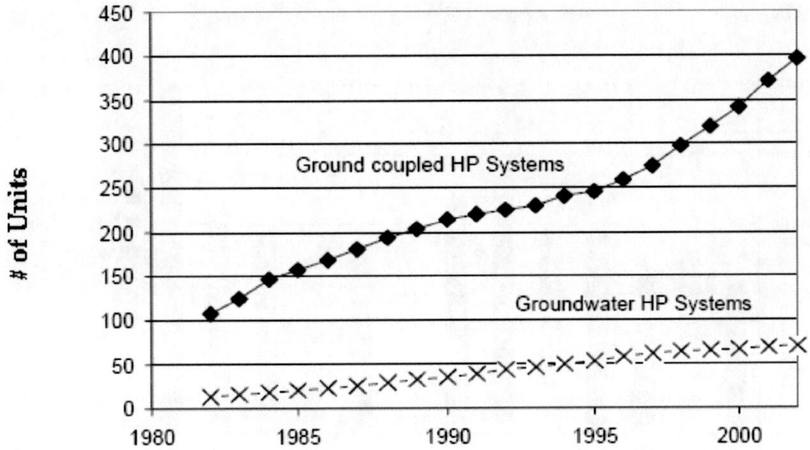

Source: Curtis 2005

Figure 2-12. Installed Capacity (MWt) of GSHPs in Switzerland (1980 – 2001)

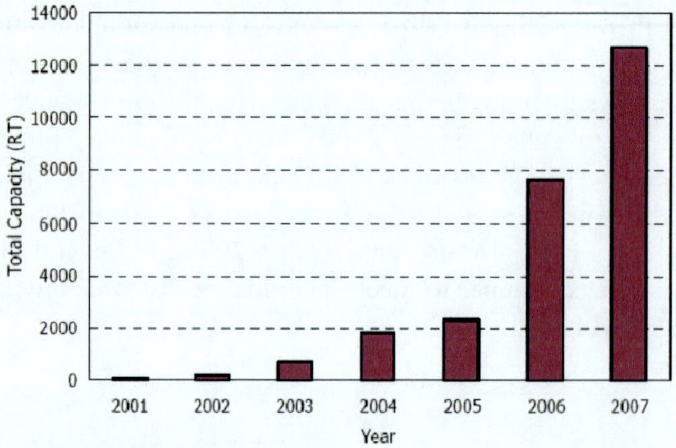

Source: Park, 2008 (IEA HPC 2008, S9,P6)

Figure 2-13. Total capacity of GSHP supply in Korea

The Korean GSHP market has shown tremendous growth since 2001 (Figure 2-13). This growth has been fueled by legislation passed in 2005 by the Korean government that required new public buildings to incorporate alternative and renewable energy sources.

U.S.

The United States has the largest worldwide GSHP installed base at approximately 1,000,000 units. Annual sales are currently estimated to be approximately 60,000 units per year, representing 245,000 tons of capacity (EIA 2006). The energy consumption of the U.S. GSHP market is estimated to be 25.5 trillion Btu in primary energy, which is five times what it was in 1990.

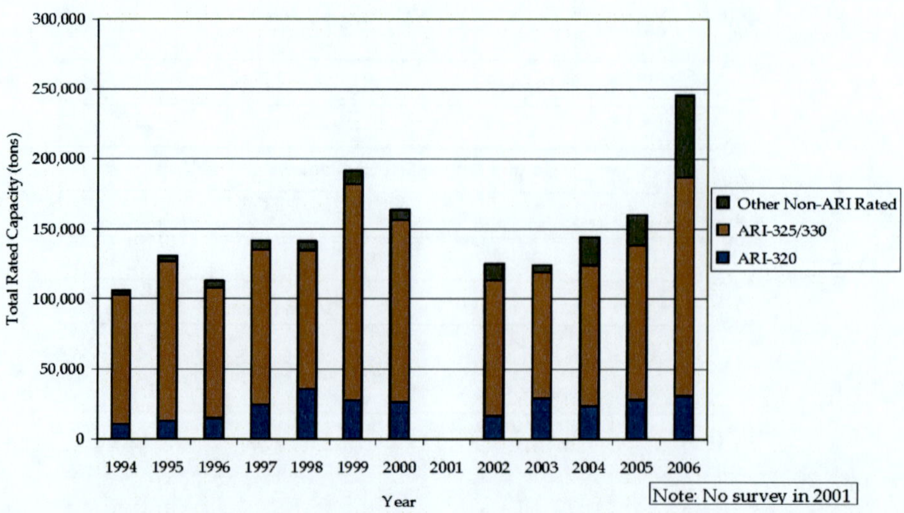

Source: EIA Survey of Geothermal Heat Pump Shipments 2006, Table 3.2 (2008).

Figure 2-14. Capacity of GSHP Shipments by Model Type[4]

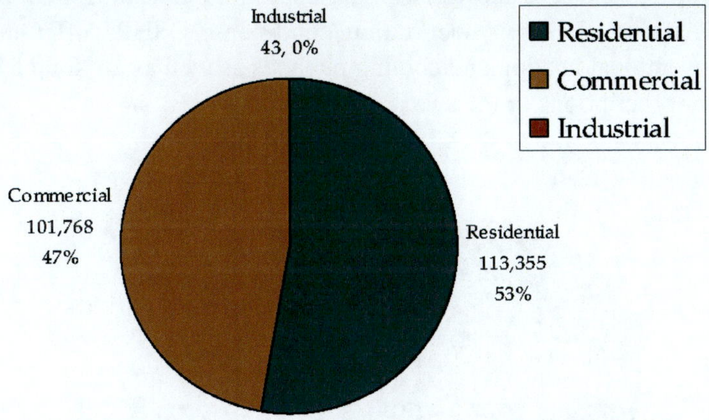

Source: EIA Survey of Geothermal Heat Pump Shipments 2006, Table 3.2 (2008)

Figure 2-15. GSHP Shipments by Sector in tons (2006)

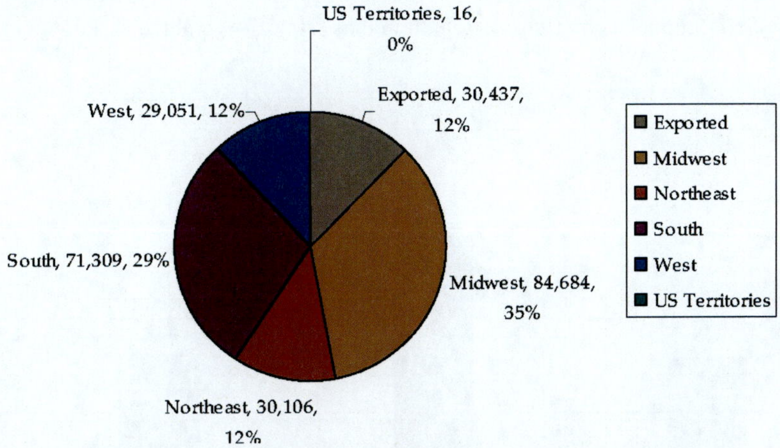

Source: EIA Survey of Geothermal Heat Pump Shipments 2006, Table 3.2 (2008)

Figure 2-16. GSHP Shipments by Census Region in tons (2006)

Figure 2-14 shows the annual shipments in terms of tons of capacity since 1994. In 2006, shipments reached just under 245,000 tons.

Figure 2-15 through Figure 2-18 give a sense for how the GSHP market is segmented. Just over half of shipments in 2006 were for residential applications, while the remaining shipments were commercial. The retrofit market for schools has seen substantial growth in recent years. Over 600 schools have GSHP systems installed, especially schools located in Texas.

As shown in Figure 2-16, GSHPs have a presence in all census regions although the market has historically been dominated by the Midwestern and southern states. Figure 2-17 shows the segmentation by census region, weighted by the population of each census region. The Midwest and South are home to the major GSHP manufacturers and have more personnel trained in GSHP installation and maintenance than other regions.

GSHP rated efficiencies are shown in Figure 2-18. With a heating COP of about 4.0, and a cooling SEER of up to 19.4 under rating conditions, GSHPs offer a high-efficiency alternative to conventional heating and cooling methods as well as air-source heat pumps. See footnote above for descriptions of the subcategories.

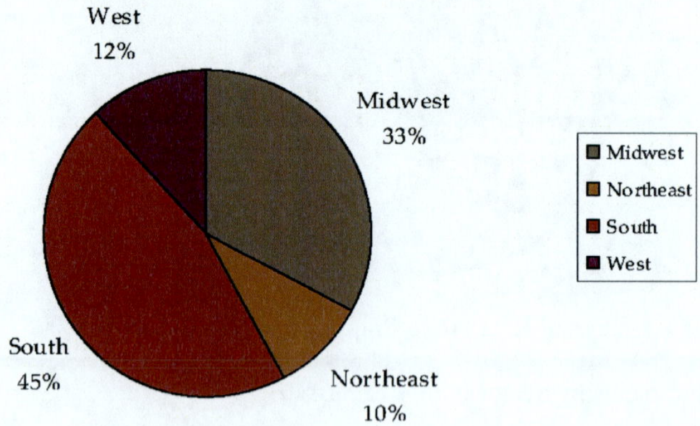

Source: EIA Survey of Geothermal Heat Pump Shipments 2006, Table 3.2 (2008)

Figure 2-17. GSHP Shipments by Census Region in tons (2006) - Weighted by Population

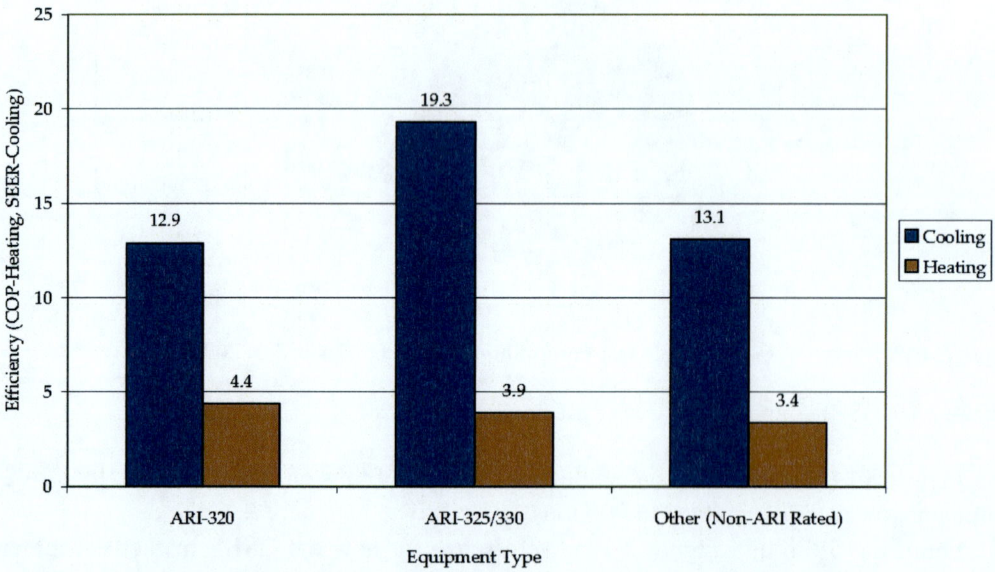

Source: EIA Survey of Geothermal Heat Pump Shipments 2006, Table 3.2 (2008)

Figure 2-18. US Average GSHP Efficiency Ratings (2006)

3. NATIONAL PRIMARY-ENERGY -SAVINGS POTENTIAL FOR GSHPS

Our approach to projecting the national primary-energy savings[5] of GSHPs is outlined below.

3.1. Approach to Projecting National Energy Savings

Our review of the available literature provided a number of analyses of the economics and energy-savings potential of GSHPs. However, we found no direct comparisons to alternative high-efficiency HVAC technologies, such as advanced ASHPs, nor sufficient documentation to use available analyses for projecting national energy savings. Therefore, we used simplified spreadsheet analyses to project the economics and energy-saving potential of GSHPs. Figure 3-1 outlines our overall approach to projecting the national energy savings for GSHPs and other advanced technologies. We first identified the cost and performance characteristics of the two energy-saving technologies considered (GSHPs and advanced ASHPs), as well as the existing space-conditioning technologies that would be displaced:

- Conventional ASHPs
- Conventional furnaces and air conditioners.

We then established geographical regions for analysis. We do this because energy savings and economics can vary significantly, depending on regional construction practices, climate conditions, and utility rates. We analyze two representative building applications (single-family residential and small commercial/institutional). We then project national energy impacts for each:

- Technology displacement option
- Scenario
- Representative building application.

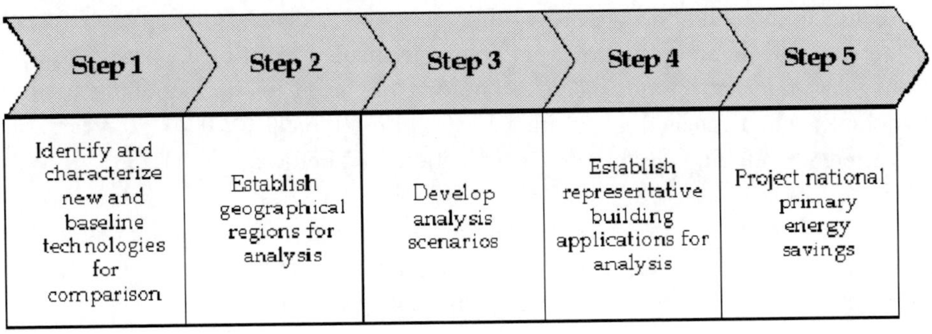

Figure 3-1. Approach to Projecting National Energy Savings

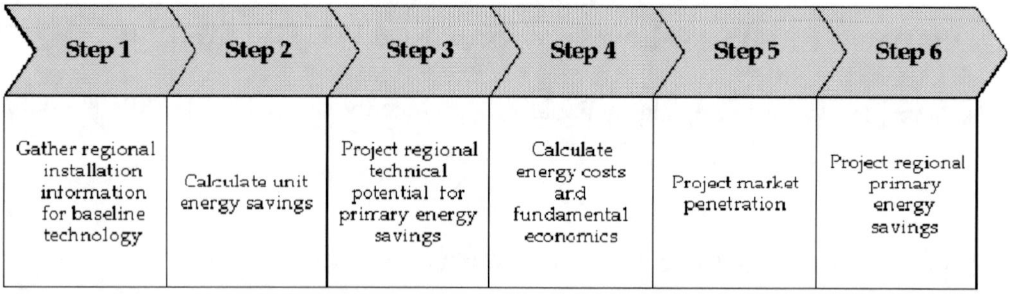

Figure 3-2. Approach to Projecting Energy Savings for a given Region, Application and Scenario

The economics and energy savings of GSHPs (or any advanced space-conditioning technology) will vary with geographic region due to variations in:

- Utility rates
- Climate conditions
- Typical construction characteristics
- Financial incentives provided by the state or local utility.

There are other geographic variables that impact the cost and performance of ground coupling (such as soil type, available land, and environmental regulations), but these characteristics can vary significantly within any reasonably sized geographic region, so it is difficult to evaluate their impacts quantitatively.

Figure 3-2 outlines our approach to projecting energy impacts for a given region, scenario, and application. We evaluate potential energy savings two ways:

- *Technical potential:* Primary energy savings that would result if 100 percent of installations of the baseline technology are replaced with the advanced technology. Technical potential places a theoretical upper bound on energy-savings potential.

- *Market potential:* Ultimate primary energy savings that one would expect based on the "fundamental" energy-savings economics. Market potential is always lower than technical potential because the higher first costs and other complexities of the advanced technology will prevent complete displacement of the baseline technology. We assume that the advanced technology has been in the marketplace sufficiently long to have reached its ultimate saturation (typically 10 to 20 years for high-efficiency building equipment—see Section 3.12 below).

The "fundamental" energy-savings economics are determined assuming that market-entry barriers have been surmounted, but that no financial incentives, such as rebates, tax credits or low-interest loans, are available. Market-entry barriers can include:

- Increased first costs, poor performance, or poor reliability specifically associated with:
    - Low manufacturing volumes
    - Immature product designs

- Inexperienced/poorly trained installers and service technicians
- Lack of awareness
- Lack of familiarity, leading to perceived risks that can, in turn, inflate costs or discourage potential end users
- Lack of supporting sales, installation, and service infrastructure.

While incentives are often available to install energy-saving technologies, looking at the unsubsidized economics gives a better sense of which advanced technology would leverage incentives most effectively.

We then project market penetrations based on generalized relationships for market penetration as a function of economic attractiveness. We adjust the results to consider the impacts of financial incentives and non-economic factors, as appropriate. Lastly, we multiply projected market penetrations by expected energy savings to project national energy savings.

Table 3-1. Competing Residential Space-Conditioning Technologies [EIA 2007]

Technology	Rated Cooling Efficiencies	Rated Heating Efficiencies	Typical Installed Cost[a]
Gas-Fired Furnace	--	**Typical: 80% AFUE; 780 kWh/yr** ENERGY STAR®: 90% AFUE; 500 kWh/yr 2007 Best Available: 96% AFUE; 275 kWh/yr	**$24.00/kBtuh** $32.70/kBtuh $44.00/kBtuh
Oil-Fired Furnace	--	**Typical: 81% AFUE; 850 kWh/yr** ENERGY STAR®: 83% AFUE; 800 kWh/yr 2007 Best Available: 95% AFUE; 650 kWh/yr	**$23.80/kBtuh** $26.20/kBtuh $50.50/kBtuh
Central A/C (Air Source)	Typical: 13 SEER ENERGY STAR®: 14 SEER Best Available: 21 SEER	--	**$814/ton** $886/ton $1714/ton
Central Heat Pump (Air Source)	Typical: 13 SEER ENERGY STAR®: 14 SEER Best Available: 17 SEER[b]	**Typical: 7.7 HSPF** ENERGY STAR®: 8.2 HSPF **2007 Best Available: 10.6 HSPF[b]**	**$1450/ton** $1570/ton **$2300/ton**
Ground-Source Heat Pump	Typical: 16 EER ENERGY STAR®: 14.1 EER Best Available: 30 EER	**Typical: 3.4 COP** ENERGY STAR®: 3.3 COP **2007 Best Available: 5.0 COP**	**$3000/ton** $2830/ton **$5250/ton**

a. Based primarily on retrofit installations, as this is the general case. Figures are mid-range values from EIA 2007. Heat-pump costs are per nominal ton of cooling capacity.

b. The "best available" was selected based on highest heating efficiency. Higher cooling efficiencies are available.

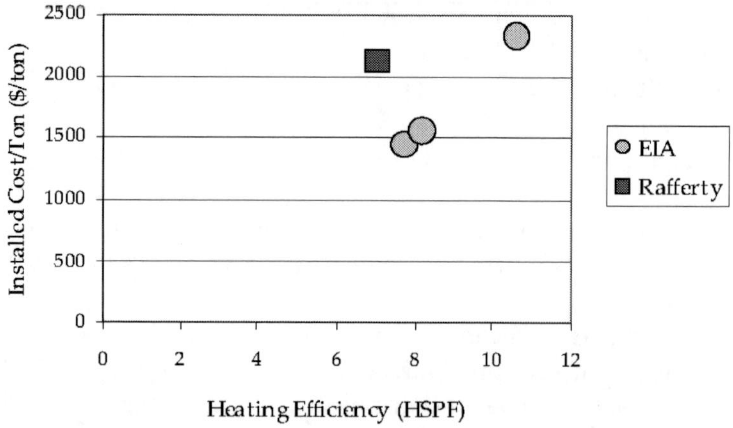

Figure 3-3. Installed-Cost Estimates for Residential ASHPs vs. Heating Efficiency

3.2. Installed-Cost and Performance Estimates

Outlined below are installed-cost and performance estimates for residential and commercial applications. Cost estimates are for retrofit applications, as this is the most common application. However, significant reductions in installation costs are possible in new construction, especially in housing developments or other planned communities.

3.2.1. Residential applications

Table 3-1 lists rated efficiencies and installed-cost estimates for a range of residential space-conditioning technologies as of 2007. Values in bold are used in this analysis.

Figure 3-3 compares EIA installed-cost estimates for the residential ASHPs to those from one other source [Rafferty 2008]. EIA costs are lower than other estimates. Upon reviewing the alternative source, the EIA estimates appeared most credible. Estimates by Kavanaugh are old [Kavanaugh 1995]. Estimates by Rafferty were adjusted from a 1995 estimate of $4400 for a 3-ton ASHP [Kavanaugh 1995}. It is not clear how Rafferty adjusted the 1995 estimate.

Kavanaugh and Rafferty also provided estimates for the combined installed cost of a gas furnace and conventional 3-ton air-conditioning system:

- Kavanaugh 1995: $4300 (in $1995)
- Rafferty 2008: $6200.

Comparing these estimates to EIA's (Table 3-1 above), we would estimate and installed cost of about $4200, based on a:

- 75-kBtuh gas furnace (capacity assumed) at $24.00/kBtuh: $1800
- 3-ton air conditioner at $814/ton: $2400
- Total: $4200.

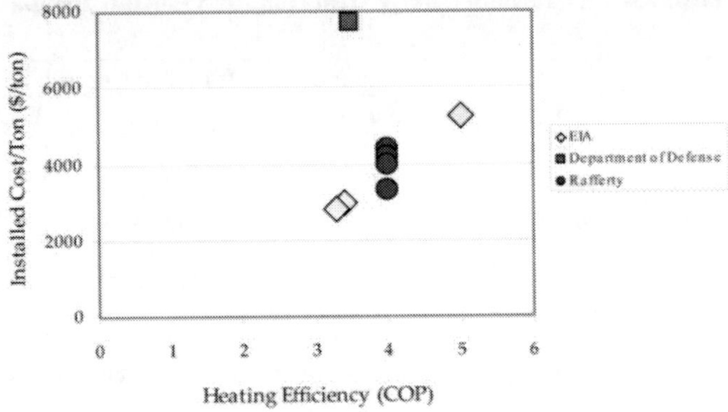

Figure 3-4. Installed -Cost Estimates for Residential GSHPs vs. Heating Efficiency

For reasons similar to those outlined above for ASHP costs, we elected not to use the Kavanaugh and Rafferty estimates.

Figure 3-4 compares installed-cost estimates for residential GSHPs (as a function of rated heating efficiency) from various sources. Most estimates are for GSHPs having 3-ton nominal cooling capacities. For reasons similar to those outlined above for ASHP costs, we elected not to use the Kavanaugh and Rafferty estimates. While DOD has a substantial installed-cost database, their average costs seemed suspiciously high given the large contracts let—many were to install hundreds of residential heat pumps [DOD 2007]. We do not know if DOD had special provisions that contributed to the costs.

Figure 3-5 shows the approximate breakdown of GSHP cost by major component. The ground loop is the single most expensive component, accounting for about 30 to 35 percent of the installed cost (depending on whether the ductwork is included). Henderson reports that ground loops for four demonstration homes in New York State cost between \$1000/ton and \$1800/ton[6] [Henderson 1998]. If we assume this is 35.5 percent of the installed cost, then the associated installed costs would range from \$2800/ton to \$5100/ton, which is not inconsistent with the installed cost we used (\$3000/ton for a typical -efficiency GSHP).

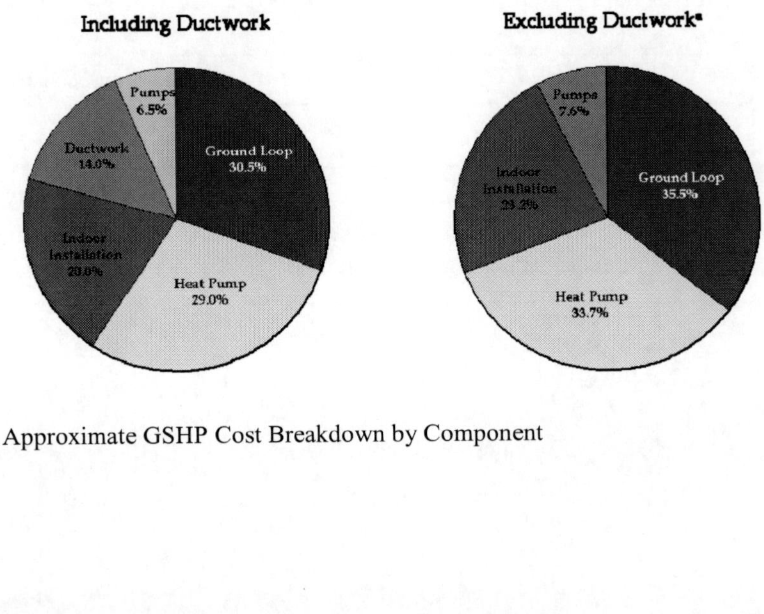

Figure 3-5. Approximate GSHP Cost Breakdown by Component

Table 3-2. Installed-Cost Estimates for Ground Loops

Type	Ground-Loop Installed Cost			
	Kavanaugh 1995 (1995 dollars)		Rafferty 2008 (2008 dollars)	
	$/ton	Relative Cost	$/ton	Relative Cost
Vertical Loop	$ 2,999	1.11	$4400/ton	1.32
Slinky	$ 2,875	1.06	$4200/ton	1.26
Horizontal Loop	$ 2,712	1.00	$4000/ton	1.20
Open Loop	-----	-----	$3300/ton	1.00

Table 3-3. Competing Commercial Space-Conditioning Technologies [EIA 2007]

Technology	Rated Cooling Efficiencies	Rated Heating Efficiencies	Typical Installed Cost[a]
Gas-Fired Furnace	--	Typical: 80% thermal High Efficiency: 82% thermal	$8.1/kBtuh $8.8/kBtuh
Oil-Fired Furnace	--	Typical: 81% thermal	$8.1/kBtuh
Roof-Top Air Conditioner	Typical: 10.1 EER High Efficiency: 12.0 EER	--	$65.6/kBtuh $85.0/kBtuh
Roof-Top Heat Pump	Typical: 10.3 EER High Efficiency: 11.7 EER	Typical: 3.2 COP High Efficiency: 3.4 COP	$73.0/kBtuh $97.0/kBtuh

a. Based primarily on retrofit installations, as this is the general case. Figures are mid-range values from EIA 2007. Heat-pump costs are per nominal ton of cooling capacity.

In reality, ground-loop costs vary, depending on type used. Estimated installed costs for various types of ground loops are shown in Table 3-2.

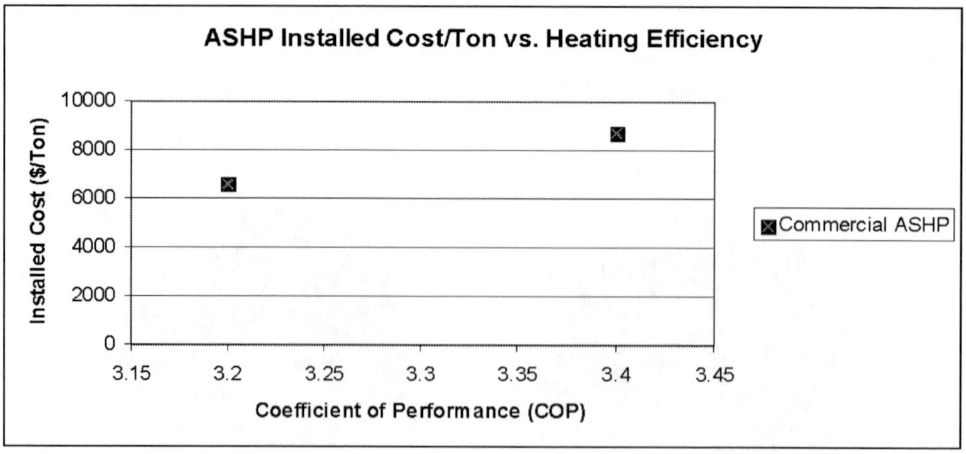

Figure 3-6. Installed-Cost Estimates for Commercial ASHPs

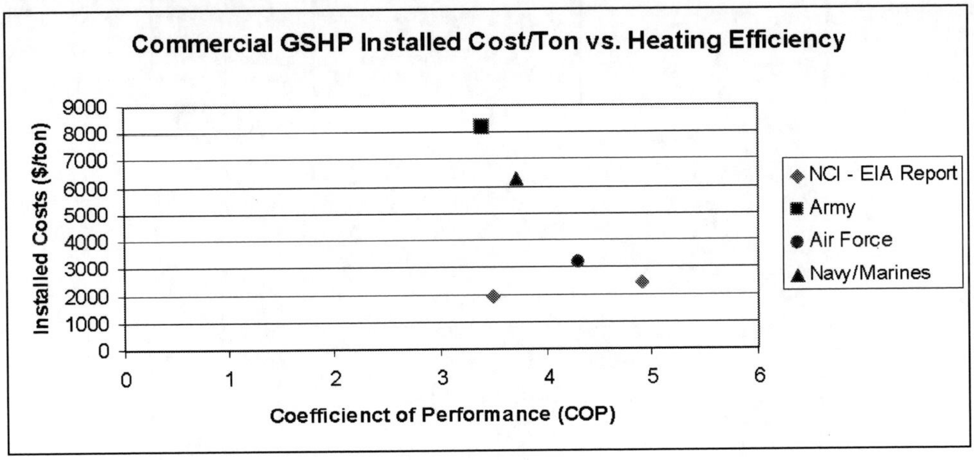

Figure 3-7. Installed-Cost Estimates for Commercial GSHPs

3.2.2. Commercial Applications

Table 3-3 lists rated efficiencies and installed-cost estimates for a range of small-commercial space-conditioning technologies as of 2007. Values in bold are used in this analysis.

Figure 3-6 shows installed-cost estimates for commercial ASHPs as a function of rated heating efficiency

Figure 3-7 compares installed-cost estimates for commercial GSHPs as a function of rated heating efficiency. ETA estimates are similar (per unit capacity) to ETA estimates for residential applications. DOD estimates, however, are substantially higher, which is surprising, given the large scope of most of the DOD projects. Again, we elected to use ETA values.

3.3. Maintenance Costs

Two investigators of commercial/institutional GSHPs report reduced maintenance costs compared to conventional equipment [Martin 2000; Cane 2000]. These investigators, however, did not account for differences in equipment age, and neither suggests that their results can be applied broadly.

For residential applications, ASHPs require periodic cleaning of the outdoor coil to maintain good performance, while GSHPs may require some maintenance to maintain the glycol solution in the ground loop. In either case, we assume that both require one annual maintenance call, and the fixed costs associated with that call are the bulk of the cost.

In the end, we assumed that maintenance costs were roughly the same for all technology options and did not account for them.

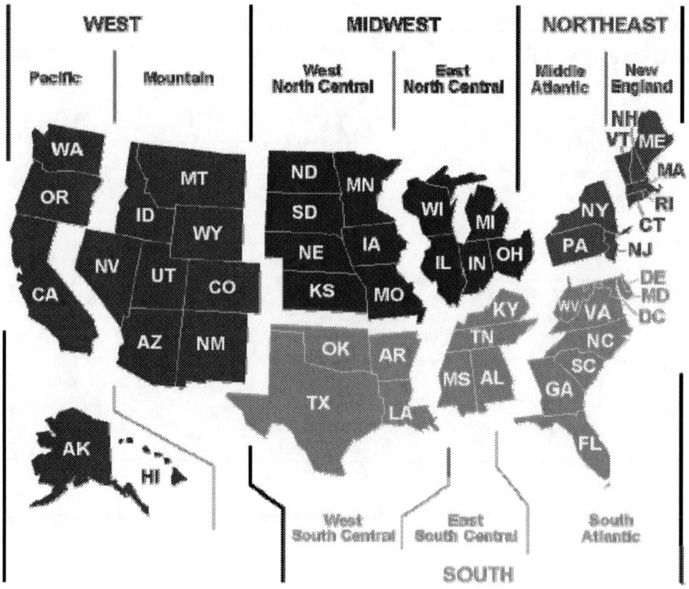

Source: EIA website

Figure 3-8. U.S. Census Regions

3.4. U.S. Regions for Analysis

Common ways to divide the U.S. geographically include:

- Climate regions/zones
- U.S. census regions
- States
- Other geopolitical divisions, such as counties or zip codes
- Utility service areas.

We used U.S. major census regions (see Figure 3-8) for our regional analysis because:

- Heat-pump installation and shipment data are available by census region
- Although they don't make ideal climate regions, one can assign approximate climate conditions to census regions
- More detailed modeling is beyond the scope of this analysis.

3.5. Scenarios for U.S. Market Projections

We selected three scenarios under which to project the national energy -saving potentials of GSHPs:

- "Carbon Tax plus R&D" Scenario: DOE invests in successful cost-reduction R&D and a carbon tax increases utility rates, including electricity, natural gas, fuel oil, and propane
- 'Successful R&D" Scenario: DOE invests in successful cost-reduction R&D, but no carbon tax is imposed
- "Business as Usual" Scenario: No successful R&D and no carbon tax imposed.

None of the three scenarios includes the impacts of financial incentives such as tax credits or utility rebates, consistent with our objective of investigating the fundamental economics of GSHPs compared to alternatives.

As illustrated in Appendix B, EIA projections for electric rates under three scenarios (high price, low price, and reference) show very little variation in projected rates through 2030 (adjusted to 2006 dollars) [AEO 2008]. EIA projections also vary little within each census region. However, the EIA does not account for the possibility of a carbon tax in their projections, even in their high-price case. Projections of the cost impacts of a carbon tax vary, as shown in Table 3-4. We assume that 2006 electric rates do not vary for the foreseeable future unless a carbon tax is imposed. We assume a carbon tax, if imposed, would increase electricity prices by 30 percent. We use 30 percent for all regions, even though carbon taxes would probably vary significantly by region.

Likewise, EIA projections show little variation in natural-gas rates over time (when adjusted for inflation). We assume that all inflation-corrected fuel prices remain at 2006 levels, if no carbon tax is implemented. If a carbon tax is implemented, we assume a 30 percent increase in all fuel prices. The impacts of an actual carbon tax would probably vary by fuel, and might vary regionally as well.

When considering potential GSHP cost reductions associated with R&D, we considered only improvements in the ground loop, assuming that improvements in other system components would apply equally well to the baseline technology (ASHPs) and, therefore, not change the overall economics of using GSHPs. As shown in Figure 3-5 above, the ground loop accounts for about 30 to 35 percent of the a typical GSHP installed cost.

Each of the three scenarios is described further below.

Table 3-4. Projected Impacts on Electricity Price if a Carbon Tax is instituted

Source	Projected Electricity Price Increase by 2020 (%)	
	Real	**Adjusted to 2006 Dollars**
Lieberman-Warner Climate Security Act of 2008 (Bill) [EIA 2008]	--	5% to 27%
Sanders-Boxer Proposal [Paltsev 2007]	73%	36%[a]
Bingaman-Specter Proposal [Paltsev 2007]	44%	13%[a]
Assumed Scenario for Analysis (see discussion)	--	30%

a. Adjusted based on an expected 27% electricity price increase (real dollars) between 2005 and 2020 [Paltsev 2007].

Under the "Carbon Tax plus R&D" Scenario, we assumed the following conditions prevail:

- No financial incentives are available for high-efficiency GSHPs or ASHPs
- GSHP R&D results in a 30-percent decrease in the average installed cost of a ground loop (in current dollars), with no performance penalty
- Any improvements in ASHP installed costs are similarly applicable to GSHPs, and vice versa, i.e., no change in installed-cost differentials except for the reduction in ground-loop costs discussed above
- A carbon tax is instituted, raising utility prices by 30 percent (in current dollars).

Under the "Successful R&D" Scenario, we assumed the following conditions prevail:

- No financial incentives are available for high-efficiency GSHPs or ASHPs
- GSHP R&D results in a 30-percent decrease in the average installed cost of a ground loop (in current dollars), with no performance penalty
- Any improvements in ASHP installed costs are similarly applicable to GSHPs, and vice versa, i.e., no change in installed-cost differentials except for the reduction in ground-loop costs discussed above
- No carbon tax is instituted, and utility prices remain steady (in current dollars) per EIA projections.

Under the "Business as Usual" Scenario, we assumed the following conditions prevail:

- No financial incentives are available for high-efficiency GSHPs or ASHPs
- Any improvements in ASHP installed costs are similarly applicable to GSHPs, and vice versa, i.e., no change in installed-cost differentials
- No carbon tax is instituted, and utility prices remain steady (in current dollars) per EIA projections.

3.6. Representative Building Applications

We selected representative building profiles for two applications (one residential and one commercial) for each of the nine census regions. While construction characteristics vary to reflect local codes and architecture, the profiles used are intended to represent comparable applications among the various regions.

3.6.1. Residential application

We selected a single-family home of about 3000 sq. ft. This is much larger than average, but probably more representative of the typical purchaser of a GSHP than an average-size home would be. Table 3-5 summarizes the characteristics of this home in the five cities for which load data were available [TIAX 2006][7].

Figure 3-9 shows the annual space-conditioning loads used for the each census region. We either selected a city from our database, or extrapolated load data by rationing heating degree days (or cooling degree days) for a city we judged to represent better the climate in that census region.

3.6.2. Commercial application

We selected a small office building of about 6500 sq. ft. for our analysis. This is a smaller commercial building than average, but will be representative of a large subset of commercial buildings. The size of building will vary slightly due to differences in northern construction methods versus southern construction methods.

Table 3-5. Characteristics of Representative Single -Family Home [TIAX 2006]

Item	Description
Construction	• Suburban, Single Family • 1 – 2 Stories • Post -2000 Construction • 2800 to 3320 sq. ft. Conditioned Space • Crawl Space
Occupancy	• 2–4 Occupants, depending on location
Locations	• Minneapolis; Washington, DC; New York City; Sacramento; Chicago[a]

a. For cities not included in original load data, we ratioed heating degree days to estimate heating loads, and cooling degree days to estimate cooling loads.

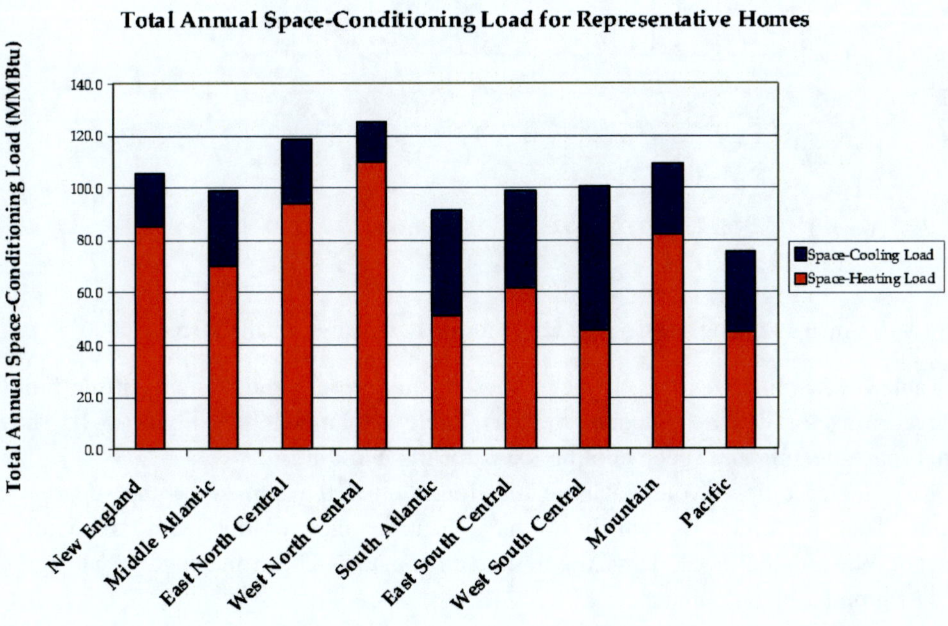

Figure 3-9. Annual Heating & Cooling Loads for Representative Single -Family Home (3000 ft^2)

Table 3-6. Characteristics of Representative Commercial Building [Huang 1999].

Item	Description
Construction	• Small Office • 1 – 2 Stories • Post -2000 Construction • ~6500 sq. ft. Conditioned Space
Occupancy	• 470 Occupants
Locations	• Minneapolis; Washington, DC; Chicago; Houston; Los Angeles[a]

a. For cities not included in original load data, we ratioed heating degree days to estimate heating loads, and cooling degree days to estimate cooling loads.

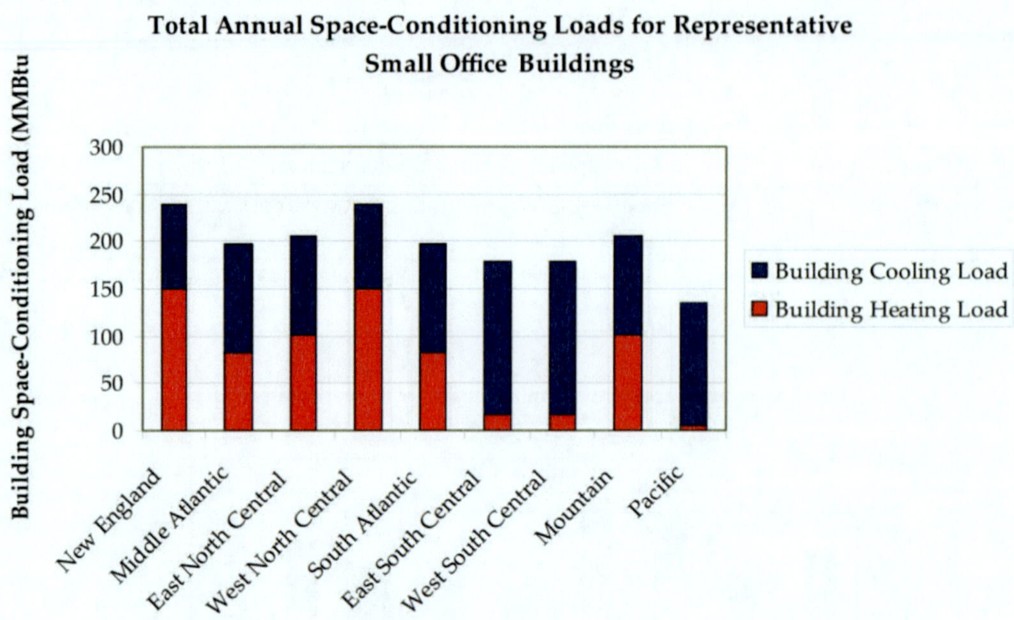

Figure 3-10. Annual Heating and Cooling Loads for Representative Small Offices (~6000 sq. ft.)

Table 3-6 summarizes the characteristics of this representative small office building in the five cities for which load data were available [Huang 1999][8]. Figure 3-10 shows the annual space-heating and space-cooling loads for this building.

We selected cities in our database that represented the climate of the separate census regions. The five cities for which we have heating and cooling loads are Minneapolis, Chicago, Washington D.C., Los Angeles, and Houston. These matched to an appropriate census region based on climate.

3.7. Baseline Technology Energy Consumption

Energy consumptions for the baseline technologies are estimated below. "Baseline technology" is the technology against which we wish to compare performance. Since new equipment generally has higher efficiencies compared to the installed base (because of appliance and equipment energy-conservation standards, or simply because of advances in product design), some energy savings will accrue simply through normal replacement cycles, without additional DOE action. So that we don't double count this portion of the energy savings associated with high-efficiency technology options, we calculate energy savings relative to typical new equipment, not the existing stock.

3.7.1. Baseline technology energy consumption—residential

We compare GSHPs and advanced ASHPs to two baseline residential technologies described in Section 3.2 above:

- Typical (conventional) ASHP
- Typical furnace and central air conditioner.

We consider three fuel types for the furnace—natural gas, fuel oil, and propane, although not intermediate results are shown for fuel oil and propane. We include parasitic electric consumption associated with furnaces.

3.7.2. Baseline technology energy consumption—commercial

We compare GSHPs and advanced ASHPs to one baseline commercial technology described in Section 3.2 above—a typical (conventional) ASHP.

3.8. Unit Energy Savings

We projected unit energy savings for both the commercial and residential representative applications. We consider both space-cooling and space-heating benefits of GSHPs, but we did not consider the option for domestic (service) water heating using the heat pump. We did not include domestic water heating because it is also an option for ASHPs and air-source air conditioners, although the later provides the benefit only during the cooling season. In any case, the energy savings associated with GSHP water heating are small compared to the space-heating savings in all but southern climates [Rafferty 2008].

3.8.1. Residential unit energy savings

Figure 3-11 and Figure 3-12 compare the primary unit energy consumptions (UECs) for the various technology options in the New England and Middle Atlantic regions. Similar charts for the other census regions are included in Appendix B.

Figure 3-13 and Figure 3-14 show the resulting unit energy savings (UES) for New England and the Middle Atlantic, respectively, for the typical ASHP baseline. Appendix B includes the charts for other regions. Figure 3 -15 and Figure 3 -16 show the resulting unit energy savings (UES) for New England and the Middle Atlantic, respectively, for the gas furnace/AC baseline. Appendix C contains the charts for other census regions.

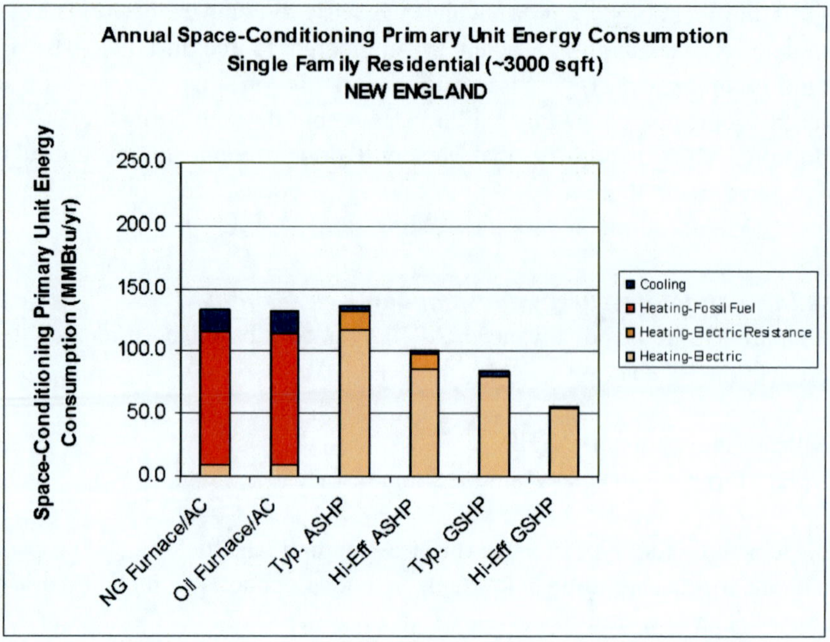

Figure 3-11. Primary Unit Energy Consumption Comparison—New England Residential

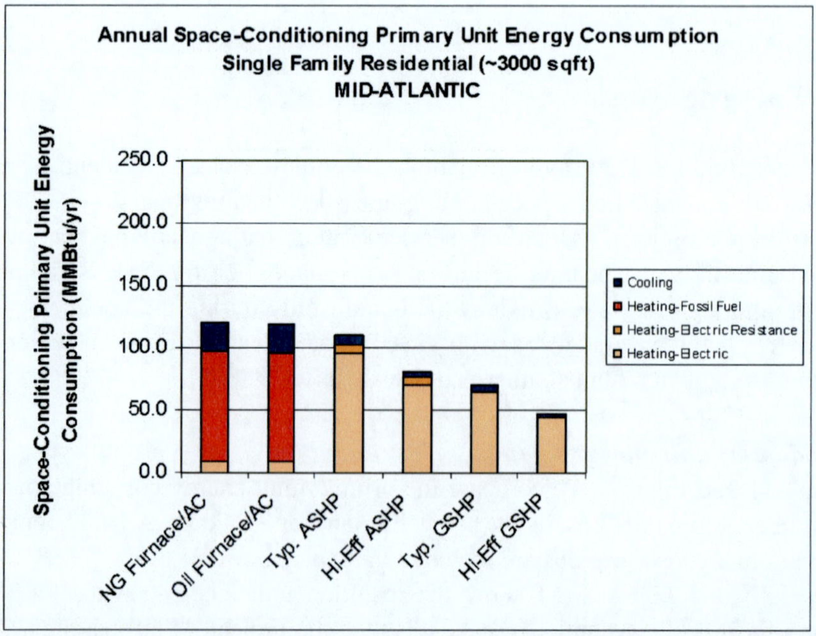

Figure 3-12. Primary Unit Energy Consumption Comparison—Middle Atlantic Residential

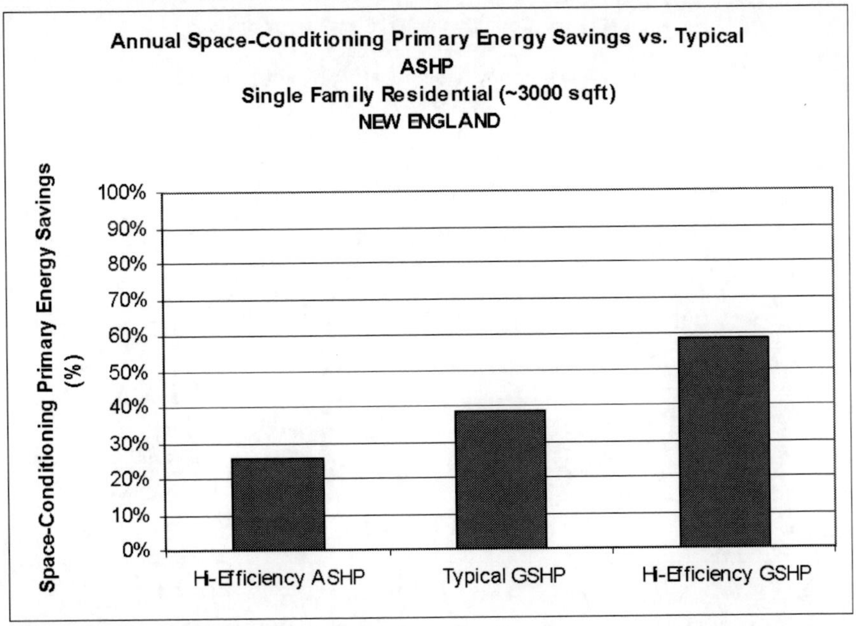

Figure 3-13. Primary Unit Energy Savings Comparison—ASHP Baseline—New England Residential

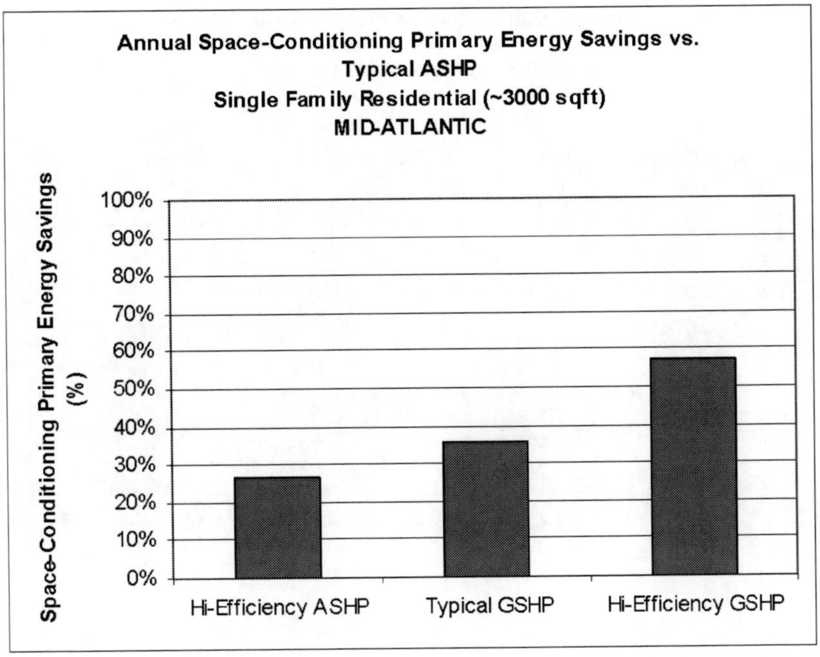

Figure 3-14. Primary Unit Energy Savings Comparison—ASHP Baseline—Middle Atlantic Residential

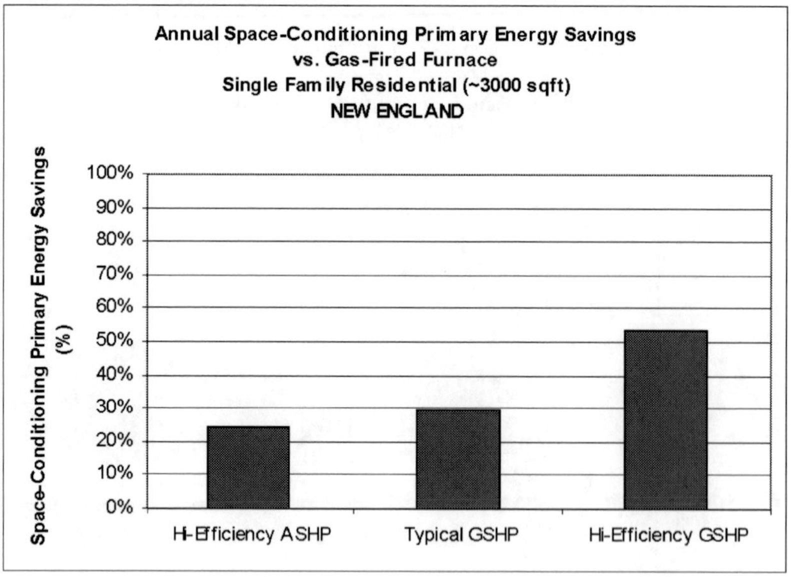

Figure 3-15. Primary Unit Energy Savings Comparison—Furnace/AC Baseline—New England Residential

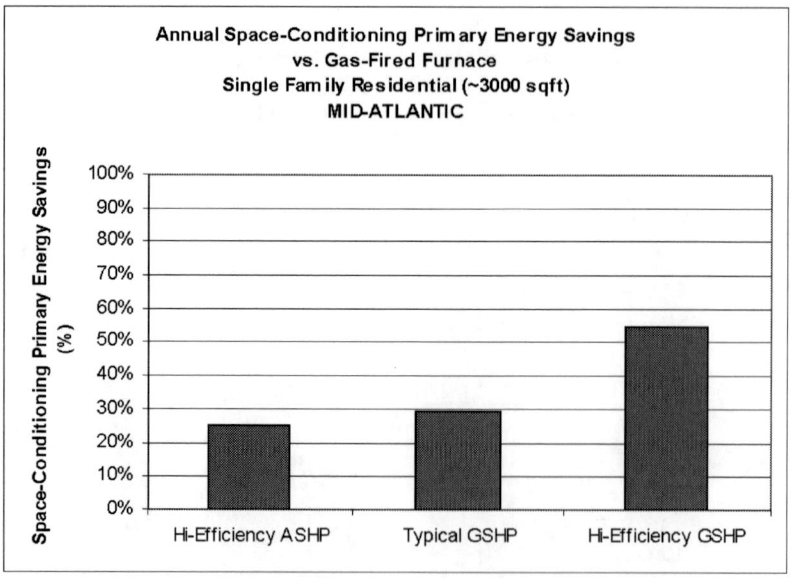

Figure 3-16. Primary Unit Energy Savings Comparison—Furnace/AC Baseline—Middle Atlantic Residential

Compared to typical-efficiency ASHPs, ranges of primary unit energy savings are (as a percent of space-conditioning energy consumption):

- Advanced ASHPs: 20 to 30 percent

- Typical-Efficiency GSHPs: 25 to 50 percent
- High-Efficiency GSHPs: 50 to 70 percent

Compared to typical-efficiency furnaces (natural gas, propane or fuel oil) and air conditioners, ranges of primary unit energy savings are (as a percent of space- conditioning energy consumption):

- Advanced ASHPs: 20 to 30 percent
- Typical-Efficiency GSHPs: 25 to 30 percent
- High-Efficiency GSHPs: 50 to 60 percent.

These ranges are slightly narrower than reported above for the typical-efficiencyASHP baseline because there is less regional variation in heating-season performance compared to ASHPs.

3.8.2. Commercial unit energy savings

Figure 3-17 and Figure 3-18 compare the primary unit energy consumptions (UECs) for the various technology options in the New England and Middle Atlantic regions. Similar charts for the other census regions are included in Appendix D.

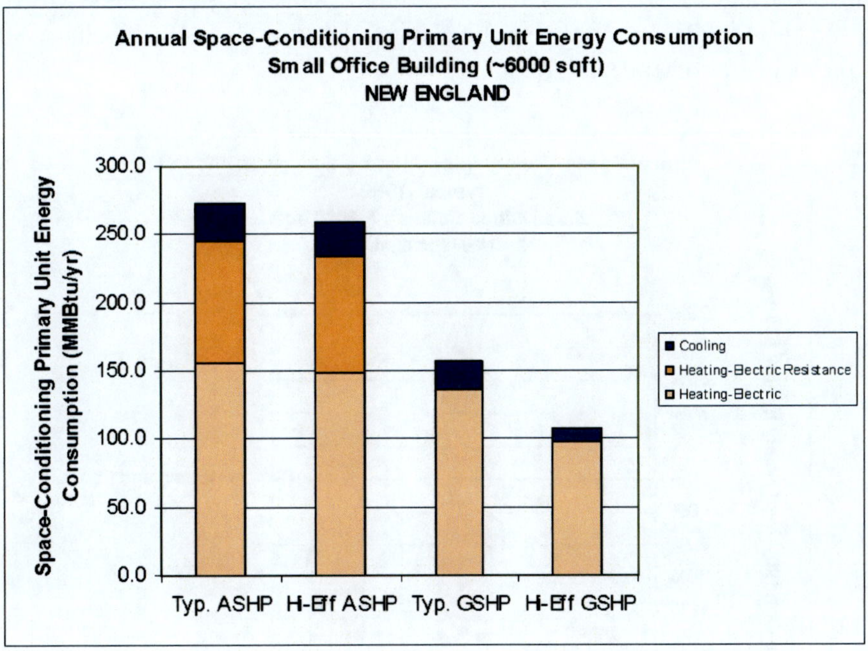

Figure 3-17. Primary Unit Energy Consumption Comparison—New England Commercial

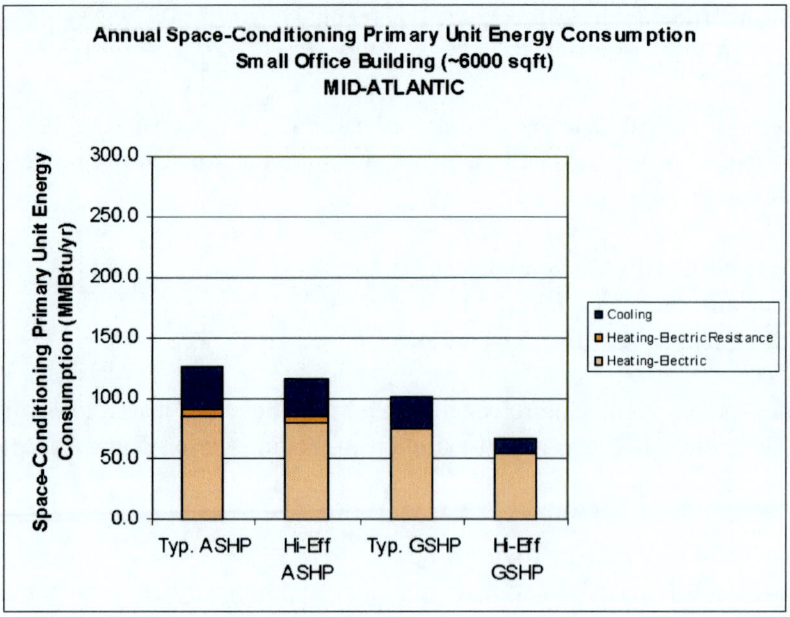

Figure 3-18. Primary Unit Energy Consumption Comparison—Middle Atlantic Commercial

Figure 3-19 and Figure 3-20 show the resulting unit energy savings (UES) for New England and the Middle Atlantic, respectively, for the typical ASHP baseline. Appendix E contains the charts for other census regions.

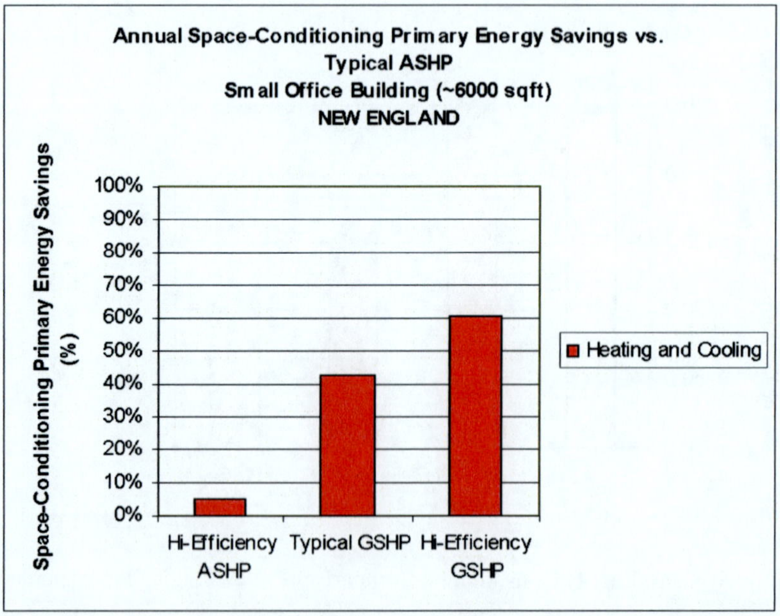

Figure 3-19. Primary Unit Energy Savings Comparison—ASHP Baseline—New England Commercial

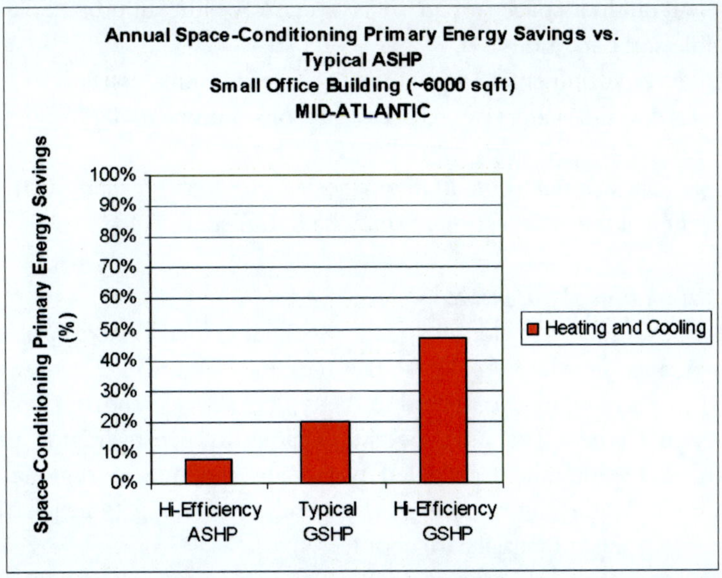

Figure 3-20. Primary Unit Energy Savings Comparison—ASHP Baseline—Middle Atlantic Commercial

Compared to typical-efficiency ASHPs, ranges of primary unit energy savings are (as a percent of space -conditioning energy consumption):

- Advanced ASHPs: 5 to 10 percent
- Typical-Efficiency GSHPs: 20 to 45 percent
- High-Efficiency GSHPs: 45 to 60 percent

In the commercial example, the "advanced" ASHP is only marginally more efficient compared to the typical ASHP, so its energy savings are less than for the residential example above. Ranges of savings for the two GSHPs are also slightly lower compared to the residential example above.

3.9. Technical Potential Primary Energy Savings

"Technical potential" refers to the theoretical primary energy savings associated with replacing 100 percent of the technically applicable baseline installations with the advanced technology. While this potential will never be achieved, it does suggest an upper limit, should the advanced technology be universally adopted. We calculate the technical potential as if all existing equipment has the efficiency of new equipment. This way, we don't double count the energy savings that will occur anyway, without further DOE action, based on normal replacement cycles.

To calculate technical potential for GSHPs rigorously, one would adjust for the number of installation sites that are not technically feasible for ground loops, such as sites having:

- Insufficient outdoor spacc (urban areas or densely built suburban areas)
- Unsuitable soil conditions
- Restrictive environmental regulations or ownership constraints on drilling or excavating, ground-water use, or use of glycol solutions underground.

We did not adjust technical potential for these factors, as they are difficult to quantify. Therefore, our technical potential estimates may be optimistic.

3.9.1. Residential technical potential

Figure 3-21 shows the estimated technical potential primary energy savings for various advanced technologies for the typical ASHP baseline. Figure 3-22 shows the technical potential primary energy savings compared to the typical furnace and air-conditioner baseline for three furnace-fuel types. Technical potential varies by census region primarily due to differences in baseline equipment installed base, but also due to regional differences in climate and seasonal heating and cooling efficiencies. Assuming typical-efficiency GSHPs are installed, total technical potentials are about:

- Typical ASHP Baseline: 0.5 Quad
- Typical Natural-Gas Furnace and Air-Conditioner Baseline: 2.1 Quad
- Typical Fuel-Oil Furnace and Air-Conditioner Baseline: 0.3 Quad
- Typical Propane Furnace and Air-Conditioner Baseline: 0.2 Quad
- Total Technical Potential Primary Energy Savings: 3.1 Quad.

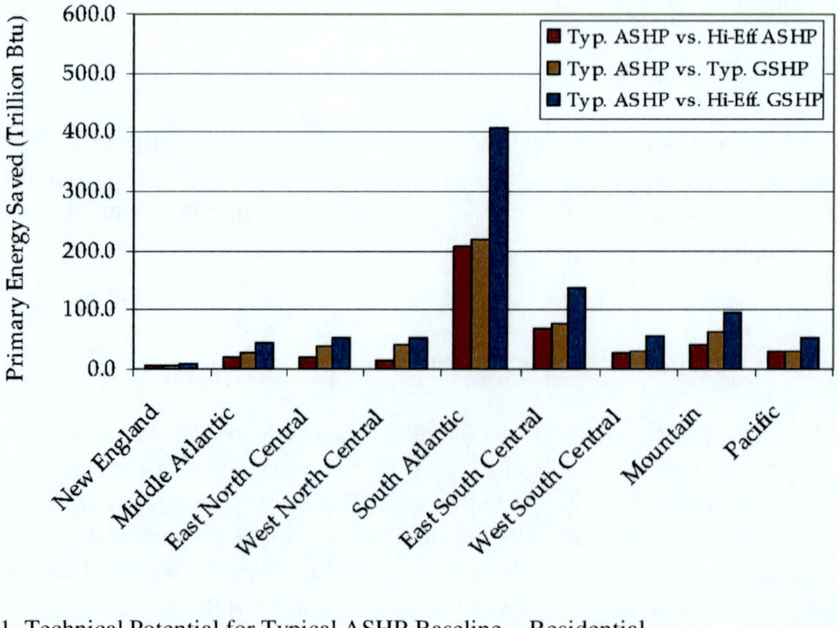

Figure 3-21. Technical Potential for Typical ASHP Baseline—Residential

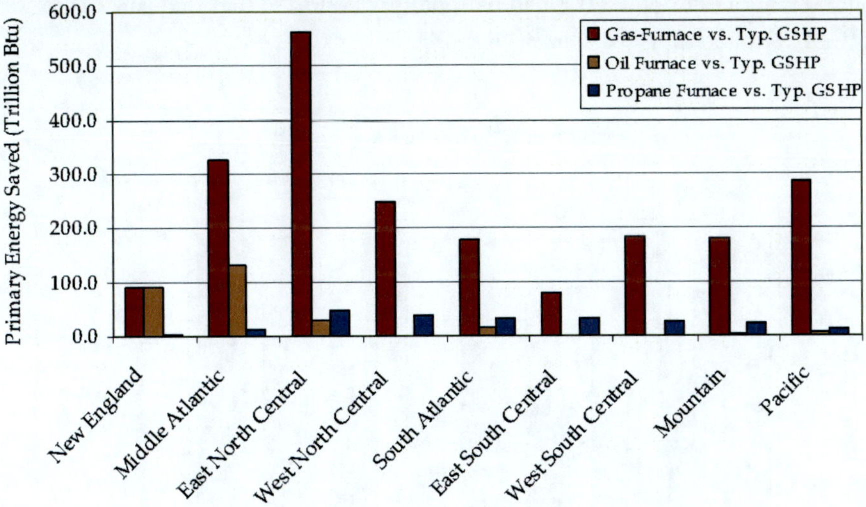

Figure 3-22. Technical Potential for Typical Furnace/AC Baseline—Residential

3.9.2. Commercial technical potential

We roughly estimated commercial technical potential using the following assumptions:

- Buildings under about 100,000 sq. ft. tend to use unitary packaged space-conditioning equipment and make up the primary target market for commercial GSHPs
- Roughly half the commercial space-conditioning energy consumption is associated with packaged unitary equipment.

The 2008 Buildings Energy Data Book reports that 2006 commercial space- conditioning primary energy consumption was (not including ventilation equipment):

- Space Heating: 2.17 Quad (1.18 Quad is natural gas)
- Space Cooling: 2.27 Quad (almost all electricity)
- Total: 4.44 Quad.

Based on the assumptions above, about half of this was consumed by the target market, or roughly 2 Quad. As noted above, typical GSHPs energy savings are between 20 and 45 percent, depending on climate. If we use 30 percent as a midrange savings, commercial GSHPs could save about 0.6 Quad if applied in all unitary space-conditioning applications.

3.9.3. School technical potential

Public schools (K-12), a subset of commercial buildings, are a noteworthy early market for GSHPs, because they often have ample space for the ground coil and often can justify longer payback periods compared to buildings in the private sector. Schools can potentially achieve 25-50% in space-conditioning energy savings using GSHPs. We estimate the technical potential energy savings associated with schools (included in the 0.6 Quad estimate in Section 3.9.2 above) to be 0.03 Quad nationally, due to the fact that schools represent only about 2.5% of commercial floor space (DOE 2008).

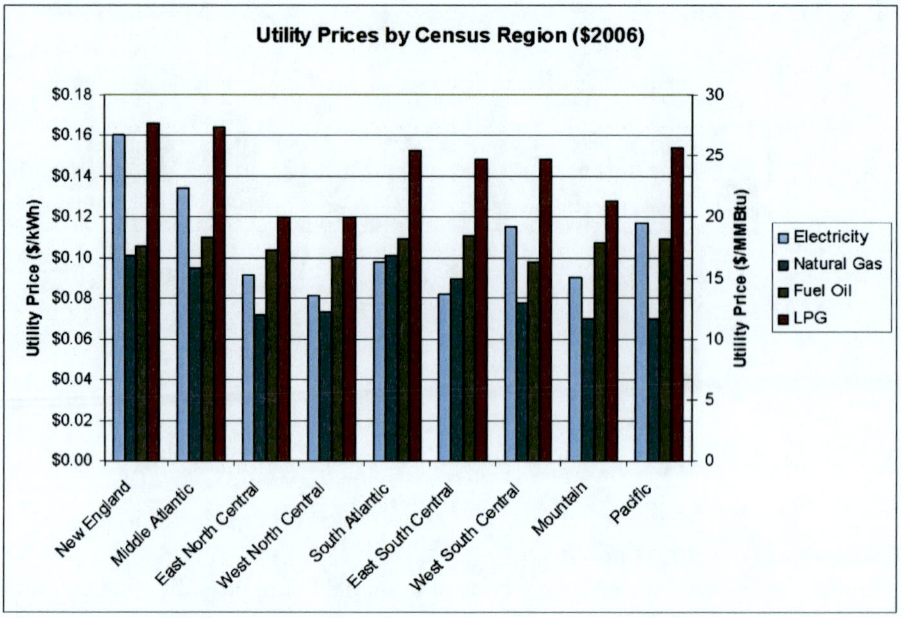

Source: Annual Energy Outlook 2008

Figure 3-23. Residential Utility Rates used in Analysis

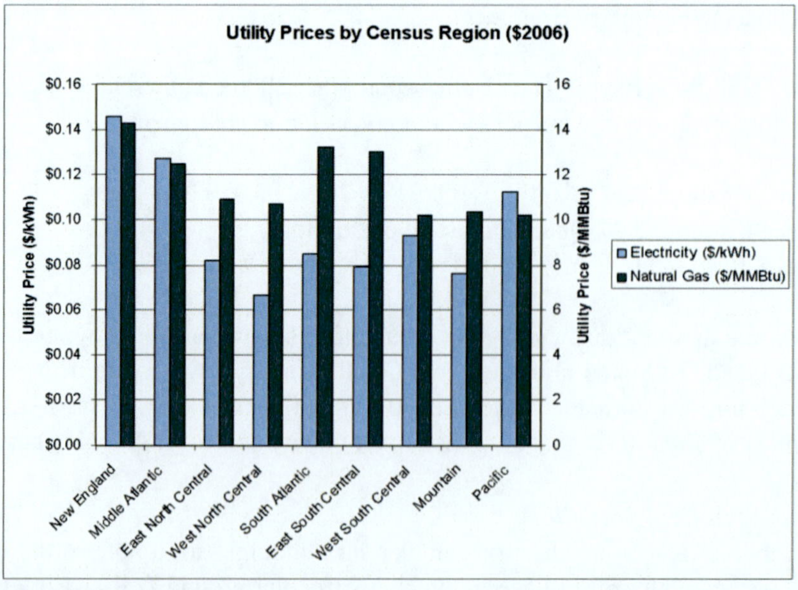

Source: Annual Energy Outlook 2008

Figure 3-24. Commercial Utility Rates used in Analysis

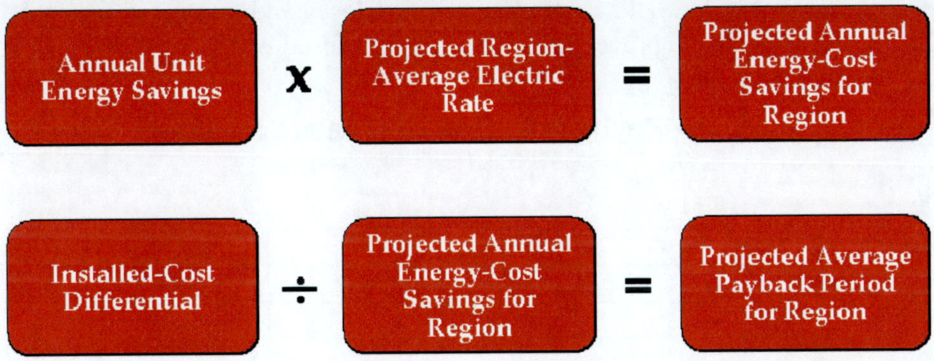

Figure 3-25. Methodology for Projecting GSHP Economics for each Region

3.10. Energy Prices

Figure 3-23 shows the residential utility prices used in this analysis. Prices are 2006 averages for each census region. As discussed in Section 3.6 above, we assumed that utility prices (corrected for inflation) do not change during the time period of our market projections unless a carbon tax is imposed.

Figure 3-24 shows the commercial utility prices used in this analysis. Prices are 2006 averages for each census region. As discussed in Section 3.6 above, we assumed that utility prices (corrected for inflation) do not change during the time period of our market projections unless a carbon tax is imposed.

3.11. Economic Analysis

We use simple payback period to represent the economic attractiveness of GSHPs and advanced ASHPs compared to alternative technologies. First, we estimate the allowable installed-cost differentials for selected payback periods. We then compare allowable installed-cost differentials to estimated differentials from above.

Projecting economics requires a regional approach. Figure 3-25 illustrates how we projected the economics (simple payback period) of GSHPs for each of the nine census regions in the nation. Annual energy costs used to calculate payback are shown in Appendices G and H for residential and commercial applications, respectively.

3.11.1. Residential fundamental economics—electric baseline technology

Figure 3-26 to Figure 3-28 show calculated simple payback periods for GSHPs and advanced ASHPs compared to typical-efficiency ASHPs.

GSHPs could achieve simple payback periods of roughly 5 to 10 years in the northeast and Midwest if R&D efforts lower ground-loop installed costs and/or a significant carbon tax is imposed. The economics of advanced ASHPs are, however, as good or better compared to GSHPs. Furthermore, our analysis does not account for the improved heating-season performance of advanced ASHPs currently being developed and introduced for cold-climate applications, which would improve the economics of advanced ASHPs.

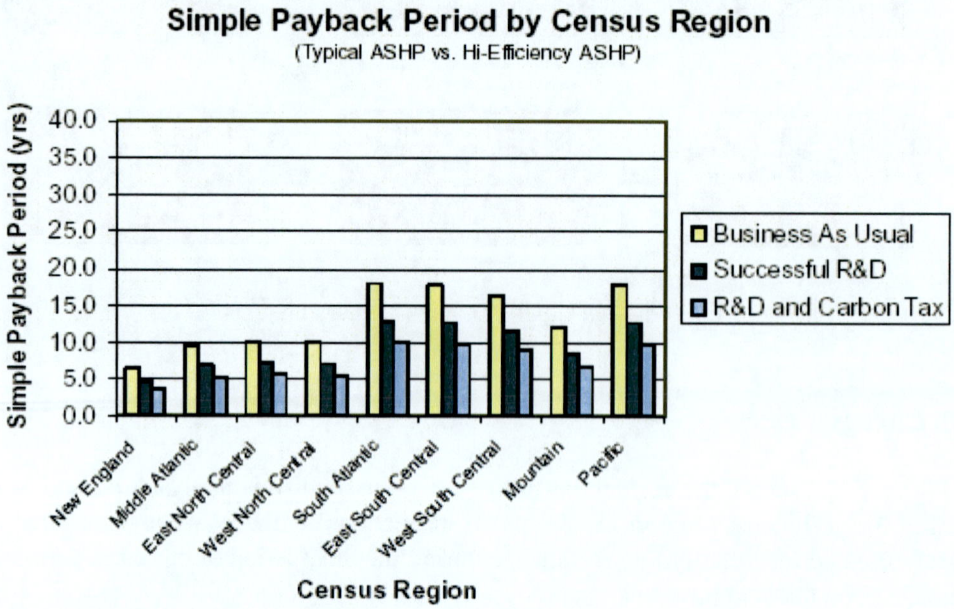

Figure 3-26. Simple Payback Periods for Representative Single -Family Home (~3000 sq. ft.)—Advanced ASHP vs. Typical -Efficiency ASHP Baseline

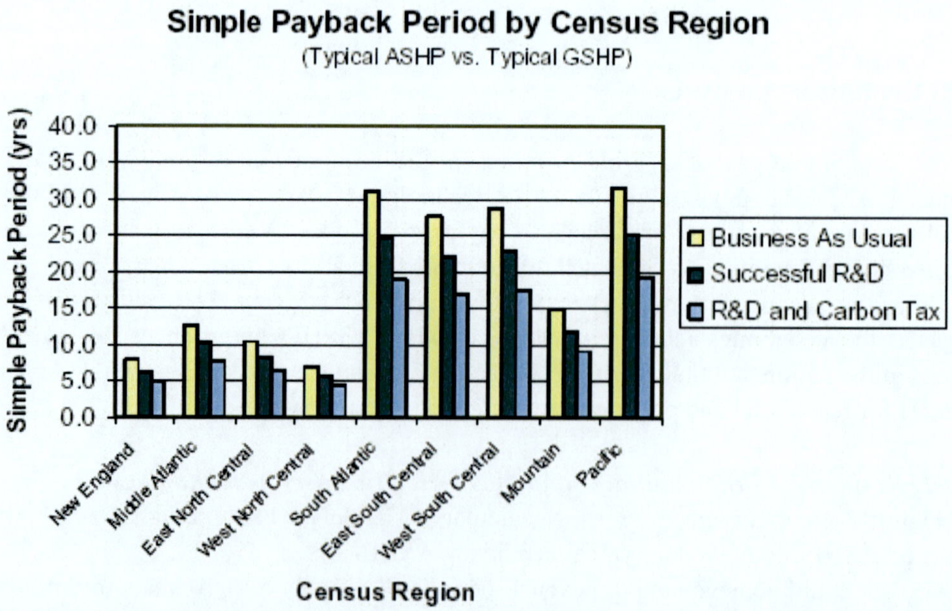

Figure 3-27. Simple Payback Periods for Representative Single-Family Home (~3000 sq. ft.) — Typical-Efficiency GSHP vs. Typical-Efficiency ASHP Baseline

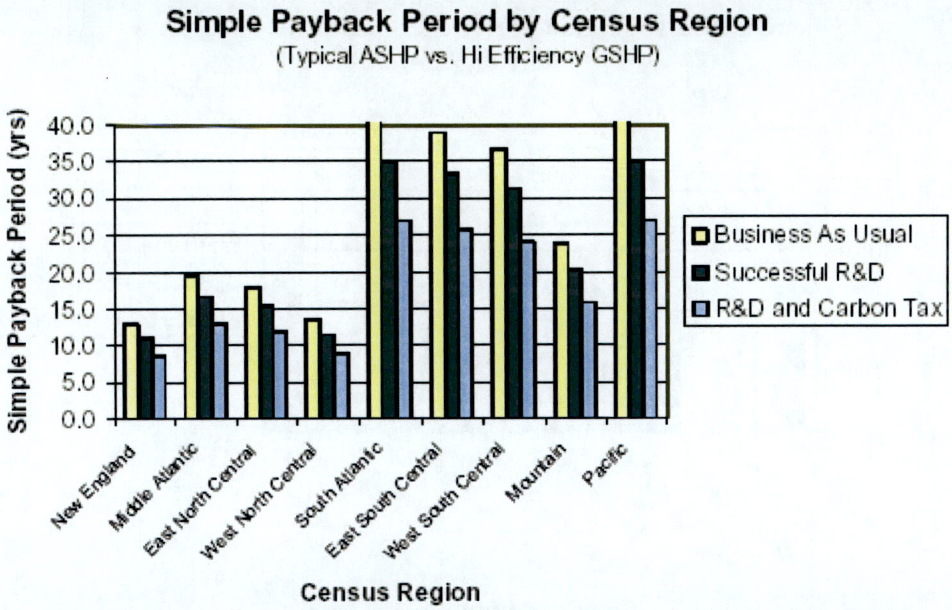

Figure 3-28. Simple Payback Periods for Representative Single-Family Home (~3000 sq. ft.) — High-Efficiency GSHP vs. Typical-Efficiency ASHP Baseline.

Both GSHPs and advanced ASHPs could see cost reductions as sales volumes increase. For example, GSHPs could benefit from streamlined business models for ground-loop installation. ASHPs could benefit from increased manufacturing economies of scale. (Advanced ASHPs currently sell at lower volumes than GSHPs, so have potential to achieve higher manufacturing economies of scale as sales volumes increase.)

3.11.2. Residential fundamental economics and market potential—natural-gas baseline technology

Figure 3-29 shows calculated simple payback periods for GSHPs and advanced ASHPs compared to typical-efficiency natural-gas furnaces and air conditioners. GSHPs could achieve payback periods of 5 to 10 years in most regions if R&D efforts reduce installed costs and/or a significant carbon tax is imposed.

3.11.3. Residential fundamental economics and market potential—fuel-oil baseline technology

Figure 3-30 shows calculated simple payback periods for typical-efficiency GSHPs compared to typical-efficiency fuel-oil furnaces and air conditioners. The results suggest that GSHPs can achieve potentially attractive payback periods in some regions compared to fuel-oil furnaces. However, fuel-oil use is generally limited to the northeast, as shown in Figure 3-31. This analysis used the 2006 average price of oil by census region from the Annual Energy Outlook to calculate energy costs (AEO 2008). Although there was a 30% rise in residential fuel oil prices between 2006 and 2008, they are expected to return to 2006 levels by the end of 2009 as shown in Figure 3-32 (STEO 2009).

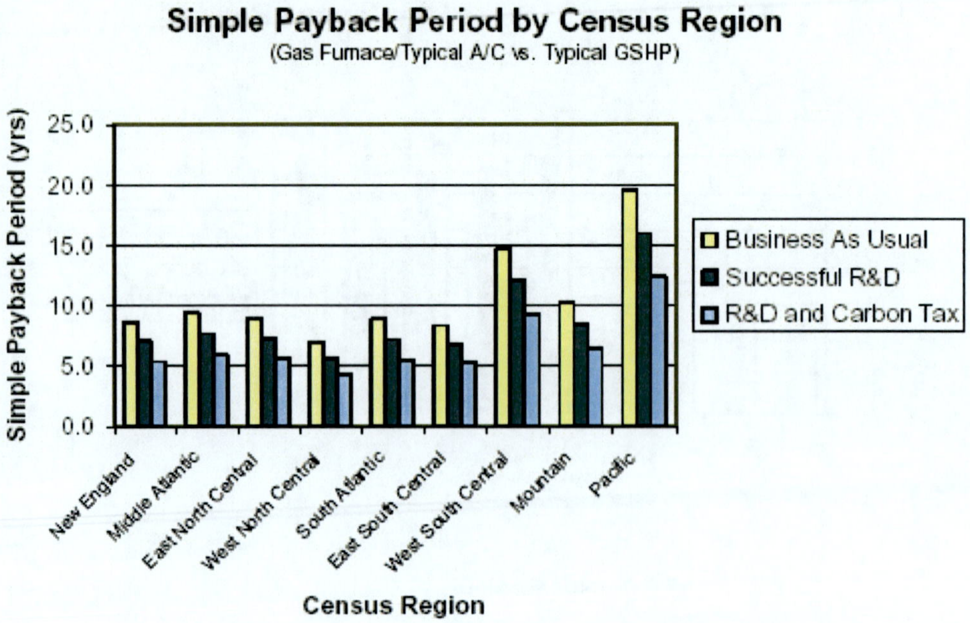

Figure 3-29. Simple Payback Periods for Representative Single -Family Home (~3000 sq. ft.) — Typical-Efficiency GSHP vs. Typical-Efficiency Natural -Gas Furnace/AC Baseline.

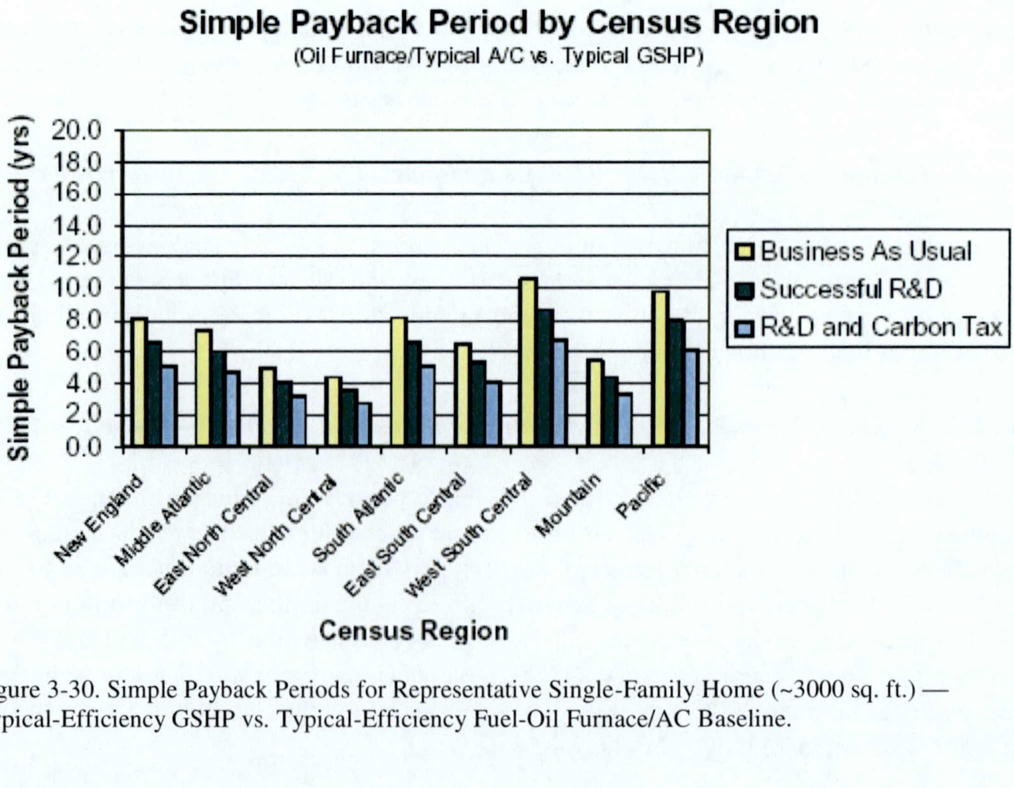

Figure 3-30. Simple Payback Periods for Representative Single-Family Home (~3000 sq. ft.) — Typical-Efficiency GSHP vs. Typical-Efficiency Fuel-Oil Furnace/AC Baseline.

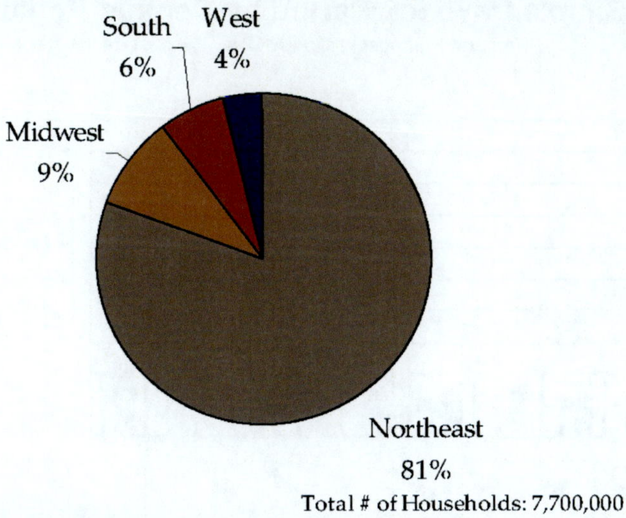

Source: RECS 2005

Figure 3-31. Number of Households with Fuel-Oil as Main Fuel Source

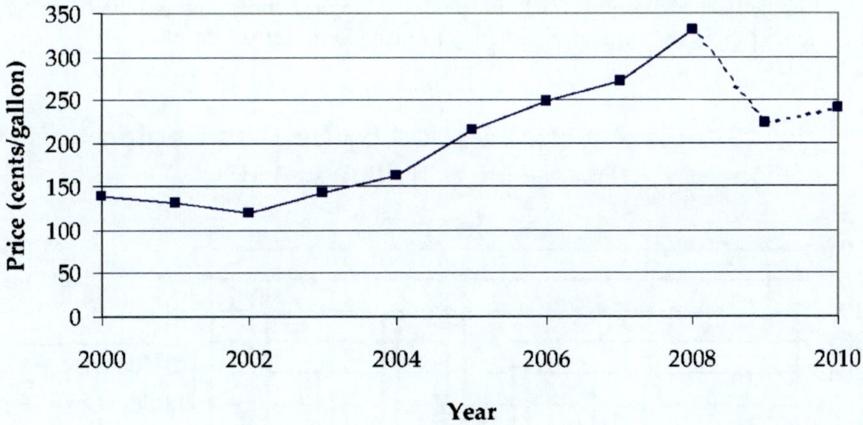

Source: EIA Short Term Energy Outlook 2009; data for 2009 -2010 are projected.

Figure 3-32. U.S. Energy Nominal Price for Heating Oil (Retail Price Including Taxes)

3.11.4. Residential fundamental economics and market potential—propane baseline technology

Figure 3-33 shows calculated simple payback periods for typical-efficiency GSHPs compared to typical-efficiency propane furnaces and air conditioners. The results suggest that GSHPs can offer fairly attractive payback periods in all regions compared to propane furnaces. However, propane use for space heating is very limited [RECS 2005].

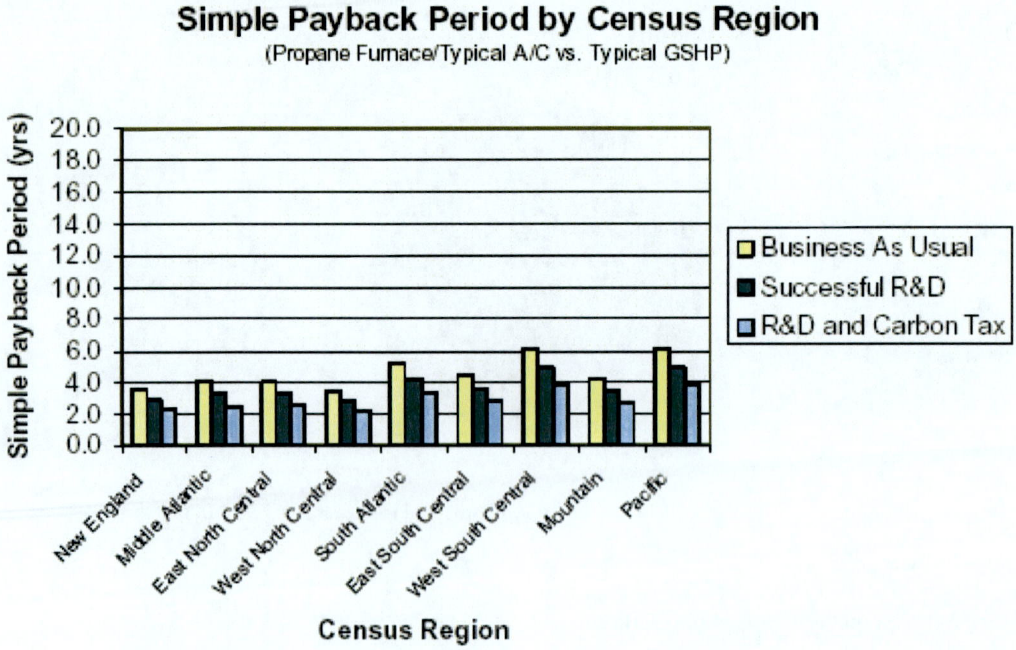

Figure 3-33. Simple Payback Periods for Representative Single-Family Home (~3000 sq. ft.) — Typical-Efficiency GSHP vs. Typical -Efficiency Propane Furnace/AC Baseline

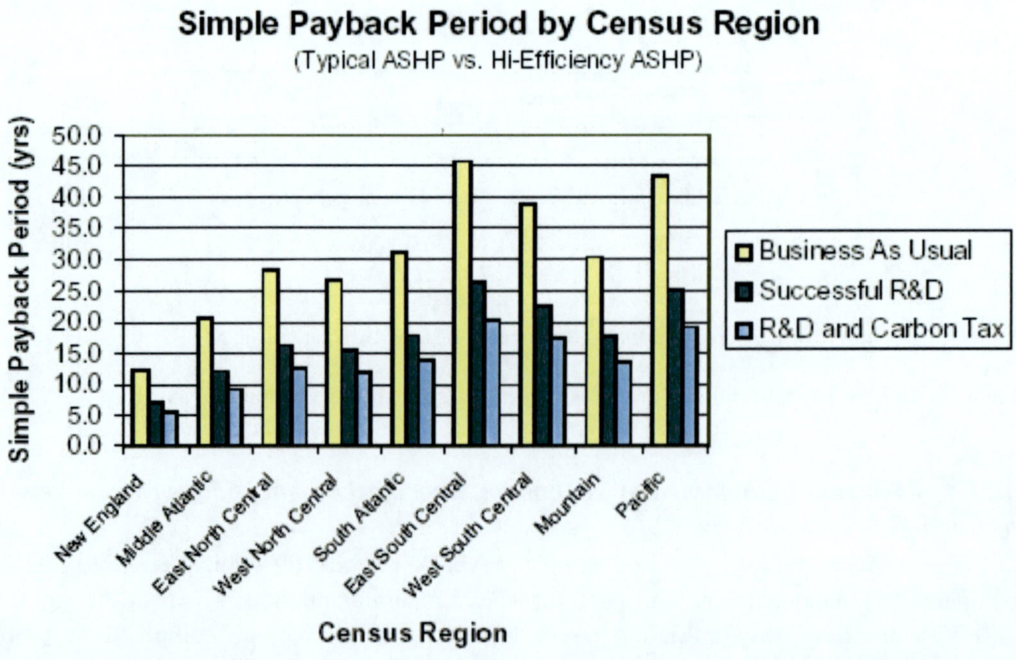

Figure 3-34. Simple Payback Periods for Representative Small Office (~6000 sq. ft.) — Advanced ASHP vs. Typical -Efficiency ASHP

3.11.5. Commercial fundamental economics—electric baseline technology

Figure 3-34 to Figure 3-36 show calculated simple payback periods for the small office example. Payback periods generally exceed ten years, except in New England. At least three factors contribute to the longer payback periods compared to the residential example.

- We used average commercial electric rates in our analysis, which are lower than residential rates
- ETA cost estimates used show installed costs per unit capacity for commercial products to be similar to those for residential applications
- Space-conditioning loads for commercial buildings are generally less weighted to the heating season than loads for single-family residences, and GSHPs provide a disproportionate benefit during the heating season.

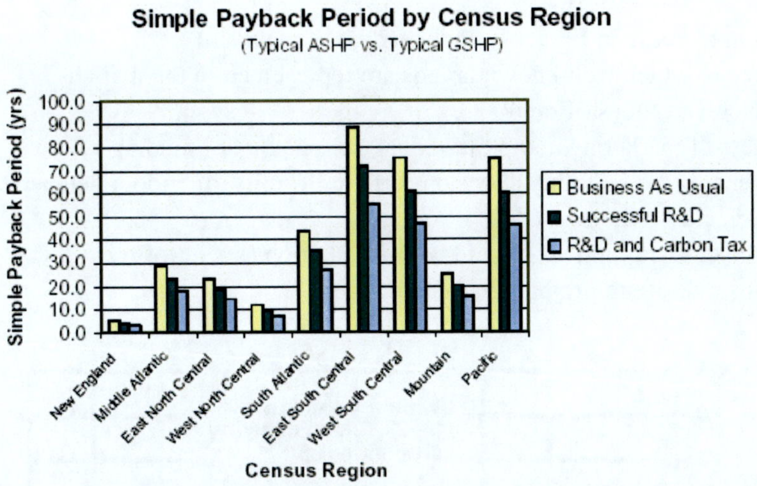

Figure 3-35. Simple Payback Periods for Representative Small Office (~6000 sq. ft.)—Typical-Efficiency GSHP vs. Typical-Efficiency ASHP

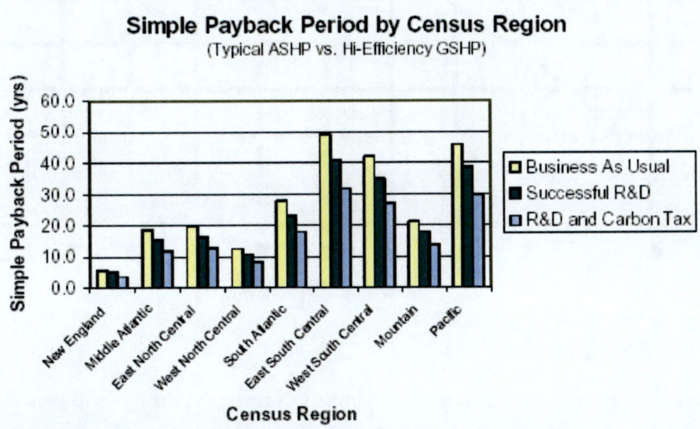

Figure 3-36. Simple Payback Periods for Representative Small Office (~6000 sq. ft.)—HighEfficiency GSHP vs. Typical-Efficiency ASHP

3.12. Projected Market Penetration

Many relationships have been developed that project market penetration of energy-saving technologies as a function of an economic parameter—usually simple payback period. One such market-penetration relationship that is commonly used in DOE analyses was developed by Arthur D. Little, Inc. (see Figure 3-37). The market-penetration curves suggest that payback periods of five years or longer lead to ultimate market penetrations under 10 percent. We consider a five-year payback as a "threshold" payback for widespread market adoption of energy-saving technologies. Payback periods of ten years suggest that applications are limited to niche markets.

Of course, actual market penetration will depend on many factors that are not specifically accounted for by simple payback period. For space-conditioning equipment, these factors may include:

- Percent increase in first cost (independent of payback)
- Degree to which well known brands are represented in the market
- Product warranties offered
- Success of marketing and promotional campaigns or branding
- Non-energy benefits such as comfort (uniformity of indoor temperature, humidity control, etc.) or noise
- Degree of disruption associated with installation (for retrofits)
- End-user desire to project a "Green image".

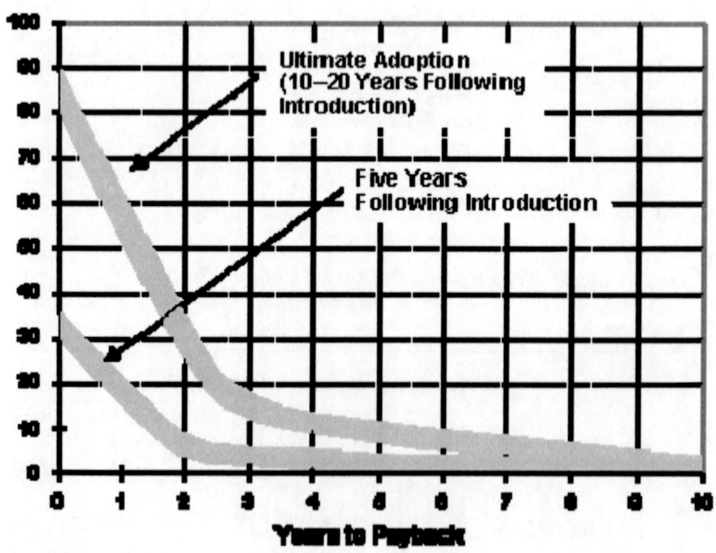

Source: ADL 1995

Figure 3-37. Market-Penetration Relationship for Energy-Saving Building Equipment

Therefore, these relationships can only be used as a general guideline for projecting market penetrations. The impact of subsidies such as utility rebates or tax credits is to shift the curves to lower payback periods.

3.13. Projected National Primary Energy Savings

As illustrated in Figure 3-38, we project national primary energy savings for each of the three scenarios based on:

- Market -penetration projections (from Section 3.7 above)
- Unit energy savings, as a percent of baseline unit energy (from Section 3.9 above)
- Regional energy consumption for the baseline technology (from Section 3.8 above).

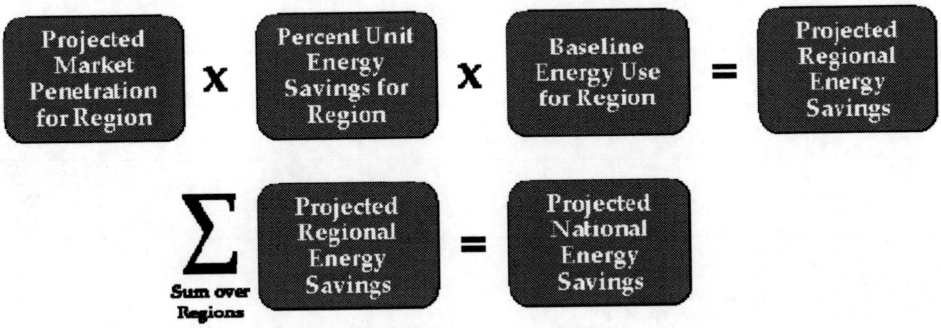

Figure 3-38. Methodology for Projecting National Primary Energy Savings

Table 3-7. Ultimate Market Penetrations and Primary Energy Savings for Typical ASHP Baseline — Residential

Scenario	Advanced Technology	New England		Middle Atlantic		National	
		Market Penetration	Primary Energy Savings, TBtu	Market Penetration	Primary Energy Savings, TBtu	Market Penetration	Primary Energy Savings, TBtu
Business as Usual	Advanced ASHP	7.5%	0.6	2.6%	0.9	0.5%	3
	Typical GSHP	5.1%	0.6	--	--	1%	6
	Hi-Eff. GSHP	--	--	--	--	--	--
Success-ful R&D	Advanced ASHP	10.6%	0.9	7.0%	2.6	3%	11
	Typical GSHP	7.8%	0.9	1.4%	0.7	3%	11
	Hi-Eff. GSHP	0.1%	--	--	--	--	--
R&D plus Carbon Tax	Advanced ASHP	12.4%	1.0	9.7%	2.6	7%	28
	Typical GSHP	10.2%	1.2	1.4%	0.7	3%	11
	Hi-Eff. GSHP	4.3%%	0.8	--	--	1%	4

Table 3-8. Ultimate Market Penetrations and Primary Energy Savings for Typical Natural -Gas Furnace/AC Baseline—Residential

Scenario	Advanced Technology	New England		Middle Atlantic		National	
		Market Penetration	Primary Energy Savings, TBtu	Market Penetration	Primary Energy Savings, TBtu	Market Penetration	Primary Energy Savings, TBtu
Business as Usual	Advanced ASHP	8.5%	6.3	8.1%	22.7	6%	130
	Typical GSHP	3.9%	3.5	2.7%	8.7	3%	60
	Hi-Eff. GSHP	--	--	--	--	--	--
Successful R&D	Typical GSHP	6.5%	5.9	5.5%	18.0	5%	100
R&D plus Carbon Tax	Typical GSHP	9.3%	8.4	8.5%	27.7	8%	160

3.13.1. Residential applications

Table 3-7 lists the calculated ultimate market penetrations and primary energy savings for the typical ASHP baseline for New England, Middle Atlantic, and the nation as a whole for each scenario. In the northeast, advanced ASHPs achieve slightly higher market penetrations compared to typical GSHPs, but don't always achieve greater primary energy savings. Calculated national energy savings are small in all cases—well under 0.1 Quad even in our most optimistic scenario.

Table 3-8 lists the calculated ultimate market penetrations and primary energy savings for the typical natural-gas furnace/AC for New England, Middle Atlantic, and the nation as a whole for each scenario. In this case, the dominant markets are not in the Northeast but instead in the Midwest. Compared to typical GSHPs, advanced ASHPs achieve both higher market penetrations and higher energy savings. Due to the relatively high prices of natural gas, advanced heat pumps (both GSHPs and ASHPs) show potential for significantly more impact compared to the typical ASHP baseline above—on the order of 0.1 Quad.

Table 3-9. Ultimate Market Penetrations and Primary Energy Savings for Typical Fuel - Oil Furnace/AC Baseline—Residential

Scenario	Advanced Technology	New England		Middle Atlantic		National	
		Market Penetration	Primary Energy Savings, TBtu	Market Penetration	Primary Energy Savings, TBtu	Market Penetration	Primary Energy Savings, TBtu
Business as Usual	Typical GSHP	4.8%	4.4	6.0%	8.0	6%	17
Successful R&D	Typical GSHP	7.3%	6.7	8.3%	10.9	8%	23
R&D plus Carbon Tax	Typical GSHP	9.8%	9.0	10.6%	14.0	11%	29

Table 3-9 lists the calculated ultimate market penetrations and primary energy savings for the typical fuel-oil furnace/AC for New England, Middle Atlantic, and the nation as a whole for each scenario. In this case, the dominant markets are clearly in the Northeast—the only region in which fuel oil is commonly used. While potentially an interesting niche market, the projected national energy savings is small—well under 0.1 Quad.

Table 3-10 lists the calculated ultimate market penetrations and primary energy savings for the typical propane furnace/AC baseline for New England, Middle Atlantic, and the nation as a whole for each scenario. In this case, the dominant markets are in the Midwest. While potentially an interesting niche market, the projected national energy savings is small—well under 0.1 Quad.

3.13.2. Commercial applications

Table 3-11 lists the calculated ultimate market penetrations and primary energy savings for the typical ASHP baseline for New England for each scenario.

Payback periods on the order of five years result in projected market penetrations of around 10 percent. National energy savings can approach 0.1 Quad.

Table 3-10. Ultimate Market Penetrations and Primary Energy Savings for Typical Propane Furnace/AC Baseline—Residential.

Scenario	Advanced Technology	New England		Middle Atlantic		National	
		Market Penetration	Primary Energy Savings, TBtu	Market Penetration	Primary Energy Savings, TBtu	Market Penetration	Primary Energy Savings, TBtu
Business as Usual	Typical GSHP	12.4%	0.5	11.7%	1.3	11%	25
Successful R&D	Typical GSHP	13.5%	0.5	12.9%	1.4	12%	28
R&D plus Carbon Tax	Typical GSHP	23.7%	0.9	16.4%	1.8	16%	37

Table 3-11. Ultimate Market Penetrations and Primary Energy Savings for Typical ASHP Baseline—Commercial

Scenario	Advanced Technology	New England	
		Market Penetration	Primary Energy Savings, TBtu
Business as Usual	Advanced ASHP	--	--
	Typical GSHP	9%	54
	Hi-Eff. GSHP	9%	73
Successful R&D	Advanced ASHP	7%	5
	Typical GSHP	11%	64
	Hi-Eff. GSHP	10%	86
R&D plus Carbon Tax	Advanced ASHP	9%	7
	Typical GSHP	13%	74
	Hi-Eff. GSHP	12%	100

4. Other Benefits of GSHPs (Not Included in Economic Analysis)

There are a number of additional benefits associated with GSHPs relative to advanced ASHPs that are not reflected in our economic/energy-impact analysis. These include:

- Actual economics may be substantially better if variable-electric rate structures are considered, such as:
  - Commercial: Demand charges and time-of-use rates penalize peak power consumption, which is substantially lower with GSHPs than with ASHPs or standard air conditioners
  - Residential: Current/upcoming time-dependent rate structures will make energy savings at peak times (e.g. summer afternoons) very valuable.
- Reduction in peak electric demand from GSHPs is significantly greater than for ASHPs, which would benefit electric utilities and could be used as a justification for offering substantial subsidies.
- Noise reduction (no outdoor fan)
- For GSHPs using secondary loops (or open loops), the refrigeration system can be factory packaged, leading to much lower chances of refrigerant leaks, improper charging, or refrigerant-system contamination
- Life of the ground loop will most likely exceed the life of an ASHP outdoor unit. Once a GSHP has been installed, replacing the system (excluding the ground loop, which should last almost indefinitely) may be less expensive than replacing an ASHP
- Lower temperature lift should improve compressor reliability and life
- No unsightly outdoor unit (that also takes up space)
- No requirement to clean an outdoor, air-cooled heat exchanger
- No shipping size/weight restrictions that limit the outdoor-unit coil size/efficiency in an ASHP
- Successful branding of GSHPs as a renewable energy technology may encourage greater adoption and greater availability of incentives.

5. Key Barriers to GSHP's in the U.S.

The key barriers to increased use of GSHPs in the U.S. are discussed below. This includes technological, market, institutional, regulatory, and other barriers.

5.1. Technological Barriers

There are several key technological barriers to widespread adoption of GSHPs, including:

- The need for a ground loop adds significant complexity, cost, and risk:

  - Adds site-specific design considerations, which are particularly significant for single-family residential applications. Geological conditions can vary significantly even within a given neighborhood [Proffer 2008]
  - Site-evaluation costs can be high.
  - Creates risks and uncertainties in cost estimating. It is difficult for installers to provide quotes, unless prices are inflated to cover uncertainties/risks
- Generally requires installation-specific design and engineering of the ground loop
- Pumping parasitics can be high if the system is not properly designed
- Seasonal variations in ground temperature in the vicinity of ground loop keep temperature lifts higher than in theory, limiting efficiency gains
- GSHPs can be difficult and costly to install in retrofit applications

Other technological barriers include:

- Direct-exchange systems (refrigerant circuit in direct contact with the ground), while less popular today compared to secondary-loop alternatives, pose unique challenges, including:
- May be difficult to ensure adequate refrigerant-oil return
- Increased difficulty in maintaining refrigeration-loop integrity and cleanliness
- High cost of copper or aluminum refrigeration tubing/piping
- High refrigerant cost
- System repair and maintenance challenges (i.e., more difficult to recover charge and re-charge system)
- Detecting charge loss or repairing leaks can be problematic

Market barriers

GSHPs also face several market challenges:

- High installation costs result in poor payback compared to ASHPs, and limit energy savings compared to ultra -high -efficiency ASHPs, which costs less to install
- Space constraints in many urban areas
- Limited production volumes lead to higher costs
- Operating cost is dependent upon electricity price (high in NE)
- Advances in ASHPs are "raising the bar" (high -efficiency, cold climate)
- Longer project duration for installing a GSHP relative to an ASHP or furnace (which can be completed in less than one day), along with the excavation mess, is a disincentive for some customers.

5.2. Institutional, Regulatory, and Other Barriers

GSHPs face additional barriers, including:

- Environmental regulations in some regions restrict re -injection of ground water.
- Potential for glycol leaks can be a barrier

- Low market awareness among consumers
- Limited number of qualified, trained installers
- Need codes to ensure proper design and installation of ground loop and pump selection (pump parasitics issue)

6. Applicability to Zero -Energy Homes and Buildings

Advanced ASHPs are currently available in efficiency ranges suitable for zero-energy homes (ZEH). Advanced ASHPs are available in capacity ranges of 1.5-2 tons of cooling capacity, which is the range required for ZEHs. GSHPs in this capacity range, if available at all, are likely to be very expensive compared to ASHPs due to the fixed costs associated with ground-loop installation. On the other side, ZEHs can utilize lower-capacity GSHPs, which will reduce the first- cost barrier compared to conventional homes.

Space-conditioning loads for zero-energy buildings (ZEBs) will generally be less weighted to the heating season than loads for conventional commercial buildings. This results because internal heat loads (which are generally fixed)become a greater percentage of the overall space-cooling load when increased insulation and reduced infiltration lower overall space-conditioning loads. Since GSHPs provide a disproportionate benefit during the heating season, ZEBs using GSHPs may not see the same percentage reduction in space- conditioning loads as do conventional buildings. On the other side, ZEBS will be able to utilize smaller-capacity heat pumps, which will lower the first-cost barrier.

7. Summary/Conclusions

A summary, observations and conclusions follow.

7.1. Summary

We conducted this investigation to:

- Summarize the status of GSHP technology and market penetration globally
- Estimate the energy saving potential of GSHPs in the U.S.
- Identify and describe the key market barriers that are inhibiting wider market adoption of GSHPs
- Recommend initiatives that can be implemented or facilitated by the DOE to accelerate market adoption.

We used information/data obtained from:

- Available literature related to GSHPs
- Interviews with selected industry experts

- A spreadsheet analysis to evaluate the energy savings and economics by U.S. census region for representative residential and commercial applications.

Our energy-savings and economics analysis compares two high-efficiency technologies (GSHPs and advanced ASHPs) to two standard-efficiency baseline systems (conventional ASHPs and furnaces with air conditioners). We used general relationships between economics and market penetration to project ultimate market penetrations of GSHPs and associated national primary energy savings.

7.2. General Observations/Conclusions

Some general observations and conclusions about the potential for GSHPs are summarized below.

7.2.1. Fundamental economics and market potential

Based on this analysis, both GSHPs and advanced ASHPs show potential for significant unit energy savings (per-installation savings) when displacing typical- efficiency ASHPs. While GSHPs generally have efficiency advantages, advanced ASHPs tend to be somewhat more economical (as measured by simple payback period). While the GSHP market may continue to expand for many years, GSHPs are unlikely to capture a major share of the heat-pump market.

7.2.2. Potential for national energy savings

The technical potential primary energy savings associated with GSHPs is shown in Table 8-1.

Table 8-1. Potential National Primary Energy Savings

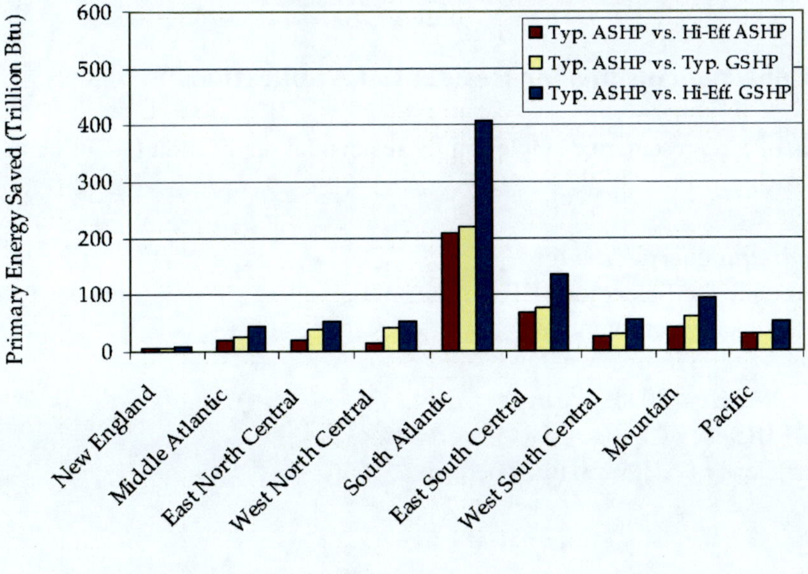

7.2.3. Key Benefits of GSHPs

In addition to high energy efficiency, GSHPs offer two key benefits:

- Can have factory-packaged refrigeration loop
- Reduces peak electric demand.

7.2.4. Key barriers to widespread application of GSHPs

GSHPs face three key barriers:

- Cost and difficulty of evaluating the suitability of individual installation sites.
- Generally requires installation-specific design and engineering of the ground loop
- Space requirements for ground coupling can be problematic in densely built areas

7.2.5. Key lessons learned from global GSHP experience

Beyond the U.S., the GSHP market is booming in Europe and starting to accelerate in Asia. Europe has a more extensive supply of published market data than the U.S., making it easier for policy-makers to analyze the trends and for manufacturers to plan for the future. Although the Asian market has only recently blossomed, China, Japan, and Korea all participate in the IEA Heat Pump Conferences and have made important contributions to the research effort. It is important to continue the international cooperation as the industry continues to mature.

The European heat pump market suffered a major blow in the late 1980's and early 1990's after booming initially during the oil crisis of the 1970's. The lack of skilled installers and poor equipment quality damaged the reputation of the technology. It is in the best interest of the GSHP industry that the U.S. maintain high quality and service through periods of growth.

Since the late 1990's, the European industry has been reborn with the help of various government and utility programs as well as rising energy prices. Programs and policies vary by country. Sweden has been most successful at penetrating the new and retrofit markets.

7.3. Observations/Conclusions for Residential Applications

Our analysis of a representative single-family residential application (about 3000 sq. ft.) suggests the following.

7.3.1. Residential unit energy savings

Compared to typical-efficiency ASHPs, ranges of unit energy savings are (as a percent of space-conditioning energy consumption):

- Advanced ASHPs: 20 to 30 percent
- Typical-Efficiency GSHPs: 25 to 50 percent
- High-Efficiency GSHPs: 50 to 70 percent

Compared to typical-efficiency furnaces (natural gas, propane or fuel oil) and air conditioners, ranges of unit energy savings are (as a percent of space- conditioning energy consumption):

- Advanced ASHPs: 20 to 30 percent
- Typical-Efficiency GSHPs: 25 to 30 percent
- High-Efficiency GSHPs: 50 to 60 percent

These ranges are narrower than reported above for the typical-efficiency-ASHP baseline because there is less regional variation in heating-season performance compared to ASHPs.

7.3.2. Residential fundamental economics and market potential—electric baseline technology

Table 7-1 shows approximate ranges of simple payback periods for GSHPs and advanced ASHPs compared to typical-efficiency ASHPs.

Generalized market-penetration curves suggest that payback periods of five years or longer lead to ultimate market penetrations under 10 percent. Payback periods of ten years suggest that applications are limited to niche markets. GSHPs approach threshold economics for widespread adoption in the northeast and Midwest if R&D efforts lower ground-loop installed costs and/or a significant carbon tax is imposed on electricity. The economics of advanced ASHPs are, however, as good or better compared to GSHPs. And, our analysis does not account for the improved heating-season performance of advanced ASHPS currently being developed and introduced for cold-climate applications, which would improve the economics of advanced ASHPs.

Projected payback periods are sensitive to installed-cost estimates. However, using installed-cost estimates from another researcher would result in even longer payback periods.

Table 7-1. Approximate Simple Payback Periods for Representative Single-Family Home (~3000 sq. ft.)—Typical-Efficiency ASHP Baseline

Scenario	Advanced Technology	Payback Range by Major Census Region (Years)			
		Northeast	Midwest	South	West
Business as Usual	Advanced ASHP	6-10	8-12	15-20	10-20
	Typical GSHP	8-15	7-11	25-35	15-35
	High-Efficiency GSHP	10-20	10-20	35-40	25-40
Successful R&D	Advanced ASHP	Same as "Business as Usual"[a]			
	Typical GSHP	6-10	6-8	20-25	10-25
	High-Efficiency GSHP	10-20	10-15	30-35	20-35
R&D plus Carbon Tax	Advanced ASHP	4-5	5-6	9-10	7-10
	Typical GSHP	5-8	4-6	15-20	9-20
	High-Efficiency GSHP	8-15	9-12	25-30	15-30

a. Successful R&D assumed to impact the GSHP ground loop only, so this scenario won't change the economics of the advanced ASHP.

Table 7-2. Approximate Simple Payback Periods for Representative Single-Family Home (~3000 sq. ft.)—Typical-Efficiency Natural-Gas Furnace and Air-Conditioner Baseline

Scenario	Advanced Technology	Payback Range by Major Census Region (Years)			
		Northeast	Midwest	South	West
Business as Usual	Advanced ASHP	5 - 6	4 - 6	5 - 9	7-12
	Typical GSHP	8-10	7 - 9	8-15	10 - 20
	High-Efficiency GSHP	12 -15	12 -15	15 - 20	15 - 30
Successful R&D	Typical GSHP	7- 8	5- 7	7 -12	8 -15
R&D plus Carbon Tax	Typical GSHP	5 -10	4- 6	5- 9	6 -12

7.3.3. Residential fundamental economics and market potential—natural-gas baseline technology

Table 7-2 shows approximate ranges of simple payback periods for GSHPs and advanced ASHPs compared to two typical-efficiency natural-gas furnaces and air conditioners.

GSHPs approach threshold economics for widespread adoption only if R&D efforts reduce installed costs and/or a significant carbon tax is imposed. The economics of advanced ASHPs are, however, are generally better compared to GSHPs. As stated above, our analysis does not account for the improved heating-season performance of advanced ASHPS currently being developed and introduced for cold-climate applications, which would improve the economics of advanced ASHPs. Our analysis also does not incorporate the potential economic advantages of lower peak demand from GSHPs, which is very valuable to utilities and can be passed on to consumers as residential variable rate pricing becomes more widespread.

7.3.4. Residential fundamental economics and market potential—fuel-oil baseline technology

Table 7-3 shows approximate ranges of simple payback periods for typical - efficiency GSHPs compared to typical-efficiency fuel-oil furnaces and air conditioners.

The results suggest that GSHPs can be economically competitive in the northeast compared to fuel-oil furnaces.

Table 7-3. Approximate Simple Payback Periods for Representative Single-Family Home (~3000 sq. ft.)—Typical-Efficiency Fuel-Oil Furnace and Air-Conditioner Baseline

Scenario	Advanced Technology	Payback Range by Major Census Region (Years) Northeast[a]
Business as Usual		7-8
Successful R&D	Typical GSHP	6-7
R&D plus Carbon Tax		4-5

a. Only the Northeast is listed because other regions use very little fuel oil.

7.3.5. Residential fundamental economics and market potential—propane baseline technology

Table 7-4 shows approximate ranges of simple payback periods for typical- efficiency GSHPs compared to typical-efficiency propane furnaces and air conditioners.

The results suggest that GSHPs can be economically competitive in all regions compared to propane furnaces. However, propane use for space heating is limited to 6% of households nationally (see Figure 7-1 for regional distribution of propane-heated households).

7.3.6. Potential national primary energy savings—residential

While the technical potential energy savings for residential applications of GSHPs is about 3.1 Quad, our economic analysis suggests that could achieve a national primary energy savings of about 0.1 Quad. While homes having propane and fuel-oil furnaces may offer an attractive target market for residential GSHPs, they are not likely to provide significant national energy impacts.

Table 7-4. Approximate Simple Payback Periods for Representative Single-Family Home (~3000 sq. ft.)—Typical-Efficiency Propane Furnace and Air-Conditioner Baseline

Scenario	Advanced Technology	Payback Range by Major Census Region (Years)			
		Northeast	Midwest	South	West
Business as Usual	Typical GSHP	3 - 4	3 - 4	4 - 6	4 - 6
Successful R&D		~3	2.5 - 3	3 - 5	3 - 5
R&D plus Carbon Tax		2 - 2.5	2 - 2.5	3 - 4	2.5 - 4

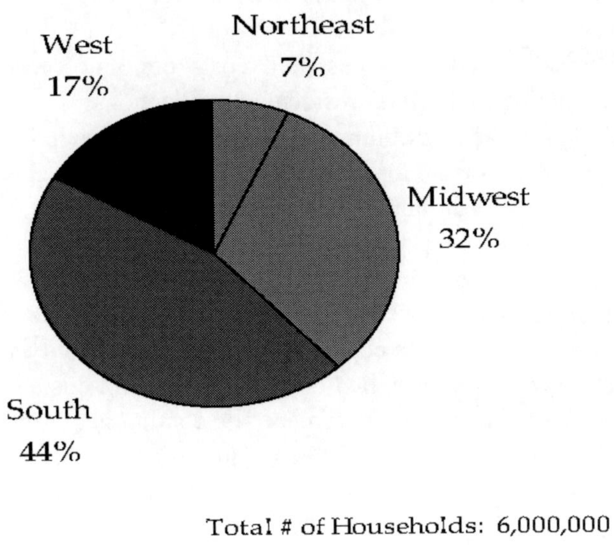

Source: RECS 2005

Figure 7-1. Distribution of U.S. Households using Propane as the Main Space-Heating Fuel

7.3.7. Zero-Energy Homes

Advanced ASHPs are currently available in efficiency ranges suitable for zero-energy homes (ZEH). Advanced ASHPs are available in capacity ranges of 1.5-2 tons of cooling capacity, which is the range required for ZEHs. GSHPs in this capacity range, if available at all, are likely to be very expensive compared to ASHPs due to the fixed costs associated with ground-loop installation. On the other side, ZEHs can utilize lower-capacity GSHPs, which will reduce the first- cost barrier compared to conventional homes.

7.4. Observations/Conclusions for Commercial/Institutional Applications

Compared to typical-efficiency ASHPs, ranges of primary unit energy savings are (as a percent of space-conditioning energy consumption):

- Advanced ASHPs: 5 to 10 percent
- Typical-Efficiency GSHPs: 20 to 45 percent
- High-Efficiency GSHPs: 45 to 60 percent

The technical potential primary energy savings in commercial buildings is roughly 0.6 Quad, of which about XXQuad is associated with public grade schools. However, the results of our analyses of a representative small office building application (about 6000 sq. ft.) generally show calculated payback periods over ten years, except in New England. At least three factors contribute to the longer payback periods compared to the residential example.

- We used average commercial electric rates in our analysis, which are lower than residential rates
- EIA cost estimates used show installed costs per unit capacity for commercial products to be similar to those for residential applications
- Space-conditioning loads for commercial buildings are generally less weighted to the heating season than loads for single-family residences, and GSHPs provide a disproportionate benefit during the heating season.

Various industry experts report, however, that real-world economics tend to be better for commercial or institutional buildings compared to residential applications. We expect that real-world pricing generally results in a cost reduction per unit capacity as capacity increases; however, we have no representative installed-cost data to confirm this.

While we did not analyze a natural-gas furnace and conditioner baseline, if the economics are similar to the typical ASHP baseline, GSHPs could ultimately achieve roughly 10 percent penetration of this market. This would result in on the order of 0.05 Quad national primary energy savings.

7.4.1. Zero-energy buildings

Space-conditioning loads for ZEBs will generally be less weighted to the heating season than loads for conventional commercial buildings. This results because internal heat loads (which are generally fixed) become a greater percentage of the overall space-cooling load when increased insulation and reduced infiltration lower overall space-conditioning loads. Since GSHPs provide a disproportionate benefit during the heating season, ZEBs using GSHPs may not see the same percentage reduction in space-conditioning loads as do conventional buildings. On the other side, ZEBS will be able to utilize smaller-capacity heat pumps, which will lower the first-cost barrier.

7.5. Observations/Conclusions for Community Applications

New-construction, planned communities may offer significantly improved economics for GSHPs. Various GSHP systems could be considered, all providing significant installed-cost reductions compared to individual installations:

- Large GSHP installation providing a district heating and cooling system to the community
- Large ground loop providing a thermal source or sink to individual water -source heat pumps in the community
- Individual GSHPS for each building/home, but installed en masse at the time of building/home construction to lower installation costs

While potentially an attractive niche market, new -construction, planned communities represent a small portion of the national building stock and, therefore, will have modest impacts on national energy consumption for quite some time.

8. Recommended Initiatives to Accelerate Market Adoption of GSHPs

We recommend that DOE support advanced heat pumps in general, rather than supporting only one type (such as GSHP). Based on our investigation, all types of heat pumps (GSHPs, ASHPs, and possibly hybrid systems) can play important roles helping DOE pursue its energy-efficiency objectives. Incentives such as federal tax credits or utility rebates can be based on energy efficiency achieved, rather than type of heat pump. R&D projects can be pursued based on the individual merit of each prospective project, rather than type of heat pump. This will require close coordination between the Geothermal Technologies Group (which is responsible for GSHPs) and Building Technologies Group (which is responsible for ASHPs). This coordination will help ensure that both types of heat pumps are developed, evaluated, and promoted in a way that ensures that apples-to-apples comparisons are made and that duplication of effort is avoided to the extent possible.

8.1. Additional Evaluations

Additional evaluations will help determine the likely impacts of R&D efforts to lower costs and to identify promotional projects that may be of interest to stakeholders. We recommend that GT and BT pursue the following evaluation activities.

8.1.1. Potential for GSHP cost reductions

Evaluate the potential for first-cost reductions for GSHPs, including potential economies of scale, alternative business models, and potential partnering relationships. Working with industry stakeholders, identify concepts to lower ground-loop installation costs then estimate their likely cost impacts. Potential concepts may include:

- Reducing the need for, and/or cost of, evaluating ground conditions (soil type/mix, thermal conductivity, water content/ground-water depth)
- For new construction, maximizing use of excavation required for the building foundation, including coupling ground loop to the foundation
- Hybrid systems using air-cooled condensers or possibly cooling towers to reduce ground-loop size while still meeting peak cooling requirements
- Additives to enhance soil conductivity in the vicinity of the ground loop
- Heat-exchanger designs, or extended surfaces that then attach to ground loops, that can be hammered into soil
- Low-cost drilling/excavation equipment, including water-jet technology.

8.1.2. Potential for ASHP cost reductions and heating-season performance improvements

Evaluate the potential for first-cost reductions for advanced ASHPs, including potential economies of scale. Include ASHPs developed specifically for cold climates that improve heating-season efficiency, as poor heating-season performance is the "Achilles' Heel" of ASHPs. While cold-climate ASHPs are currently available from a limited number of suppliers, the products are generally expensive, designs may be immature, and performance/reliability may not be sufficiently demonstrated.

8.1.3. Detailed performance and energy-benefits modeling

Since we did not conduct detailed performance modeling, our investigation does not consider:

- Potential improvements in economics due to using variable electricity rates, such as:
  - Commercial: Demand charges and time-of-use rates
  - Residential: Current/upcoming time-dependent rate structures
- Benefits of reducing peak electric demand. Understanding the peak demand reduction benefits of GSHPs is essential to justifying utility rebates that could substantially accelerate market adoption of GSHPs.

We recommend detailed performance modeling to estimate these impacts, which could significantly improve the economics of GSHPs.

In addition, as noted above, the national benefits modeling described in this report was necessarily based on many estimates and simplifications, as the project scope did not permit detailed models such as hourly load modeling. A more rigorous modeling process would better quantify the potential national benefits of GSHPs and could better target DOE activities to accelerate market adoption.

8.1.4. Schools and other government buildings

Schools and other government buildings (with assistance from energy services companies) have proven to be attractive, early market niches for GSHP installations. We did not specifically examine these opportunities in this analysis. A 1998 DOE study found GSHP systems in schools to reduce energy use for space conditioning by 25-50% compared to traditional systems, with payback periods of 2-8 years (MacMillan 2007). Our analysis of technical potential has produced similar results, and we estimate that U.S. public schools (K-12) have the potential to save 0.03 Quad collectively by using GSHPs for space-conditioning. We recommend further analysis, documentation, and publication of the energy savings, economics, reliability, comfort, installation and operational lessons learned, etc., associated with schools and other government buildings. Having this knowledge and experience available will help facilitate GSHP market growth in other building applications.

8.1.5. Promotional programs with stakeholders

Contact stakeholders to identify interest in a joint DOE promotional program. Arrange meetings with interested stakeholders to compare information, identify common interests, agree on priorities, and outline a joint collaboration effort, as appropriate. Stakeholders potentially interested in a DOE partnership to promote GSHPs may include:

- Electric utilities and the Consortium for Energy Efficiency (CEE)
- GSHP Manufacturers
- American Council for an Energy Efficient Economy (ACEEE)
- Geothermal Heat Pump Consortium, Inc.
- International Energy Agency, Heat Pump Program
- International Ground Source Heat Pump Association
- Canadian GeoExchange Coalition
- Geo-Heat Center, Oregon Institute of Technology
- South Dakota State University
- Leadership in Energy and Environmental Design.

8.2. Research and Development

Depending on the results of the additional analyses outlined above, we recommend that GT and BT consider the following R&D projects.

8.2.1. Ground-loop cost reduction

After developing and evaluating various concepts for lowering ground-loop cost, develop prototype designs for the more promising concepts. Laboratory test or field test, as appropriate.

8.2.2. Cold-climate ASHP development/cost reduction

Depending on the results of the investigation outlined in Section 9.1.2 above, perform additional development and laboratory testing to reduce cost and ensure good reliability and performance of cold-climate ASHPs.

8.3. Field Testing and Verification

We recommend that GT and BT pursue the following field -testing and performance-verification activities.

8.3.1. Ground-loop testing/evaluation

Researchers have demonstrated that the ground loop has significant impacts on ground temperature in the vicinity of the ground loop. Also, soil characteristics vary dramatically, and have significant influence on ground-loop design and performance, and even the suitability of the site for a GSHP. Further, space constraints for some installations may not permit optimal sizing of the ground loop or spacing of bore holes.

8.3.2. Rigorous performance verification and comparison to ASHP

DOE should evaluate and document the energy-savings potential of GSHPs compared to alternatives (advanced ASHPs and furnaces) through field testing and demonstrations. One option is to install both a GSHP and an ASHP (or furnace and air conditioner) in a test home and alternate use of each system. Adjust results for weather conditions and compare performance. Careful instrumentation of the ground loop is important to understand the impacts of seasonal ground-temperature variation (due to heat extraction in the winter and heat rejection in the summer).

8.3.3. Gathering data to support improved design/installation guidelines

Working with interested manufacturers and installers, DOE could encourage that

8.4. Promotion

We recommend that GT and BT pursue several advanced heat-pump promotional activities as outlined below.

8.4.1. Installation codes

DOE test procedures are used for measuring WSHP energy efficiency for ENERGY STAR® ratings.[9] Because a WSHP is factory or laboratory tested, the DOE test procedure uses a formula to estimated ground-loop pump-power requirements. Unfortunately, this does not provide assurance that the GSHP, as installed, will use an energy-efficient pump, nor a ground-loop design and glycol flow rate that ensures optimum balance between heat-transfer performance and pumping power requirements. We recommend that DOE work with state and local governments, manufacturers, and installers to develop model codes that state and local governments can utilize to ensure in-field performance is consistent with good design practice. The model codes should provide (or reference) appropriate ground-loop design and pump-selection guidelines for various installation conditions and ground-loop types. It should include functional performance testing requirements, if appropriate, to ensure that the system works as intended once installed.

8.4.2. Guidelines for selecting/designing advanced heat pumps

There are many factors to consider when selecting the appropriate heat-pump technology for a given installation, including:

- Site conditions
- Available space
- Climate
- Building type/construction
- End-user economic criteria
- End-user preferences.

Adequate tools are lacking for selecting the appropriate technology and designing the system to optimize cost and performance. DOE should work with interested stakeholders to develop, disseminate and support these tools.

8.4.3. Community-based systems

GSHPs, WSHPs, and even hybrid systems can offer significant cost and performance advantages when considered for communities. There are substantial opportunities for creative combinations with other types of community systems, such as:

- Combined Heat and Power (CHP) systems
- District heating or cooling systems, including lake-water cooling systems
- Heat recovery systems, including sources such as sewage, anaerobic digesters, or industrial waste-heat streams.

For example, WSHPs or hybrid systems installed at individual customer sites may be effective in reducing the capacity requirements for district heating and cooling systems, when a few, peak hours or days may otherwise dictate sizing requirements. Also, community-based systems provide a scale that may interest energy service companies or third-party owner/operators, helping to surmount the first-cost barrier.

8.4.4. Other promotional activities

Promotional activities should include:

- Support training for designers and installers (including drillers and excavators)

- Consider partnerships to create new business models to reduce drilling/trenching costs
- Support regional information-dissemination programs
- Work with local governments, utilities, developers, manufacturers and installers to consider community-based GSHP systems when constructing planned communities. These are especially attractive for communities that have access to lake, pond, or ocean water where, in many cases, direct cooling is possible for much or all of the cooling season.

REFERENCES

ADL, (1995). "*Fuel Cells for Building Cogeneration Applications – Cost/Performance Requirements and Markets*". Prepared for the Building Equipment Division, U.S. DOE; Arthur D. Little, Inc., January 1995.

AEO, (2008). Annual Energy Outlook 2008. Energy Information Administration, http://www.eia.doe.gov/oiaf/aeo/, June 2008.

AHRI, (2005). "*U.S. Manufacturer Domestic Shipments of Unitary Air-Source Heat Pumps*". Air-Conditioning, Heating, and Refrigeration Institute, http://ahrinet.org/ARI/util/showdoc.aspx?doc=629, 2005.

Appliance Magazine, (2008). "*31st Annual Portrait of the U.S. Appliance Industry*", Appliance Magazine, September 2008: 38.

ASHRAE, (1995). Commercial/Institutional Ground-Source Heat Pump Engineering Manual. *American Society of Heating*, Refrigerating and Air- Conditioning Engineers, Inc., 1995.

Cane, D., & Garnet, J. (2000). Update on Maintenance and Service Costs of Commercial Building Ground-Source Heat Pump Systems. *ASHRAE Journal*, January, 2000.

CBECS, (2003). Commercial Building Energy Consumption Survey 2003. Energy Information Administration, http://www.eia.doe.gov/emeu/cbecs/contents.html, 2003.

Curtis, R., Lund, J., Sanner, B., Rybach, L., & Hellstrom, G. (2005). Ground Source Heat Pumps—Geothermal energy for anyone, anywhere: current worldwide activity. In: *Proc. WGC2005*, Turkey.

DoD, (2007). "*Ground-Source Heat Pumps at Department of Defense Facilities*". Office of the Deputy Under-Secretary of Defense (Installations and Environment), January 2007.

DOE, (2002). Technical Support Document for Residential Central Air Conditioners and Heat Pumps Rulemaking: Chapter 6. U.S. Department of Energy, May 2002.

DOE, (2008). Buildings Energy Data Book. Department of Energy, http://btscoredatabook.eren.doe.gov, 2008.

Ellis, D. (2008). Field Experience with Ground-Source Heat Pumps in Affordable Low Energy Housing. In: *Proc. 9th Annual IEA Heat Pump Conference 2008.*

EIA (2000). U.S. Census Regions and Divisions Map. Energy Information Administration, http://www.eia.doe.gov/emeu/reps/maps/us_census.html, 2000.

EIA (2004). A Look at Residential Energy Consumption in 2001, Table HC3-2a. Energy Information Administration, April 2004.

EIA (2006). "*Survey of Geothermal Heat Pump Shipments 2006*". Energy Information Administration, July 2008.

EIA (2007). "*EIA - Technology Forecast Updates – Residential and Commercial Building Technologies*". Navigant Consulting, Inc., September 2007.

EIA (2008). "*Energy Market and Economic Impacts of S. 2191, the Lieberman- Warner Climate Security Act of 2007*". Energy Information Administration, April 2008.

EurObserv'ER (2008). Geothermal Energy Barometer. European Renewable Energy Council, http://www.erec.org/projects/ongoing-projects/eurobserver.html, Sept. 2007, 58-60.

Forsén, M. (2007). *Market status for ground source heat pumps within Europe*. Deliverable 2, Ground Reach, July 2007.

Forsén, M., & Nowak, T. (2008). *European Heat Pump Outlook 2008*. European Heat Pump Association, 2008.

Huang, Joe & Franconi, Ellen. (1999). "*Commercial Heating and Cooling Loads Component Analysis*." Building Technologies Department, Environmental Energy Technologies Division. Lawrence Berkeley National Laboratory. November.

Kavanaugh, S. (1995). " *Cost Containment For Ground-Source Heat Pumps*". Report for Alabama Universities- TVA Research Consortium, December 1995.

Kawazoe, S. (2005). Geothermal Power Generation and Direct Use in Japan. In: *Proc. 9th Annual IEA Heat Pump Conference 2008*.

KCPL (2008). GSHP Diagram. Kansas City Power & Light, http://www.kcpl.com/residential/images/geoheatpump.jpg.

Le Feuvre, P., Kummert, M., 2008. Ground Source Heat PUmps in the UK – Market Status and Evaluation. In: Proc. 9th Annual IEA Heat Pump Conference 2008.

Lund, J.W. (2008). Personal email communication, November, 2008.

Lund, J.W., Freeston, D.H., Boyd, & T. L. (2005a). Direct Application of Geothermal Energy: 2005 Worldwide Review. *Geothermics, 34*, 691-727.

Lund, J.W., Bloomquist, R.G., Boyd, T.L., & Renner, J. (2005b). The United States of America Country Update. In: *Proc. WGC2005*.

MacMillan, J. (2007). Ground Source Heat Pumps in Schools. *ASHRAE Journal,* September 2007, 34-38.

Martin, M., Madgett, M., & Hughes, P. (2000). Comparing Maintenance Costs of Geothermal Heat Pump Systems with Other HVAC Systems. *ASHRAE Journal*, January 2000.

Michaels (2008). "*Performance, Emissions, Economic Analysis of Minnesota Geothermal Heat Pumps*". Michaels Engineering, Final Report for Minnesota Department of Commerce, July 2008.

Paltsev, S., et al, (2007). "*Assessment of U.S. Cap-and-Trade Proposals*". MIT Joint Program on the Science and Policy of Global Change, http://web.mit.edu/globalchange/www/MITJPSPGCRpt146.pdf, April 2007.

TIAX (2006). Building Load Profiles for Micro-CHP. Research, Development, and Demonstration of Micro-CHP Systems for Residential Applications. TIAX, LLC, 2006.

NCDC (2008). "*U.S. Climate Normals*". National Climatic Data Center, National Oceanic and Atmospheric Administration, http://cdo.ncdc.noaa.gov/cgi-bin/climatenormals/climatenormals.pl, 2008.

NCI (1997). "*Tools for Market Penetration Estimates*". Second DOE OUT Analysis Workshop, Navigant Consulting, Inc. July 1997.

NRCan (2008). Overview of Commercial GSHPs. http://www.canren.gc.ca/prodserv/index.asp?CaId=169&PgId=997, December 2008.

Omer, A. (2006). Ground -source Heat Pumps Systems and Applications. *Renewable and Sustainable Energy Reviews, 12*, 2008: 344 -371.

Park, S. (2008). An Overview of the Research and the Market for Heat Pumps in Korea. In: *Proc. 9th Annual IEA Heat Pump Conference 2008*.

Pieters, H. (2007). Barriers to GSHP Market Penetration. Ground -Reach, July 2007.

Proffer, T. (2008). Personal communication with Major Geothermal, Denver, CO. November, 2008.

Rafferty, K. (2008). An Information Survival Kit for the Prospective Geothermal Heat Pump Owner. Heatspring Energy, March.

RECS (2001). Residential Energy Consumption Survey 2001. Energy Information Administration, http://www.eia.doe.gov/emeu/recs/contents.html, 2001.

RECS (2005). Residential Energy Consumption Survey 2005. Energy Information Administration, http://www.eia.doe.gov/emeu/recs/contents.html, 2005.

Sanner, B., Karytsas, C., Mendrinos, D. & Rybach, L. (2003). "Current status of ground source heat pumps and underground thermal energy storage in Europe". *Geothermics, 32*, 2003, 579-588.

STEO (2009). EIA Short Term Energy Outlook: table browser, http://www.eia.doe.gov/emeu/steo/pub/contents.html, January 2009.

Zheng, X. & Wolff, H. (2008). Geothermal Source Heat Pump in China. In: *Proc. 9th Annual IEA Heat Pump Conference 2008*.

APPENDIX A: SCOPE OF WORK

The project will begin with a kickoff teleconference with DOE to confirm the objectives, approach, and deliverables anticipated for the project. The tasks planned for the project are explained below.

Task 1. Review Status of Global GSHP Markets

In this task, we will provide an overview of the global market for GSHPs, focusing primarily on Asia and Europe. To the extent that data are available, we will provide estimates of market size in key regions and explain the types of GSHPs that are prevalent worldwide. We will also review available information on incentives that are available for GSHPs in key regions. We will examine where GSHPs have been successful and the apparent reasons for their success, as well as whether any of the lessons learned abroad may be applicable in the U.S.

Task 2. Estimate Energy Savings Potential for GSHPs in the U.S.

In task 2, we will estimate the energy savings potential of GSHPs in the U.S. under a few different scenarios, likely including optimistic, pessimistic, and "business as usual". The factors that might create these scenarios will also be discussed. A simplified cost-benefit analysis based on simple payback periods will be documented, and both new construction and retrofits will be considered.

Task 3. Examine Key Market Barriers to GSHPs in the U.S.

In task 3, we will identify and explain the most important barriers that are inhibiting growth of the GSHP market in the U.S. These may include market, technological, institutional, regulatory, or other barriers. Although we will examine these issues on a national basis, we will pay particular attention to the Northeast, where the high cost of heating oil and propane might provide an opening for GSHPs. The opportunities and barriers associated with a community loop arrangement will also be considered.

Task 4. Recommend Initiatives to Accelerate Market Adoption of GSHPs

Based on the findings of task 3, as well as any lessons learned from abroad in task 1, we will recommend initiatives that the DOE could undertake or facilitate to accelerate market adoption of GSHPs in the U.S. and particularly in the Northeast. These initiatives could be undertaken by DOE alone or in partnership with other stakeholders such as states, utilities, manufacturers, or others

Deliverables and Schedule

Our findings will be summarized in a draft final report approximately 2 months from the date of subcontract award. The revised final report will be submitted within two weeks after receipt and resolution of any comments on the draft report.

APPENDIX B: RESIDENTIAL UNIT ENERGY CONSUMPTIONS

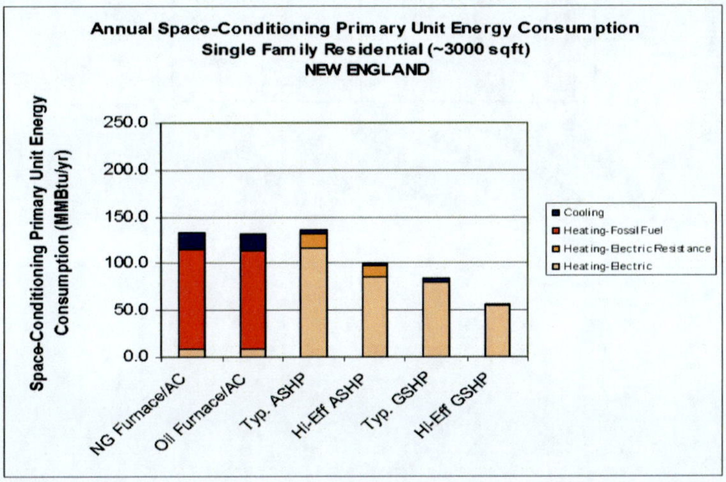

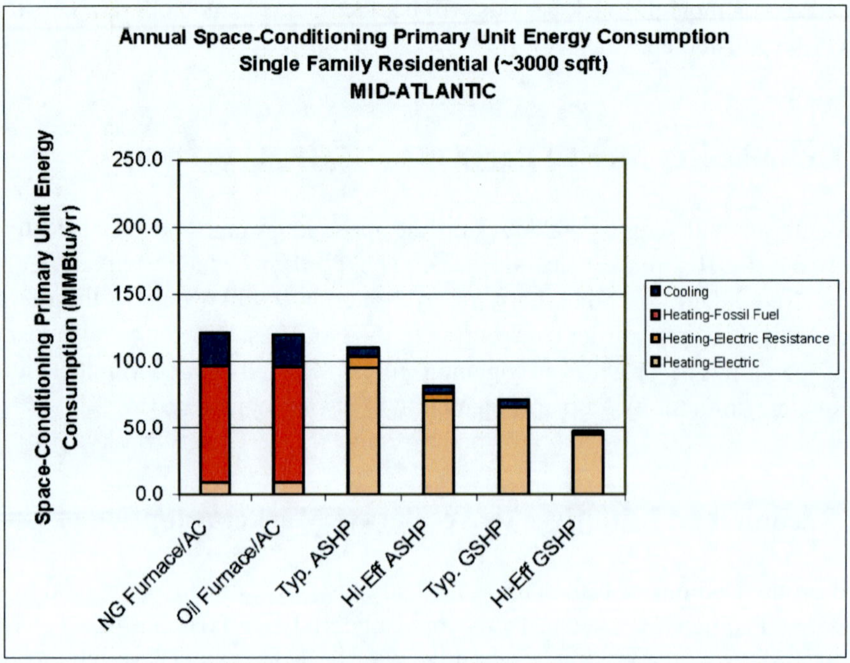
Annual Space-Conditioning Primary Unit Energy Consumption
Single Family Residential (~3000 sqft)
MID-ATLANTIC
Space-Conditioning Primary Unit Energy Consumption (MMBtu/yr)
250.0
200.0
150.0
100.0
50.0
0.0
NG Furnace/AC
Oil Furnace/AC
Typ. ASHP
Hi-Eff ASHP
Typ. GSHP
Hi-Eff GSHP
Cooling
Heating-Fossil Fuel
Heating-Electric Resistance
Heating-Electric

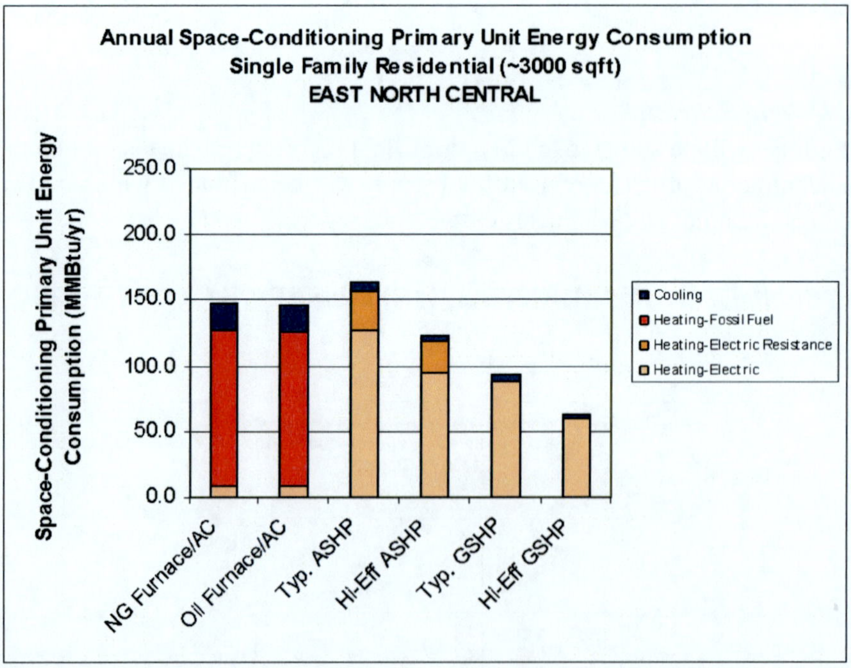
Annual Space-Conditioning Primary Unit Energy Consumption
Single Family Residential (~3000 sqft)
EAST NORTH CENTRAL
Space-Conditioning Primary Unit Energy Consumption (MMBtu/yr)
250.0
200.0
150.0
100.0
50.0
0.0
NG Furnace/AC
Oil Furnace/AC
Typ. ASHP
Hi-Eff ASHP
Typ. GSHP
Hi-Eff GSHP
Cooling
Heating-Fossil Fuel
Heating-Electric Resistance
Heating-Electric

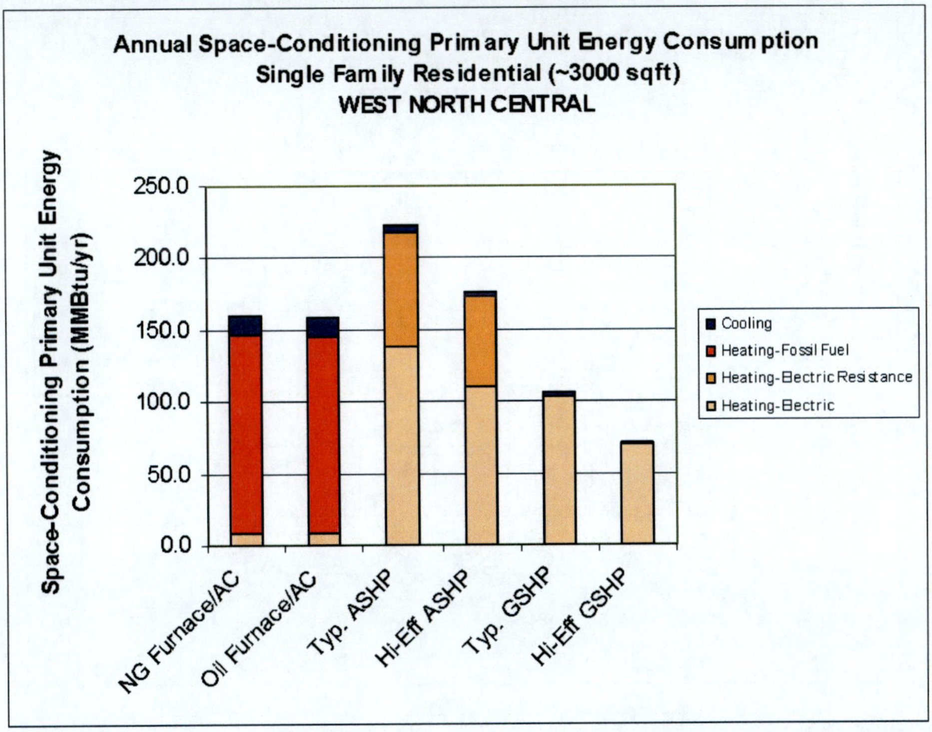
Annual Space-Conditioning Primary Unit Energy Consumption
Single Family Residential (~3000 sqft)
WEST NORTH CENTRAL
Space-Conditioning Primary Unit Energy Consumption (MMBtu/yr)
250.0
200.0
150.0
100.0
50.0
0.0
NG Furnace/AC
Oil Furnace/AC
Typ. ASHP
Hi-Eff ASHP
Typ. GSHP
Hi-Eff GSHP
Cooling
Heating-Fossil Fuel
Heating-Electric Resistance
Heating-Electric

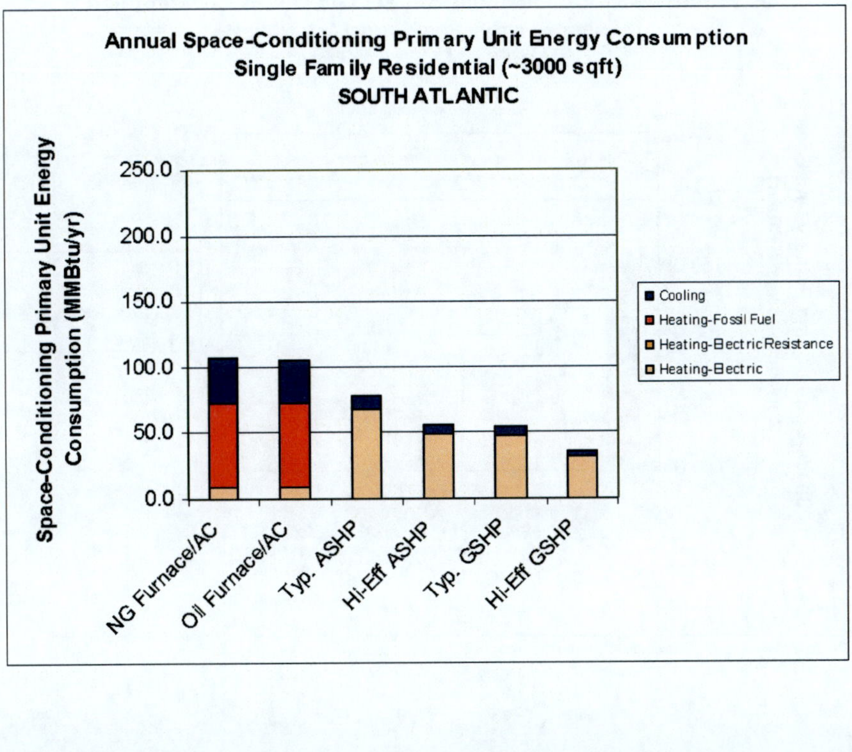
Annual Space-Conditioning Primary Unit Energy Consumption
Single Family Residential (~3000 sqft)
SOUTH ATLANTIC
Space-Conditioning Primary Unit Energy Consumption (MMBtu/yr)
250.0
200.0
150.0
100.0
50.0
0.0
NG Furnace/AC
Oil Furnace/AC
Typ. ASHP
Hi-Eff ASHP
Typ. GSHP
Hi-Eff GSHP
Cooling
Heating-Fossil Fuel
Heating-Electric Resistance
Heating-Electric

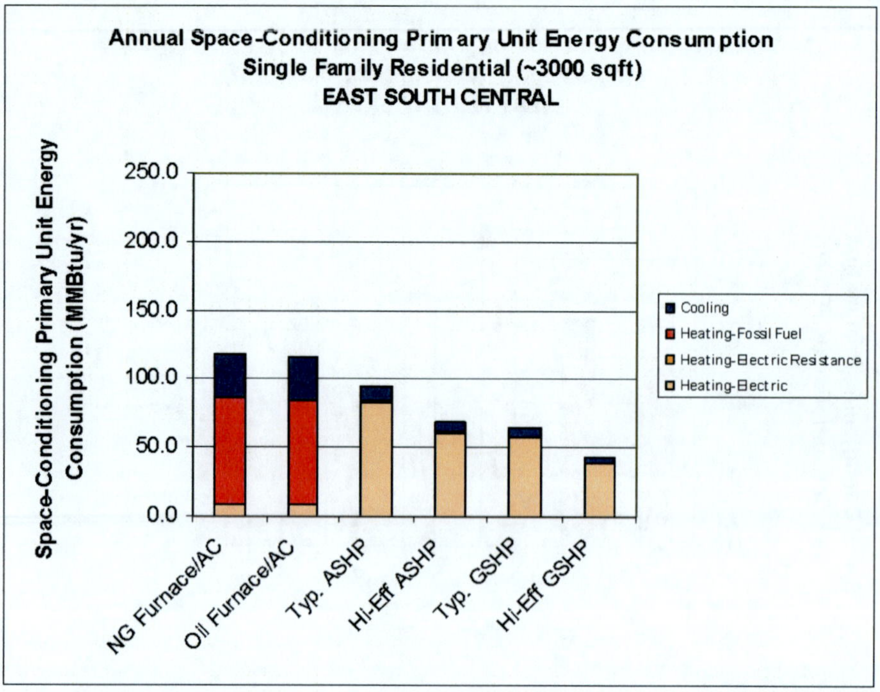
Annual Space-Conditioning Primary Unit Energy Consumption
Single Family Residential (~3000 sqft)
EAST SOUTH CENTRAL
Space-Conditioning Primary Unit Energy Consumption (MMBtu/yr)
250.0
200.0
150.0
100.0
50.0
0.0
NG Furnace/AC
Oil Furnace/AC
Typ. ASHP
Hi-Eff ASHP
Typ. GSHP
Hi-Eff GSHP
Cooling
Heating-Fossil Fuel
Heating-Electric Resistance
Heating-Electric

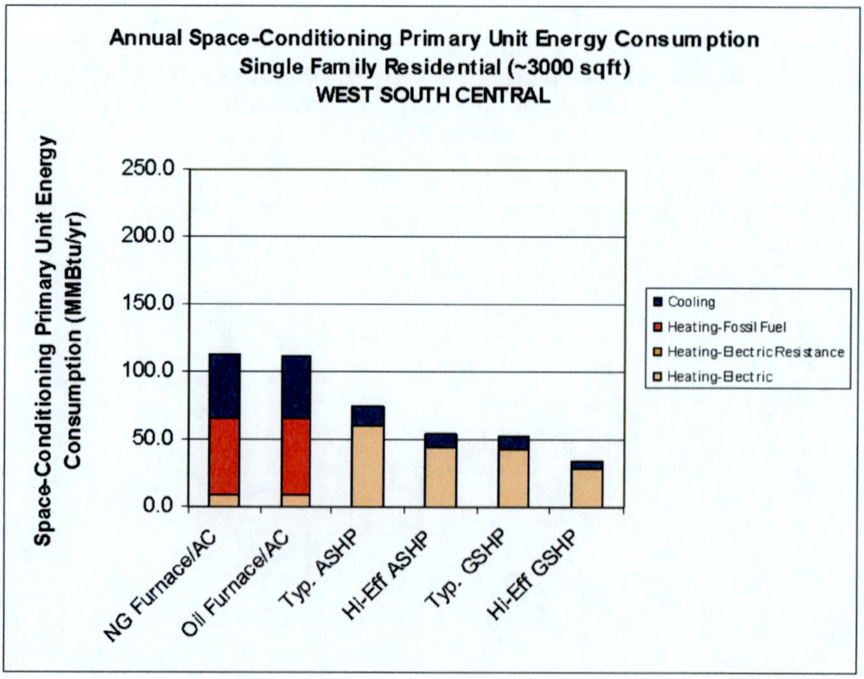
Annual Space-Conditioning Primary Unit Energy Consumption
Single Family Residential (~3000 sqft)
WEST SOUTH CENTRAL
Space-Conditioning Primary Unit Energy Consumption (MMBtu/yr)
250.0
200.0
150.0
100.0
50.0
0.0
NG Furnace/AC
Oil Furnace/AC
Typ. ASHP
Hi-Eff ASHP
Typ. GSHP
Hi-Eff GSHP
Cooling
Heating-Fossil Fuel
Heating-Electric Resistance
Heating-Electric

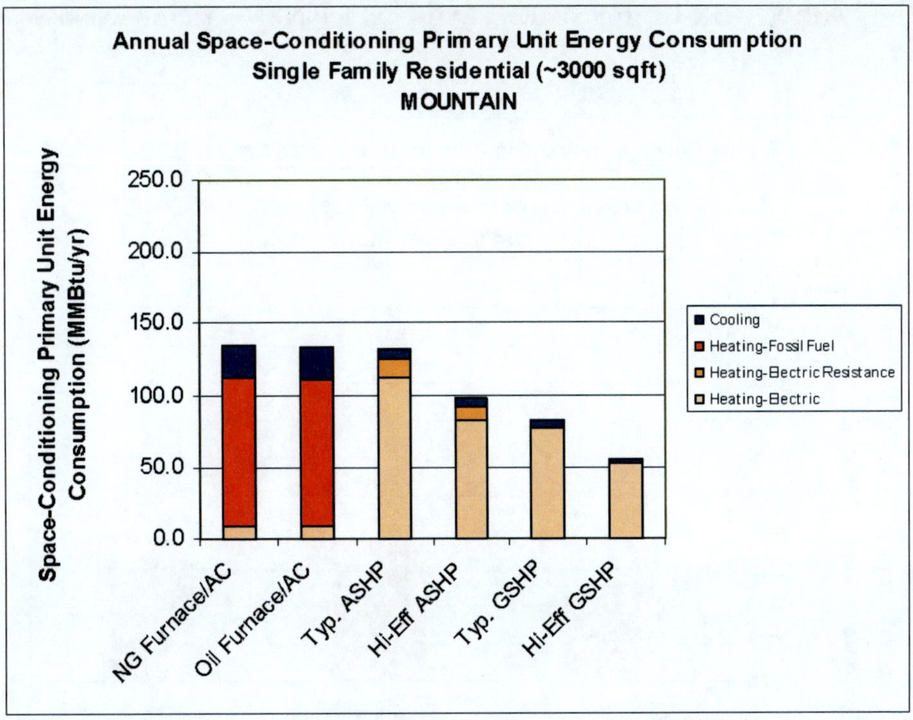
Annual Space-Conditioning Primary Unit Energy Consumption
Single Family Residential (~3000 sqft)
MOUNTAIN
Space-Conditioning Primary Unit Energy Consumption (MMBtu/yr)
250.0
200.0
150.0
100.0
50.0
0.0
NG Furnace/AC
Oil Furnace/AC
Typ. ASHP
Hi-Eff ASHP
Typ. GSHP
Hi-Eff GSHP
Cooling
Heating-Fossil Fuel
Heating-Electric Resistance
Heating-Electric

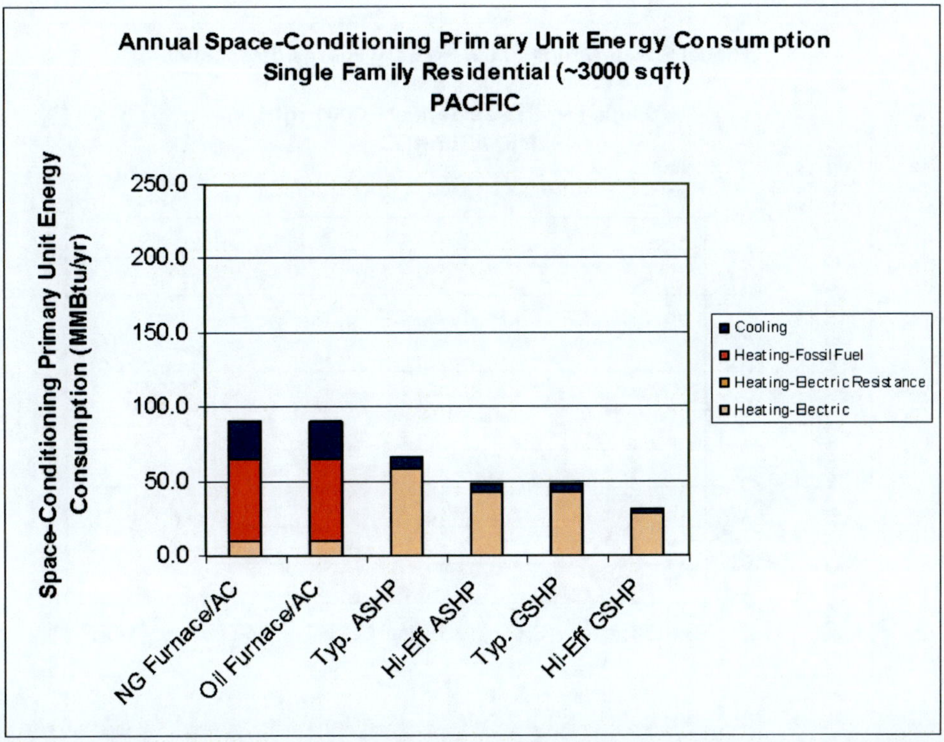
Annual Space-Conditioning Primary Unit Energy Consumption
Single Family Residential (~3000 sqft)
PACIFIC
Space-Conditioning Primary Unit Energy Consumption (MMBtu/yr)
250.0
200.0
150.0
100.0
50.0
0.0
NG Furnace/AC
Oil Furnace/AC
Typ. ASHP
Hi-Eff ASHP
Typ. GSHP
Hi-Eff GSHP
Cooling
Heating-Fossil Fuel
Heating-Electric Resistance
Heating-Electric

APPENDIX C: RESIDENTIAL UNIT ENERGY SAVINGS

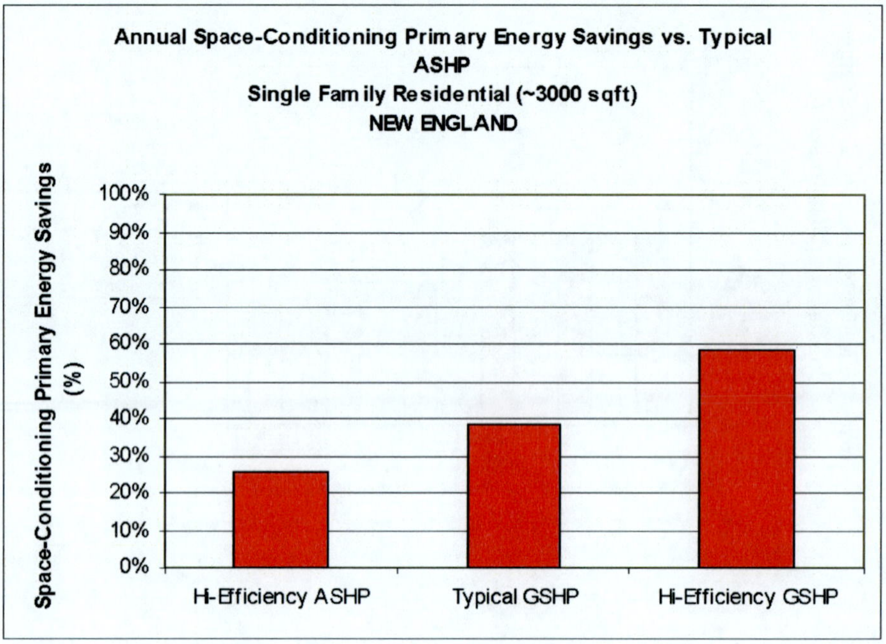

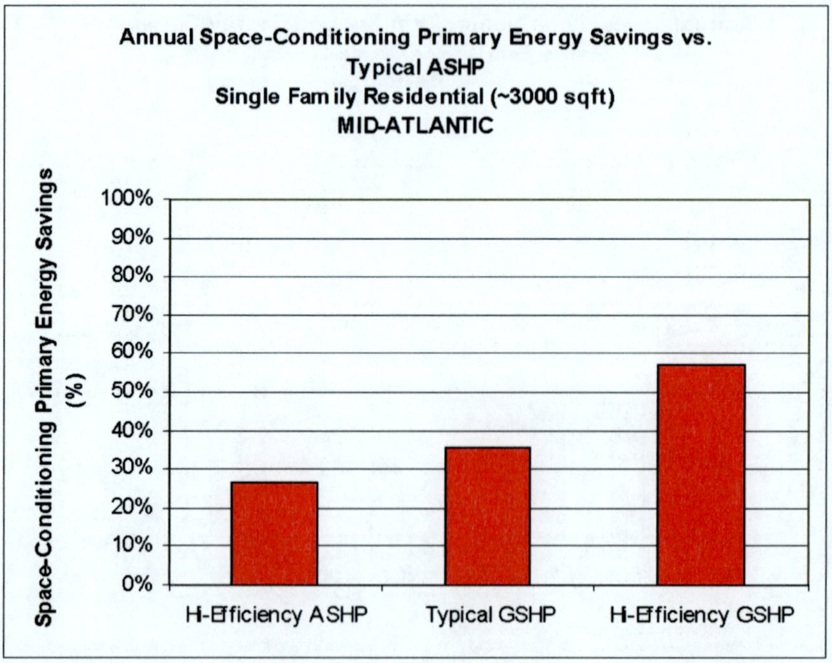

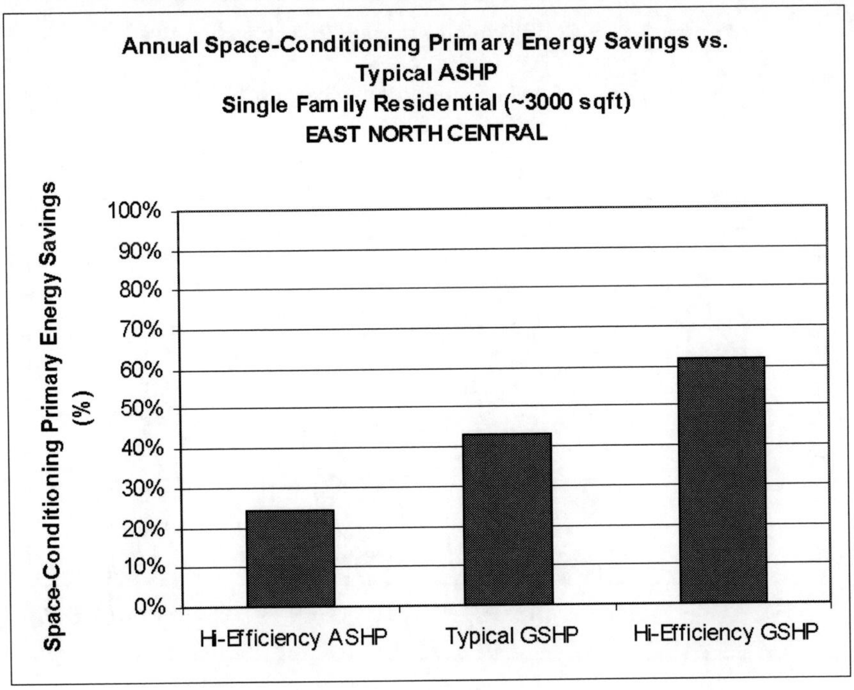
Annual Space-Conditioning Primary Energy Savings vs.
Typical ASHP
Single Family Residential (~3000 sqft)
EAST NORTH CENTRAL
Space-Conditioning Primary Energy Savings (%)
100%
90%
80%
70%
60%
50%
40%
30%
20%
10%
0%
Hi-Efficiency ASHP
Typical GSHP
Hi-Efficiency GSHP

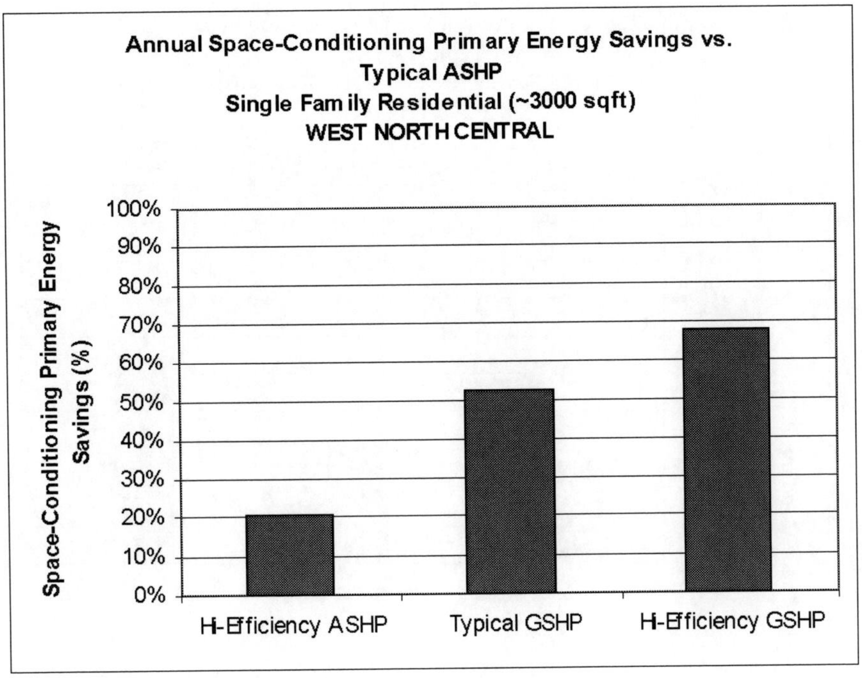
Annual Space-Conditioning Primary Energy Savings vs.
Typical ASHP
Single Family Residential (~3000 sqft)
WEST NORTH CENTRAL
Space-Conditioning Primary Energy Savings (%)
100%
90%
80%
70%
60%
50%
40%
30%
20%
10%
0%
Hi-Efficiency ASHP
Typical GSHP
Hi-Efficiency GSHP

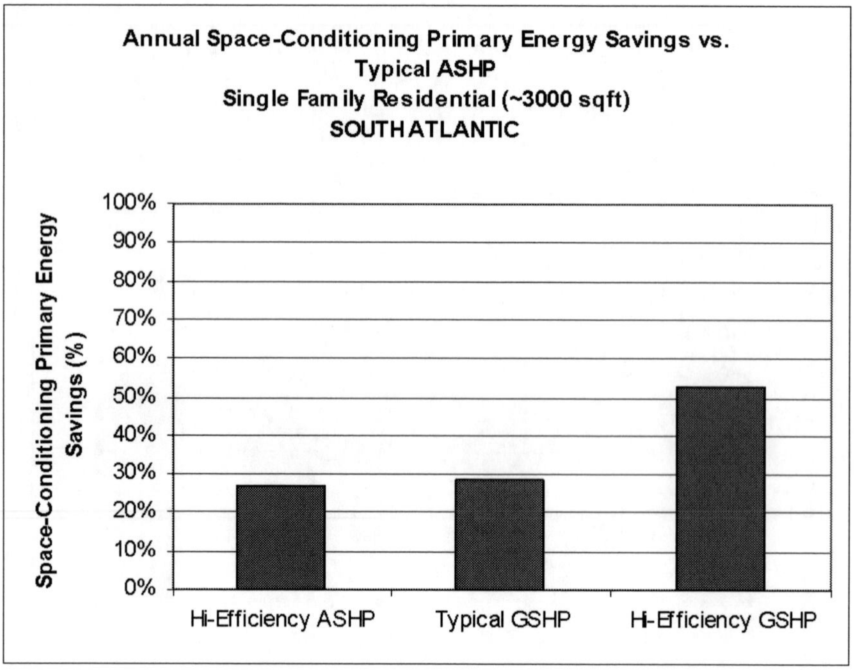
Annual Space-Conditioning Primary Energy Savings vs.
Typical ASHP
Single Family Residential (~3000 sqft)
SOUTH ATLANTIC
Space-Conditioning Primary Energy Savings (%)
100%
90%
80%
70%
60%
50%
40%
30%
20%
10%
0%
Hi-Efficiency ASHP
Typical GSHP
Hi-Efficiency GSHP

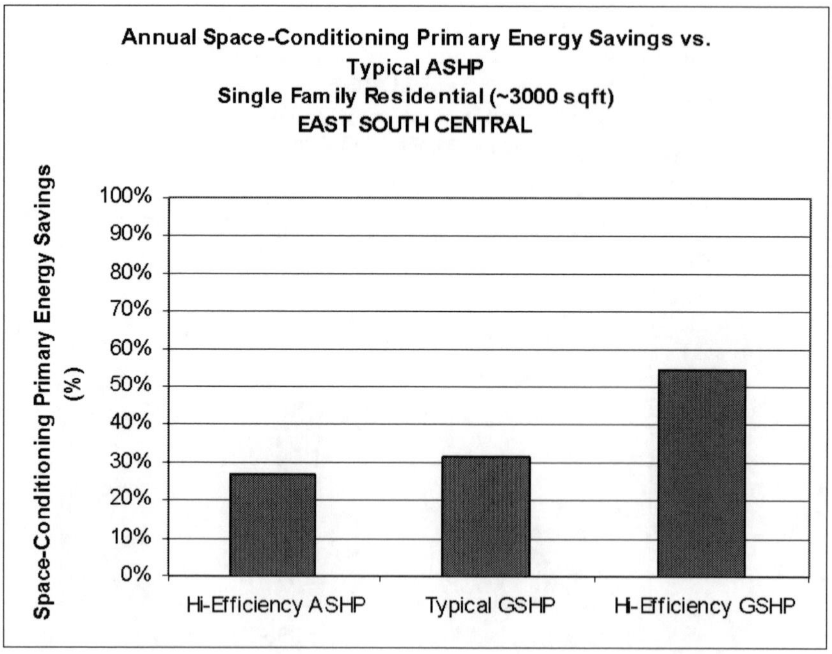
Annual Space-Conditioning Primary Energy Savings vs.
Typical ASHP
Single Family Residential (~3000 sqft)
EAST SOUTH CENTRAL
Space-Conditioning Primary Energy Savings (%)
100%
90%
80%
70%
60%
50%
40%
30%
20%
10%
0%
Hi-Efficiency ASHP
Typical GSHP
Hi-Efficiency GSHP

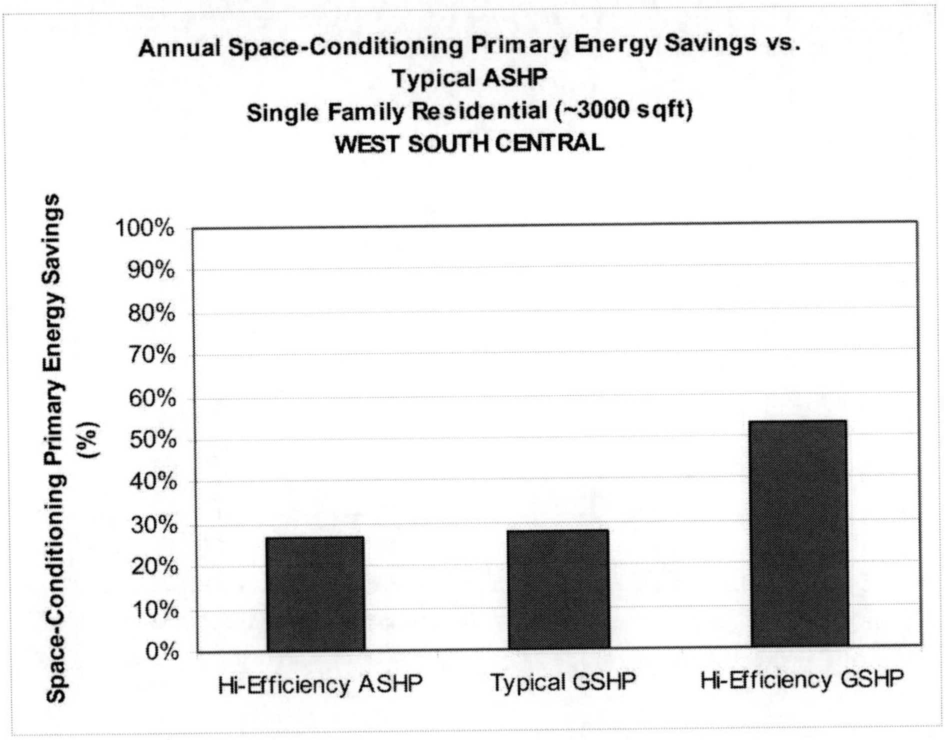
Annual Space-Conditioning Primary Energy Savings vs.
Typical ASHP
Single Family Residential (~3000 sqft)
WEST SOUTH CENTRAL
Space-Conditioning Primary Energy Savings (%)
100%
90%
80%
70%
60%
50%
40%
30%
20%
10%
0%
Hi-Efficiency ASHP
Typical GSHP
Hi-Efficiency GSHP

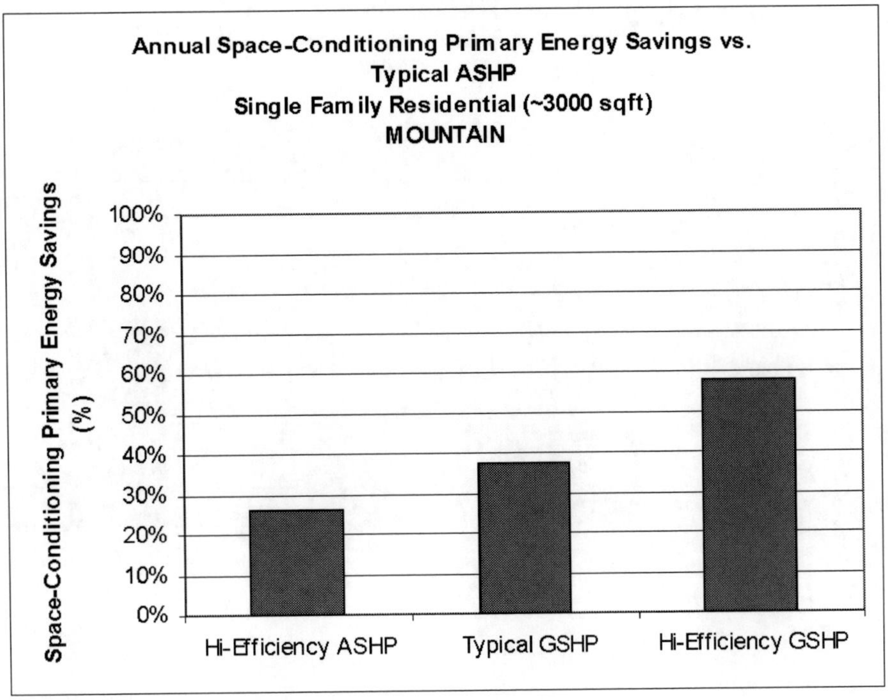
Annual Space-Conditioning Primary Energy Savings vs.
Typical ASHP
Single Family Residential (~3000 sqft)
MOUNTAIN
Space-Conditioning Primary Energy Savings (%)
100%
90%
80%
70%
60%
50%
40%
30%
20%
10%
0%
Hi-Efficiency ASHP
Typical GSHP
Hi-Efficiency GSHP

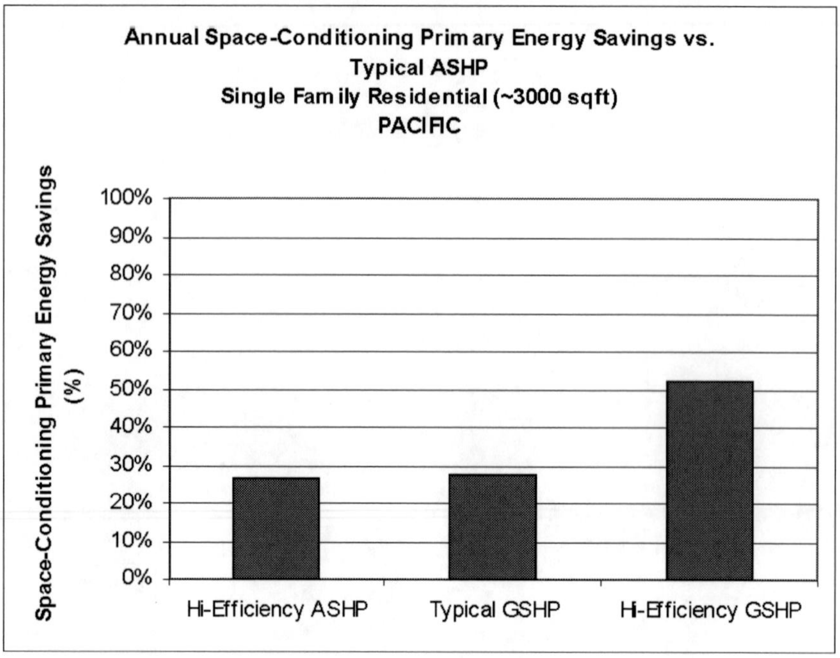
Annual Space-Conditioning Primary Energy Savings vs. Typical ASHP
Single Family Residential (~3000 sqft)
PACIFIC
Space-Conditioning Primary Energy Savings (%)
100%
90%
80%
70%
60%
50%
40%
30%
20%
10%
0%
Hi-Efficiency ASHP
Typical GSHP
Hi-Efficiency GSHP

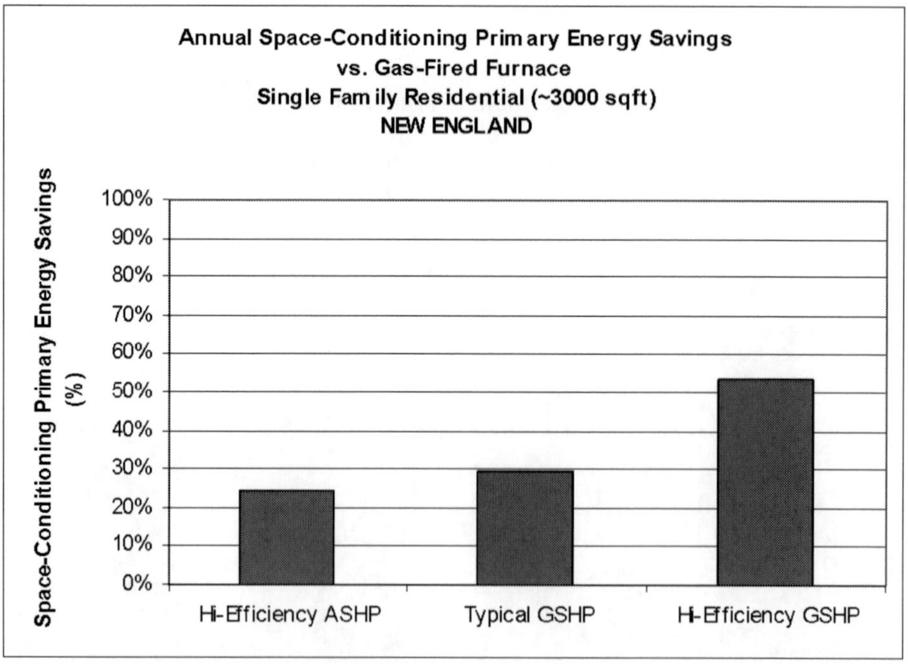
Annual Space-Conditioning Primary Energy Savings vs. Gas-Fired Furnace
Single Family Residential (~3000 sqft)
NEW ENGLAND
Space-Conditioning Primary Energy Savings (%)
100%
90%
80%
70%
60%
50%
40%
30%
20%
10%
0%
Hi-Efficiency ASHP
Typical GSHP
Hi-Efficiency GSHP

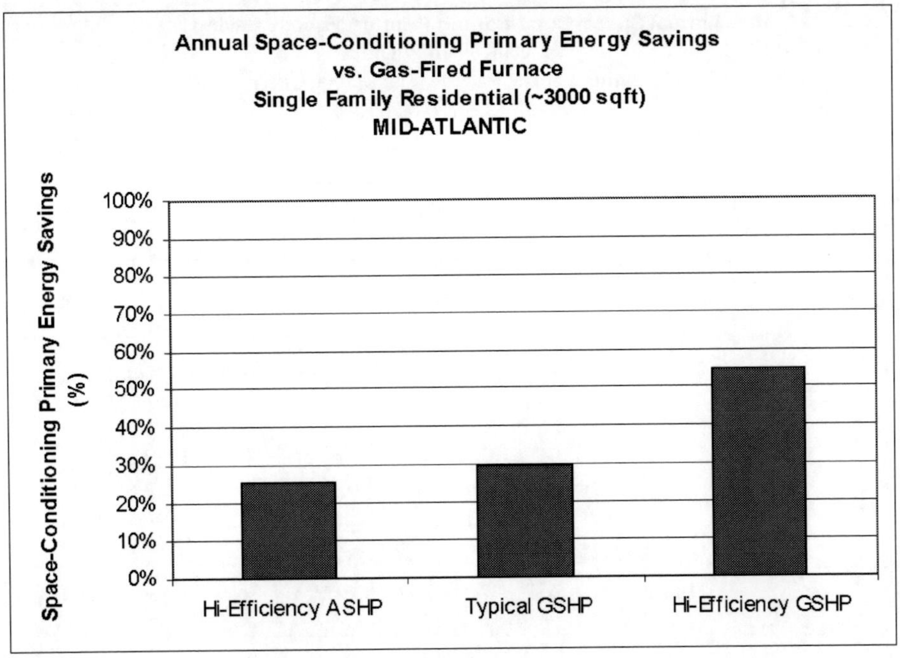
Annual Space-Conditioning Primary Energy Savings
vs. Gas-Fired Furnace
Single Family Residential (~3000 sqft)
MID-ATLANTIC
Space-Conditioning Primary Energy Savings (%)
100%
90%
80%
70%
60%
50%
40%
30%
20%
10%
0%
Hi-Efficiency ASHP
Typical GSHP
Hi-Efficiency GSHP

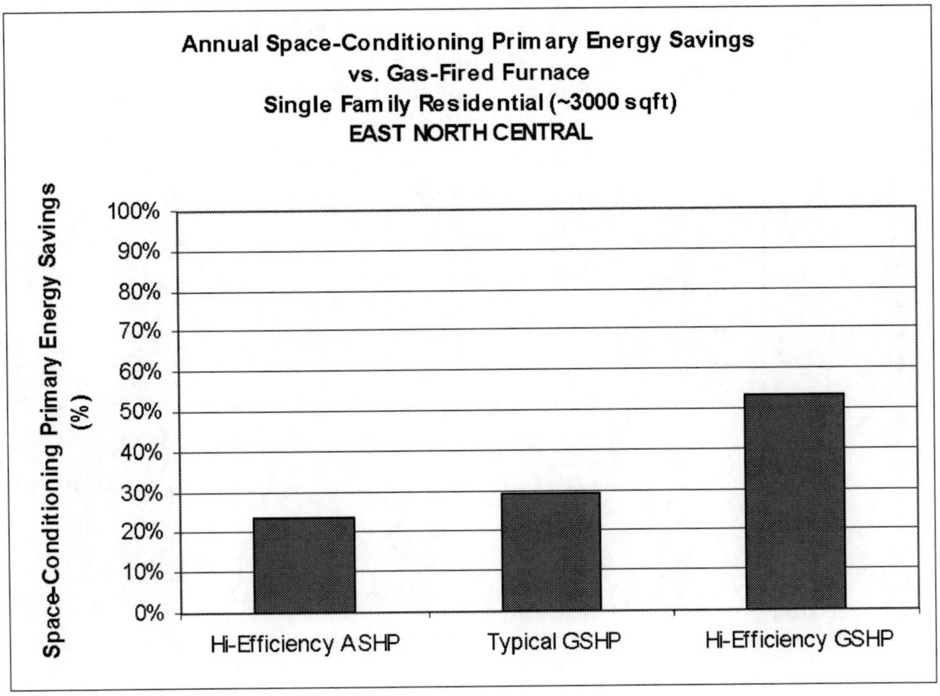
Annual Space-Conditioning Primary Energy Savings
vs. Gas-Fired Furnace
Single Family Residential (~3000 sqft)
EAST NORTH CENTRAL
Space-Conditioning Primary Energy Savings (%)
100%
90%
80%
70%
60%
50%
40%
30%
20%
10%
0%
Hi-Efficiency ASHP
Typical GSHP
Hi-Efficiency GSHP

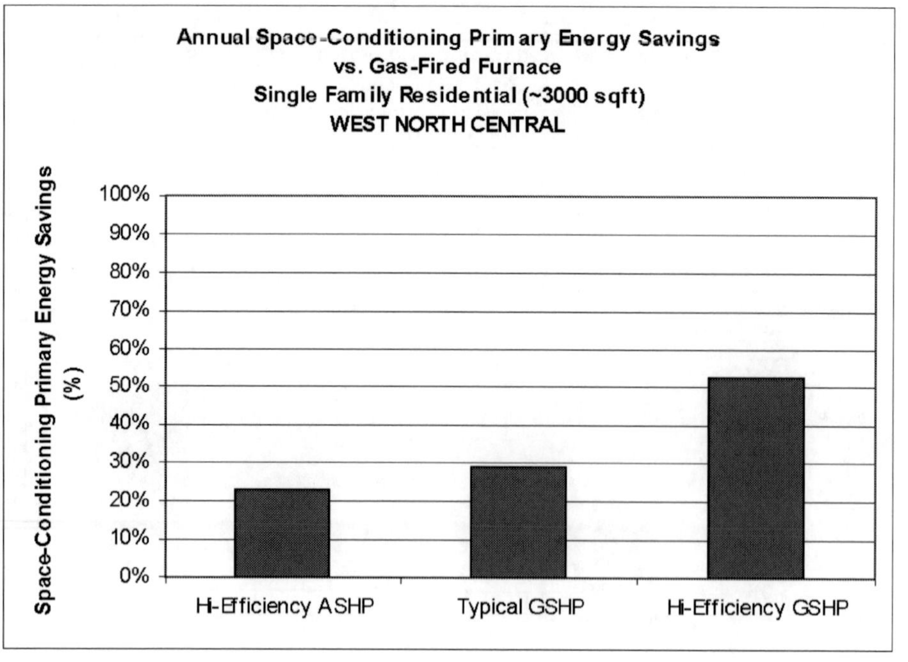
Annual Space-Conditioning Primary Energy Savings
vs. Gas-Fired Furnace
Single Family Residential (~3000 sqft)
WEST NORTH CENTRAL
Space-Conditioning Primary Energy Savings (%)
100%
90%
80%
70%
60%
50%
40%
30%
20%
10%
0%
Hi-Efficiency ASHP
Typical GSHP
Hi-Efficiency GSHP

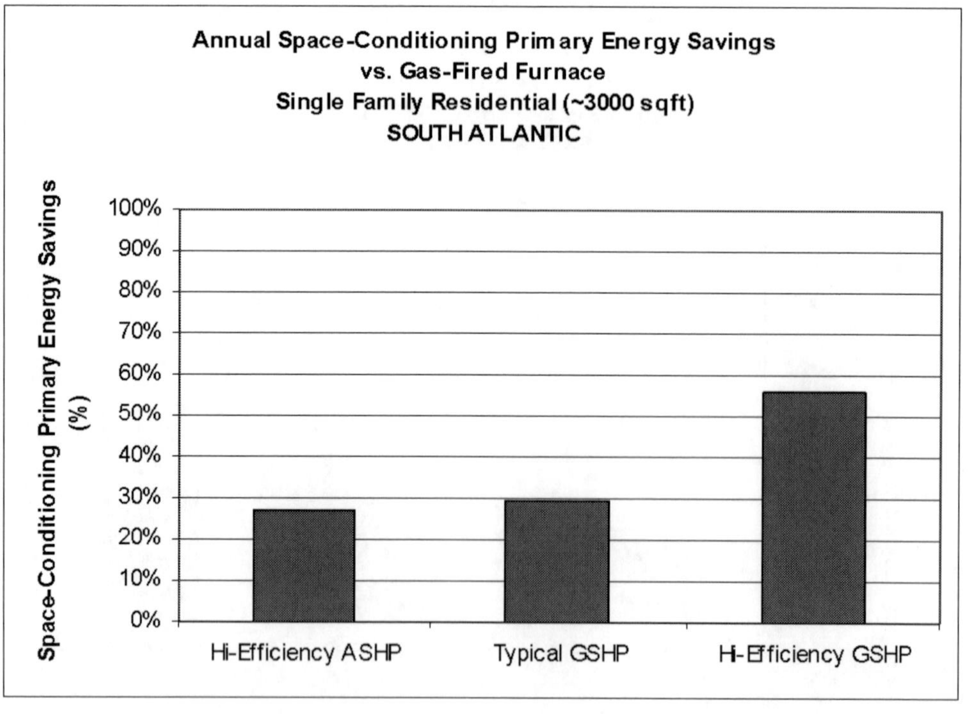
Annual Space-Conditioning Primary Energy Savings
vs. Gas-Fired Furnace
Single Family Residential (~3000 sqft)
SOUTH ATLANTIC
Space-Conditioning Primary Energy Savings (%)
100%
90%
80%
70%
60%
50%
40%
30%
20%
10%
0%
Hi-Efficiency ASHP
Typical GSHP
Hi-Efficiency GSHP

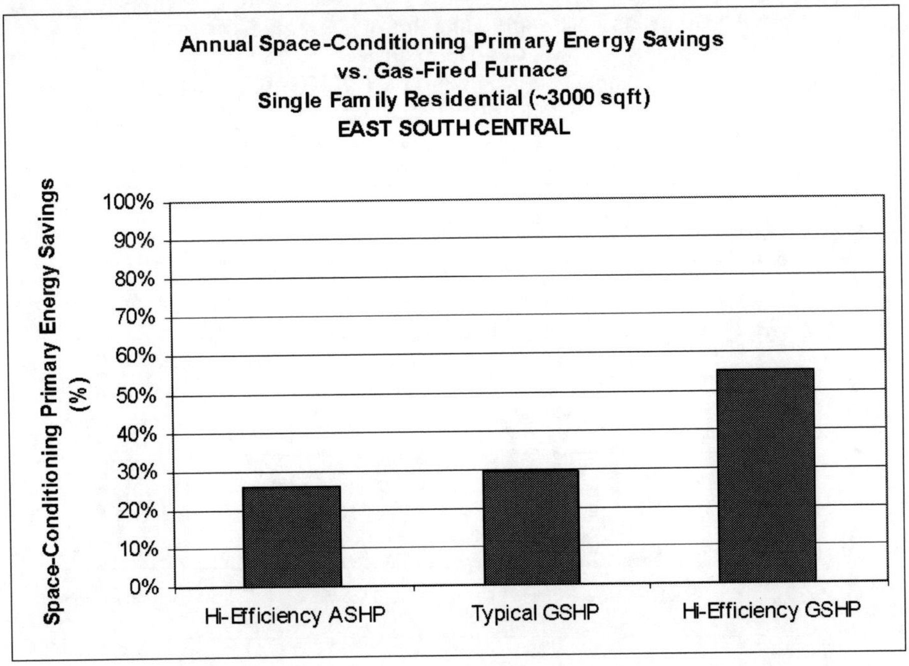
Annual Space-Conditioning Primary Energy Savings
vs. Gas-Fired Furnace
Single Family Residential (~3000 sqft)
EAST SOUTH CENTRAL
Space-Conditioning Primary Energy Savings (%)
100%
90%
80%
70%
60%
50%
40%
30%
20%
10%
0%
Hi-Efficiency ASHP
Typical GSHP
Hi-Efficiency GSHP

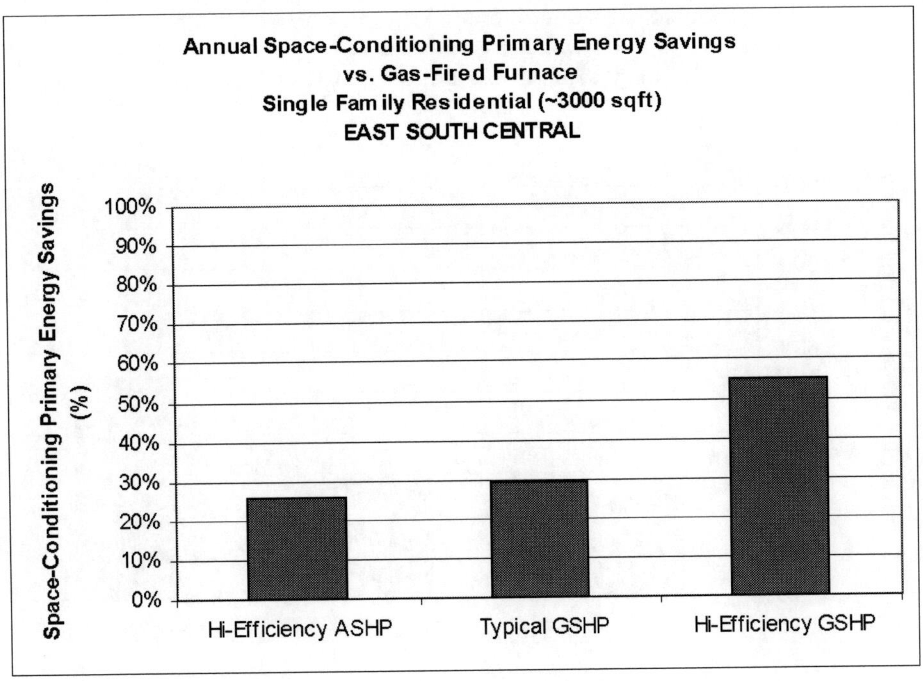
Annual Space-Conditioning Primary Energy Savings
vs. Gas-Fired Furnace
Single Family Residential (~3000 sqft)
EAST SOUTH CENTRAL
Space-Conditioning Primary Energy Savings (%)
100%
90%
80%
70%
60%
50%
40%
30%
20%
10%
0%
Hi-Efficiency ASHP
Typical GSHP
Hi-Efficiency GSHP

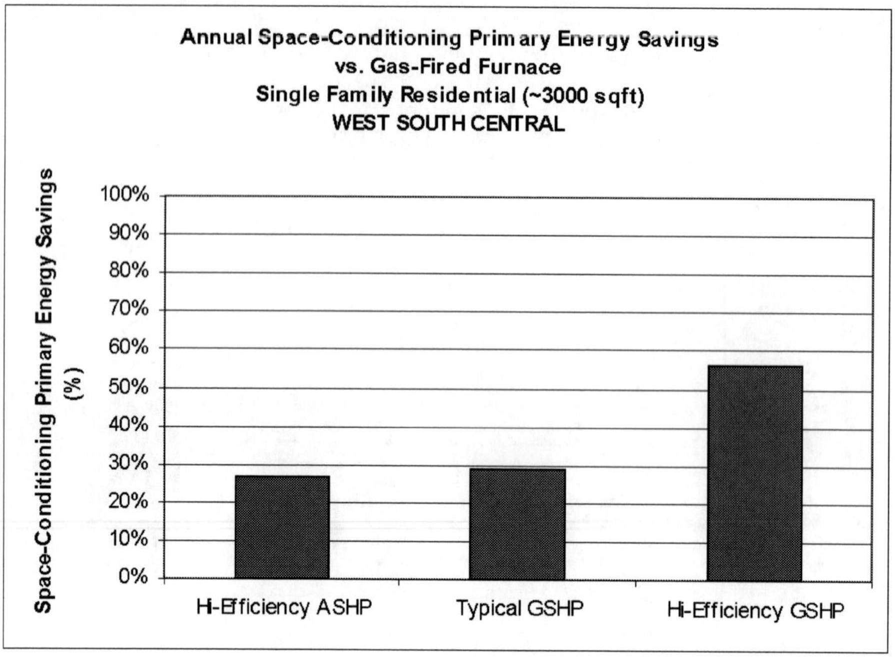
Annual Space-Conditioning Primary Energy Savings
vs. Gas-Fired Furnace
Single Family Residential (~3000 sqft)
WEST SOUTH CENTRAL
Space-Conditioning Primary Energy Savings (%)
100%
90%
80%
70%
60%
50%
40%
30%
20%
10%
0%
Hi-Efficiency ASHP
Typical GSHP
Hi-Efficiency GSHP

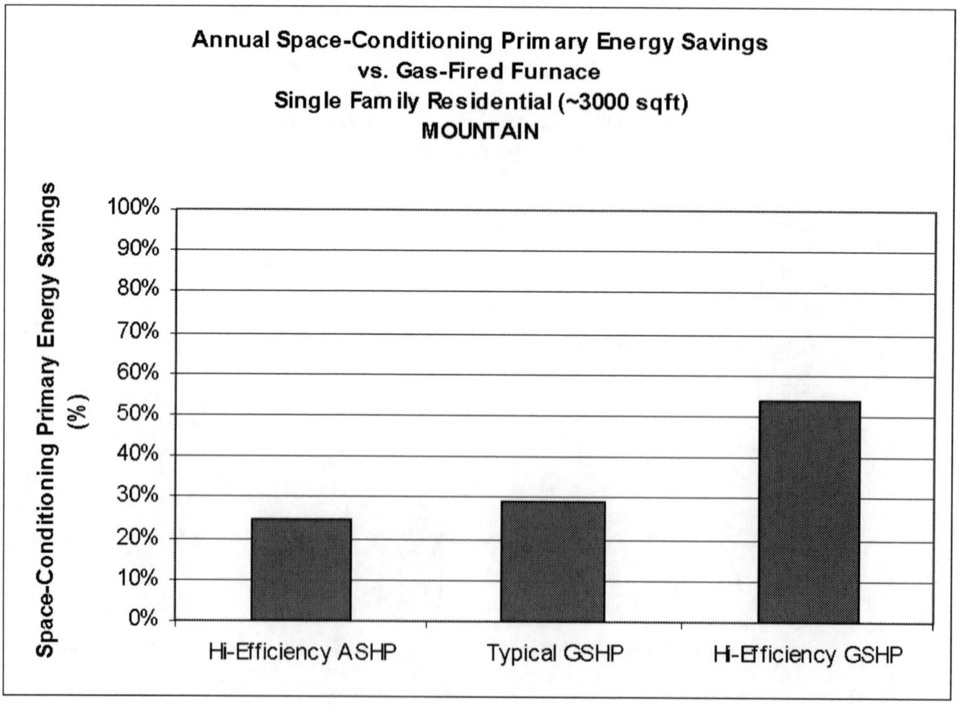
Annual Space-Conditioning Primary Energy Savings
vs. Gas-Fired Furnace
Single Family Residential (~3000 sqft)
MOUNTAIN
Space-Conditioning Primary Energy Savings (%)
100%
90%
80%
70%
60%
50%
40%
30%
20%
10%
0%
Hi-Efficiency ASHP
Typical GSHP
Hi-Efficiency GSHP

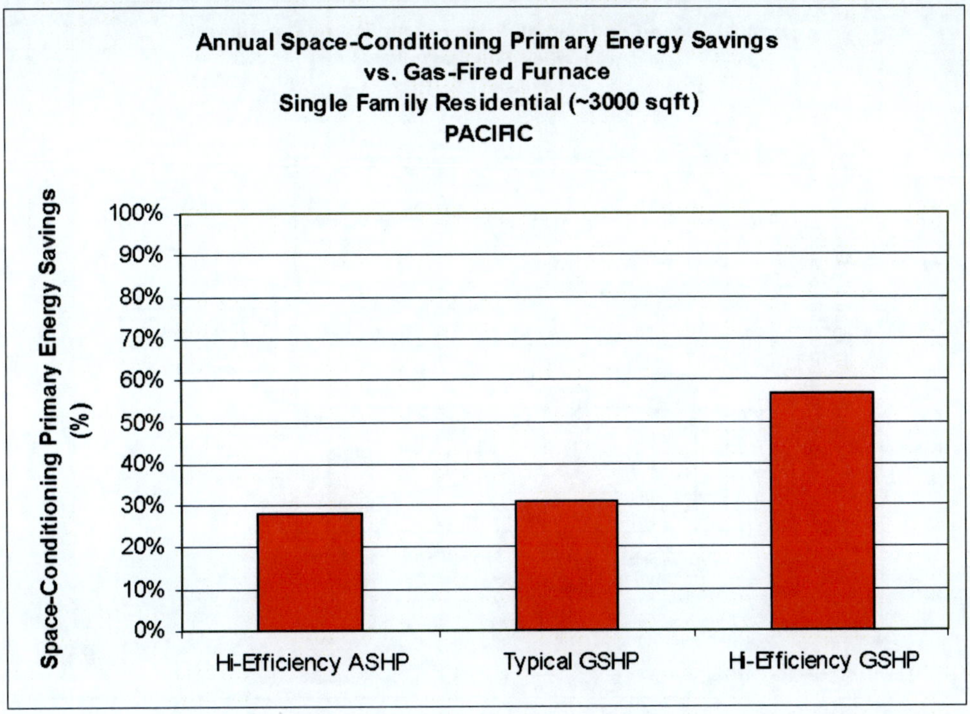

APPENDIX D: COMMERCIAL UNIT ENERGY CONSUMPTIONS

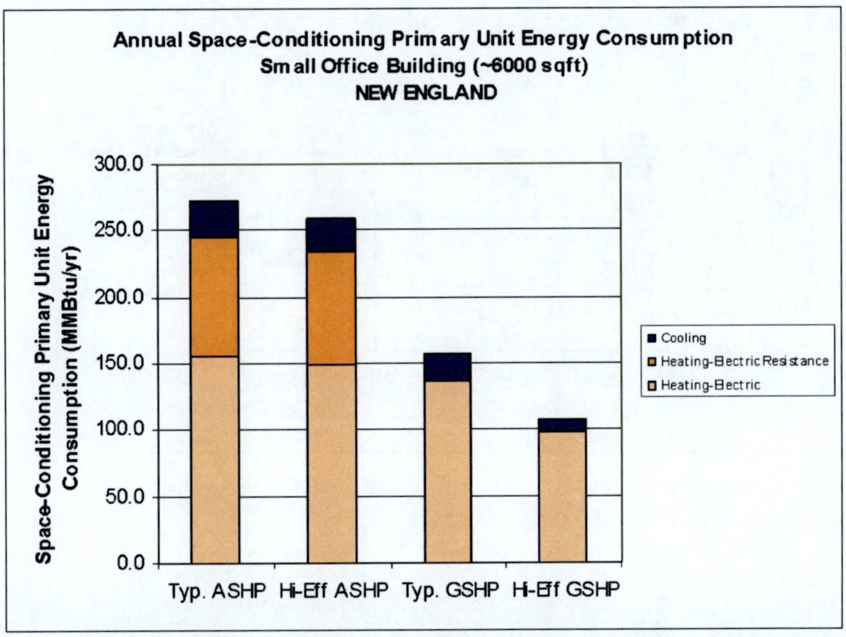

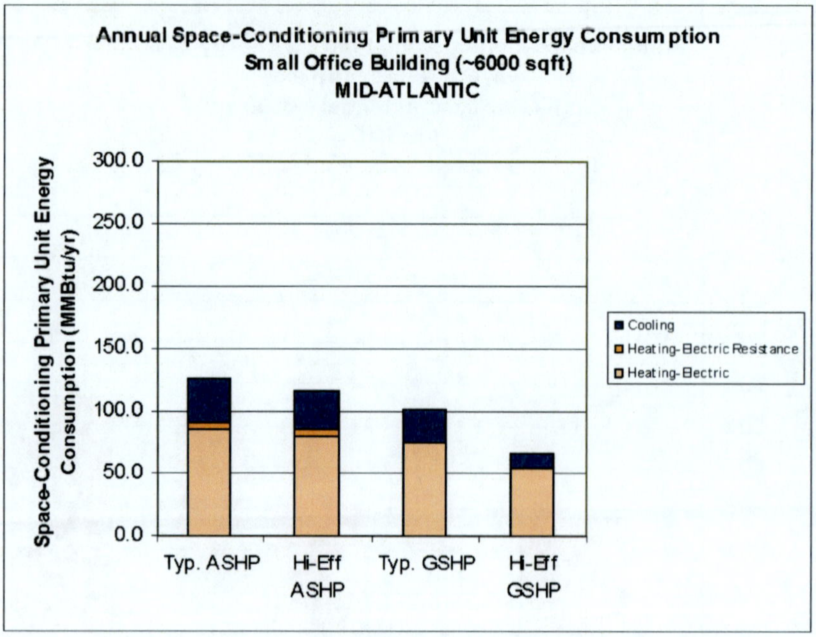
Annual Space-Conditioning Primary Unit Energy Consumption
Small Office Building (~6000 sqft)
MID-ATLANTIC
Space-Conditioning Primary Unit Energy Consumption (MMBtu/yr)
300.0
250.0
200.0
150.0
100.0
50.0
0.0
Typ. ASHP
Hi-Eff ASHP
Typ. GSHP
Hi-Eff GSHP
Cooling
Heating-Electric Resistance
Heating-Electric

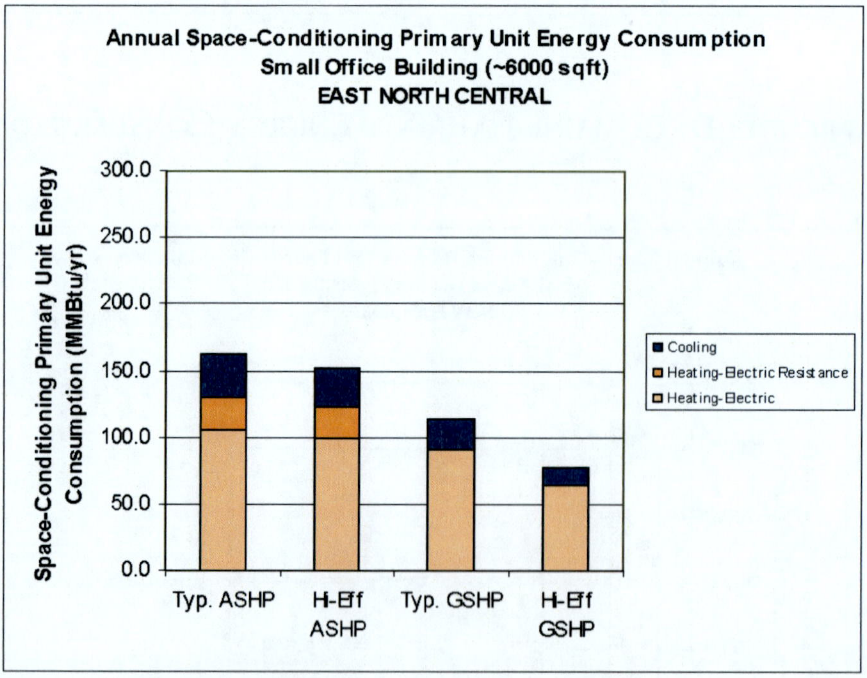
Annual Space-Conditioning Primary Unit Energy Consumption
Small Office Building (~6000 sqft)
EAST NORTH CENTRAL
Space-Conditioning Primary Unit Energy Consumption (MMBtu/yr)
300.0
250.0
200.0
150.0
100.0
50.0
0.0
Typ. ASHP
Hi-Eff ASHP
Typ. GSHP
Hi-Eff GSHP
Cooling
Heating-Electric Resistance
Heating-Electric

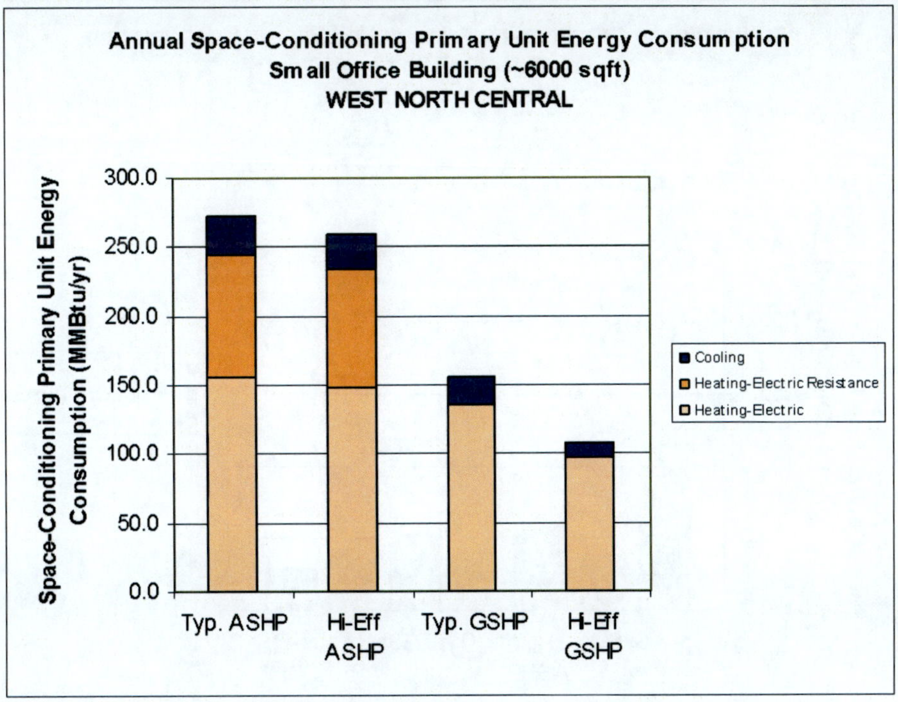
Annual Space-Conditioning Primary Unit Energy Consumption
Small Office Building (~6000 sqft)
WEST NORTH CENTRAL
Space-Conditioning Primary Unit Energy Consumption (MMBtu/yr)
300.0
250.0
200.0
150.0
100.0
50.0
0.0
Typ. ASHP
Hi-Eff ASHP
Typ. GSHP
Hi-Eff GSHP
Cooling
Heating-Electric Resistance
Heating-Electric

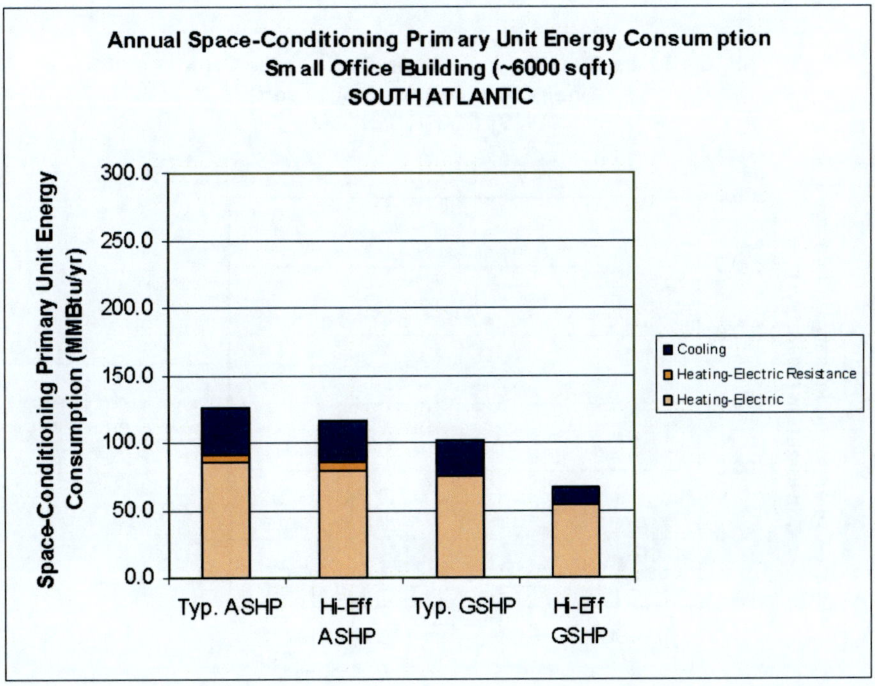
Annual Space-Conditioning Primary Unit Energy Consumption
Small Office Building (~6000 sqft)
SOUTH ATLANTIC
Space-Conditioning Primary Unit Energy Consumption (MMBtu/yr)
300.0
250.0
200.0
150.0
100.0
50.0
0.0
Typ. ASHP
Hi-Eff ASHP
Typ. GSHP
Hi-Eff GSHP
Cooling
Heating-Electric Resistance
Heating-Electric

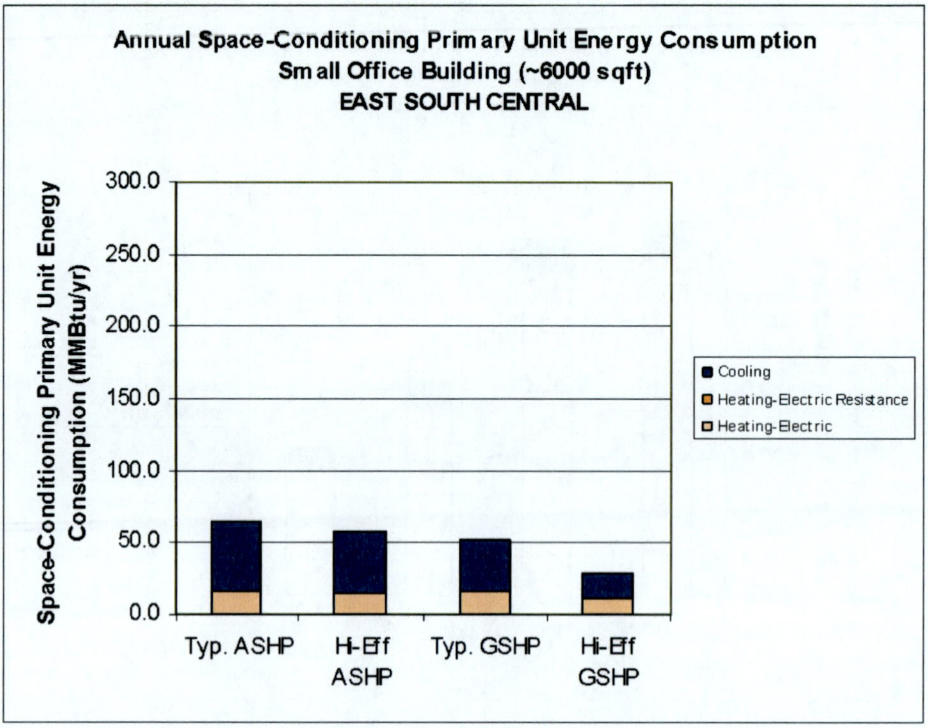
Annual Space-Conditioning Primary Unit Energy Consumption
Small Office Building (~6000 sqft)
EAST SOUTH CENTRAL
Space-Conditioning Primary Unit Energy Consumption (MMBtu/yr)
300.0
250.0
200.0
150.0
100.0
50.0
0.0
Typ. ASHP
Hi-Eff ASHP
Typ. GSHP
Hi-Eff GSHP
Cooling
Heating-Electric Resistance
Heating-Electric

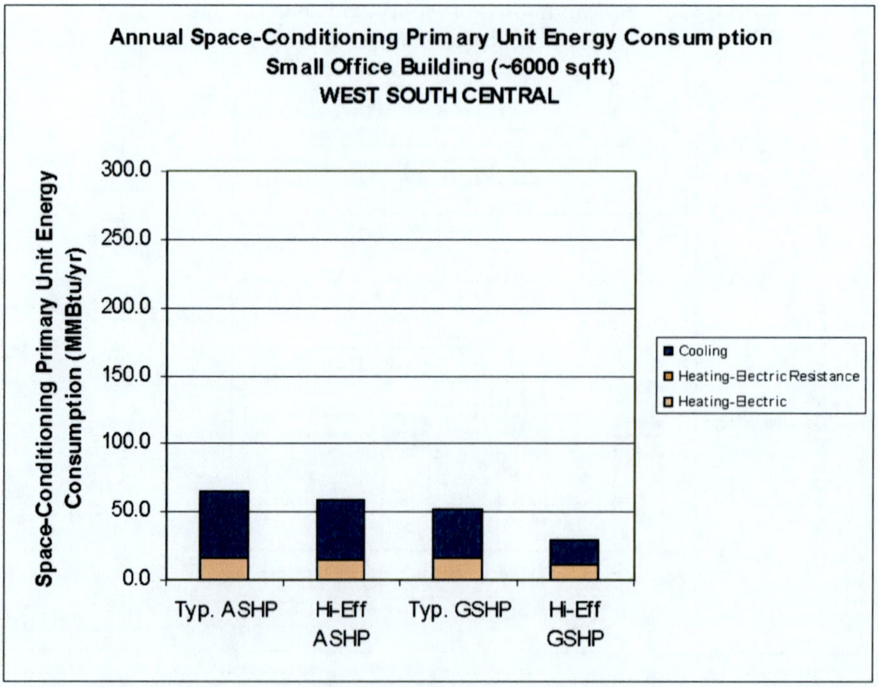
Annual Space-Conditioning Primary Unit Energy Consumption
Small Office Building (~6000 sqft)
WEST SOUTH CENTRAL
Space-Conditioning Primary Unit Energy Consumption (MMBtu/yr)
300.0
250.0
200.0
150.0
100.0
50.0
0.0
Typ. ASHP
Hi-Eff ASHP
Typ. GSHP
Hi-Eff GSHP
Cooling
Heating-Electric Resistance
Heating-Electric

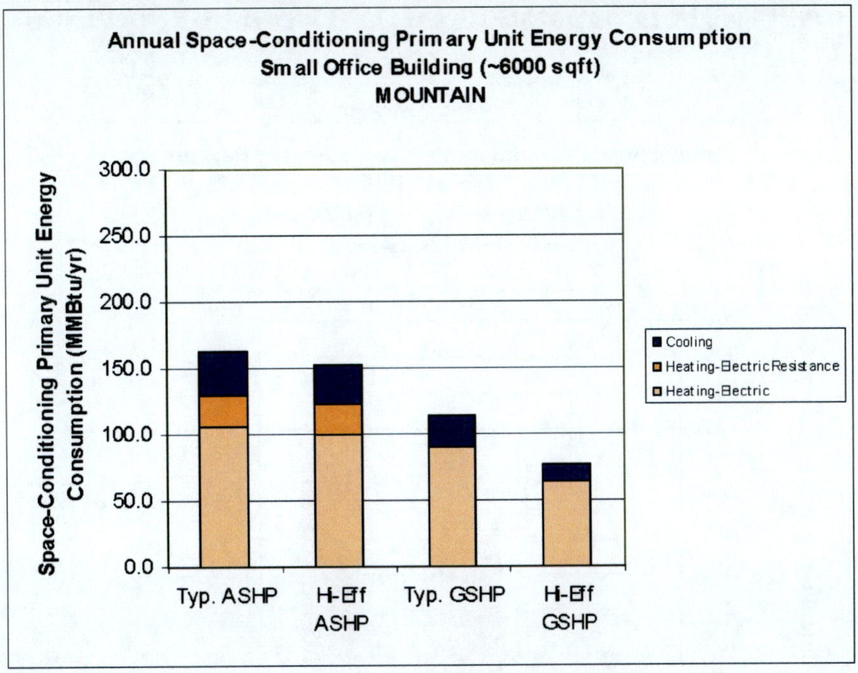
Annual Space-Conditioning Primary Unit Energy Consumption
Small Office Building (~6000 sqft)
MOUNTAIN
Space-Conditioning Primary Unit Energy Consumption (MMBtu/yr)
300.0
250.0
200.0
150.0
100.0
50.0
0.0
Typ. ASHP
Hi-Eff ASHP
Typ. GSHP
Hi-Eff GSHP
Cooling
Heating-Electric Resistance
Heating-Electric

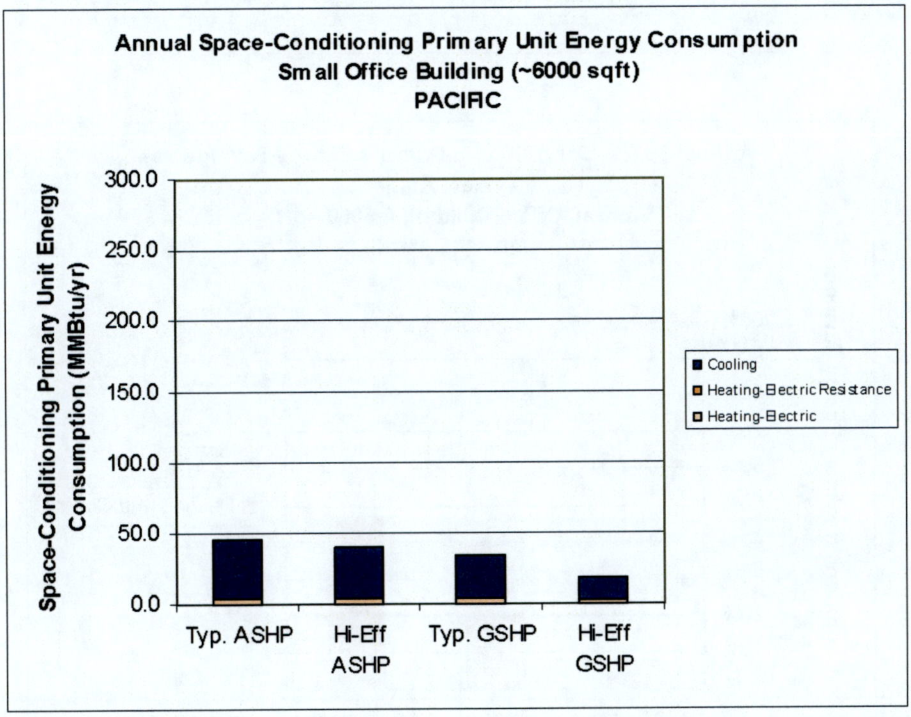
Annual Space-Conditioning Primary Unit Energy Consumption
Small Office Building (~6000 sqft)
PACIFIC
Space-Conditioning Primary Unit Energy Consumption (MMBtu/yr)
300.0
250.0
200.0
150.0
100.0
50.0
0.0
Typ. ASHP
Hi-Eff ASHP
Typ. GSHP
Hi-Eff GSHP
Cooling
Heating-Electric Resistance
Heating-Electric

APPENDIX E: COMMERCIAL UNIT ENERGY SAVINGS

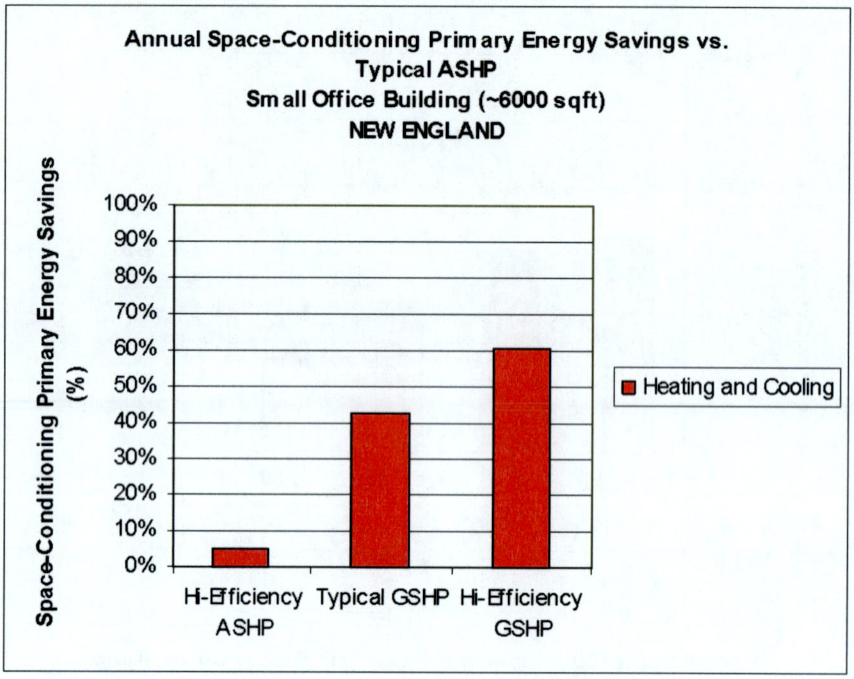

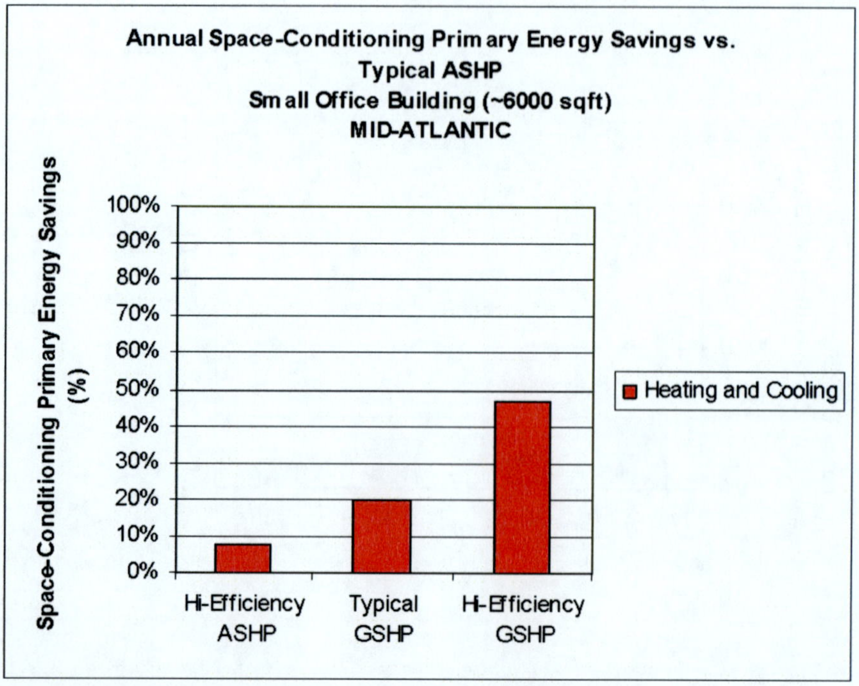

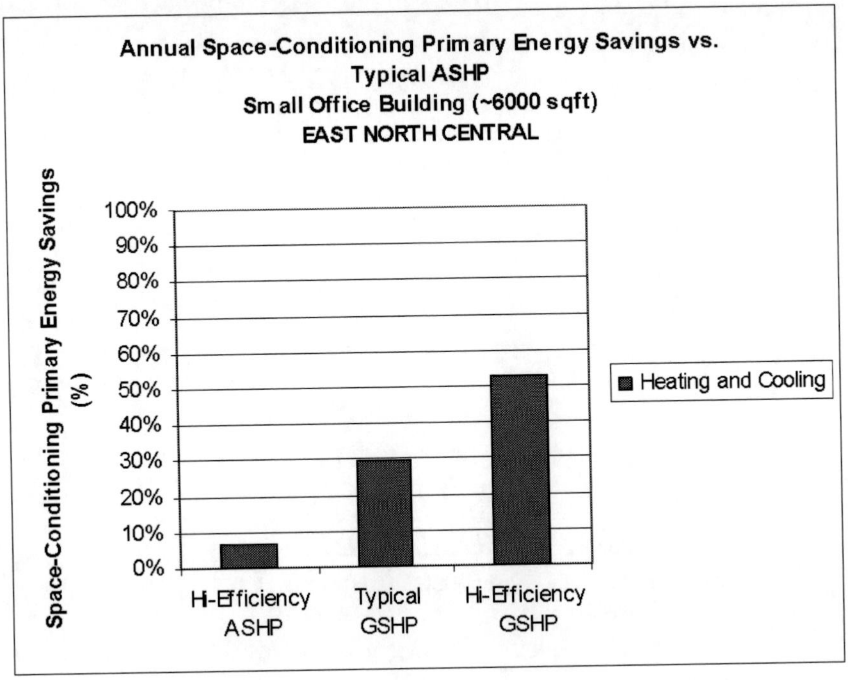
Annual Space-Conditioning Primary Energy Savings vs.
Typical ASHP
Small Office Building (~6000 sqft)
EAST NORTH CENTRAL
Space-Conditioning Primary Energy Savings (%)
100%
90%
80%
70%
60%
50%
40%
30%
20%
10%
0%
Hi-Efficiency ASHP
Typical GSHP
Hi-Efficiency GSHP
Heating and Cooling

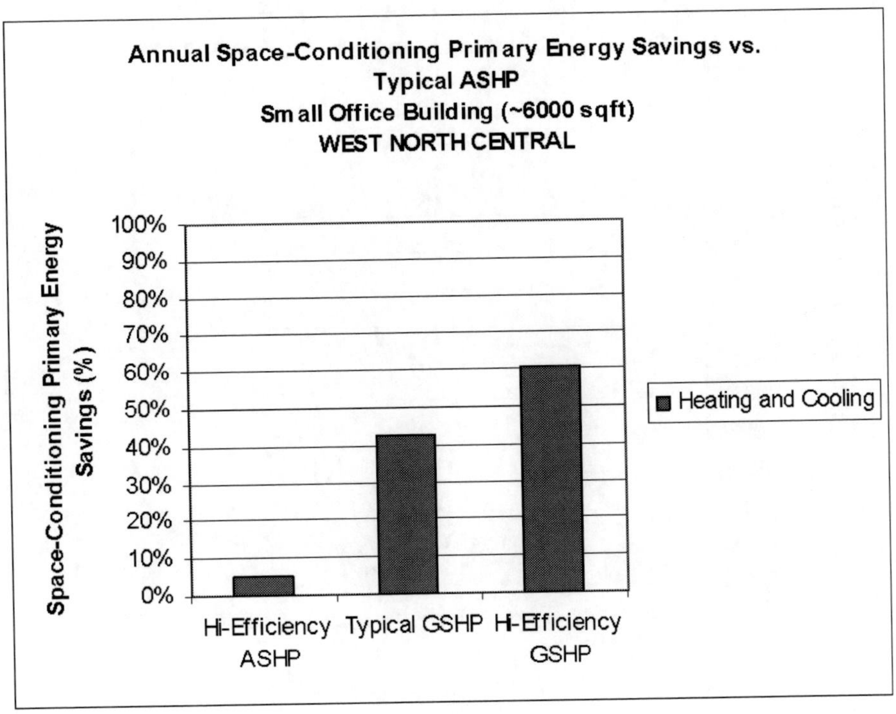
Annual Space-Conditioning Primary Energy Savings vs.
Typical ASHP
Small Office Building (~6000 sqft)
WEST NORTH CENTRAL
Space-Conditioning Primary Energy Savings (%)
100%
90%
80%
70%
60%
50%
40%
30%
20%
10%
0%
Hi-Efficiency ASHP
Typical GSHP
Hi-Efficiency GSHP
Heating and Cooling

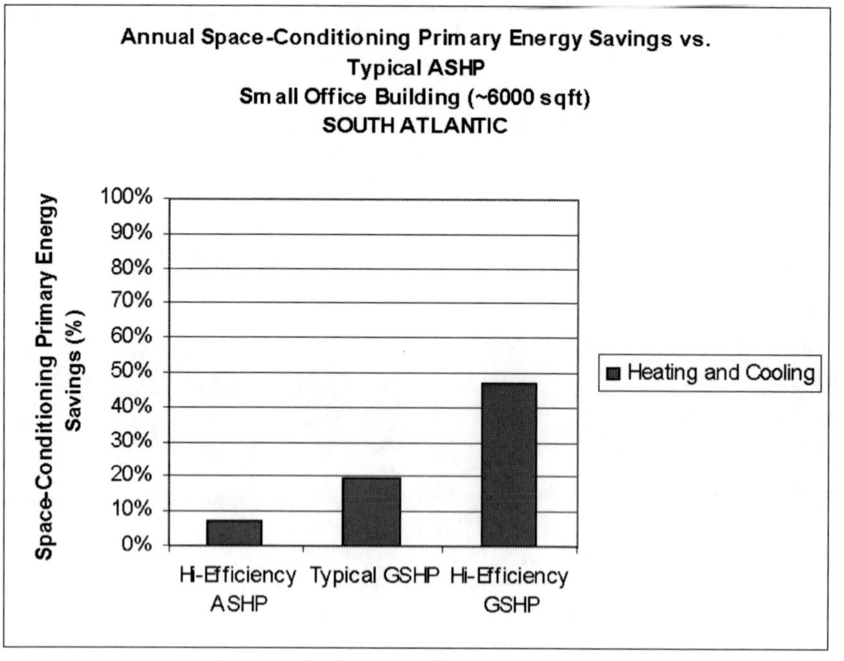

Annual Space-Conditioning Primary Energy Savings vs. Typical ASHP
Small Office Building (~6000 sqft)
SOUTH ATLANTIC
Space-Conditioning Primary Energy Savings (%)
100%
90%
80%
70%
60%
50%
40%
30%
20%
10%
0%
Hi-Efficiency ASHP
Typical GSHP
Hi-Efficiency GSHP
Heating and Cooling

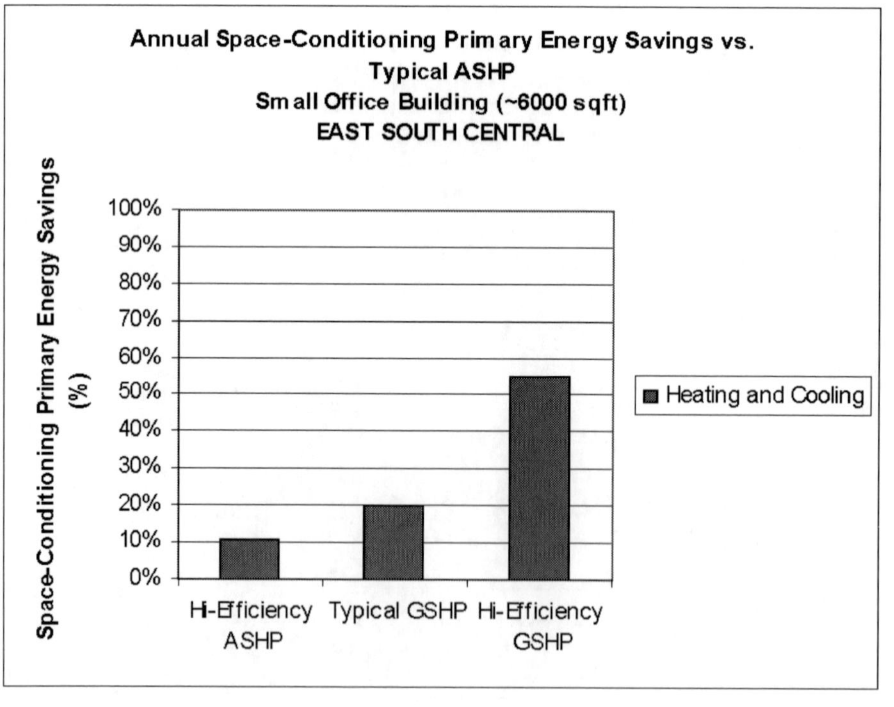

Annual Space-Conditioning Primary Energy Savings vs. Typical ASHP
Small Office Building (~6000 sqft)
EAST SOUTH CENTRAL
Space-Conditioning Primary Energy Savings (%)
100%
90%
80%
70%
60%
50%
40%
30%
20%
10%
0%
Hi-Efficiency ASHP
Typical GSHP
Hi-Efficiency GSHP
Heating and Cooling

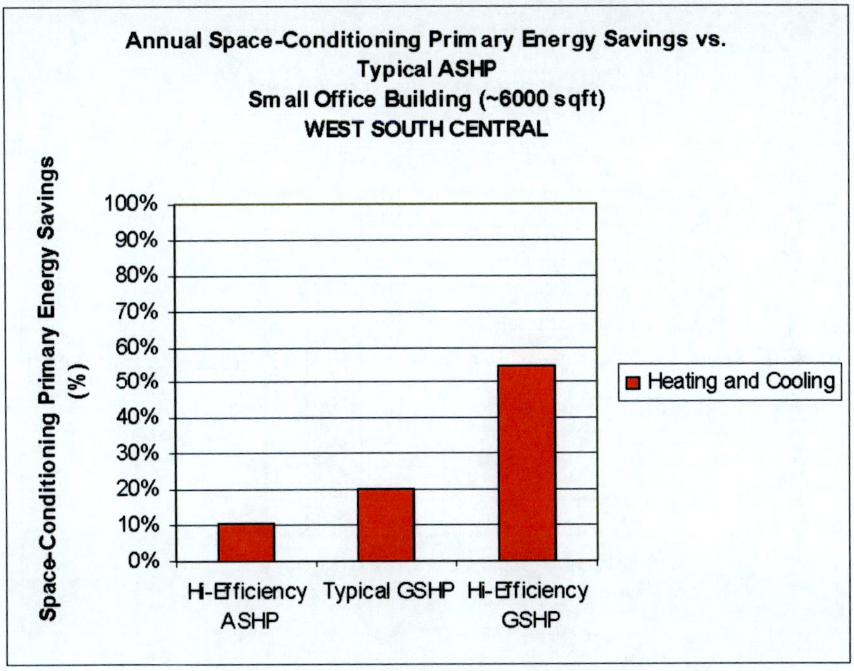
Annual Space-Conditioning Primary Energy Savings vs.
Typical ASHP
Small Office Building (~6000 sqft)
WEST SOUTH CENTRAL
Space-Conditioning Primary Energy Savings (%)
100%
90%
80%
70%
60%
50%
40%
30%
20%
10%
0%
Hi-Efficiency ASHP
Typical GSHP
Hi-Efficiency GSHP
Heating and Cooling

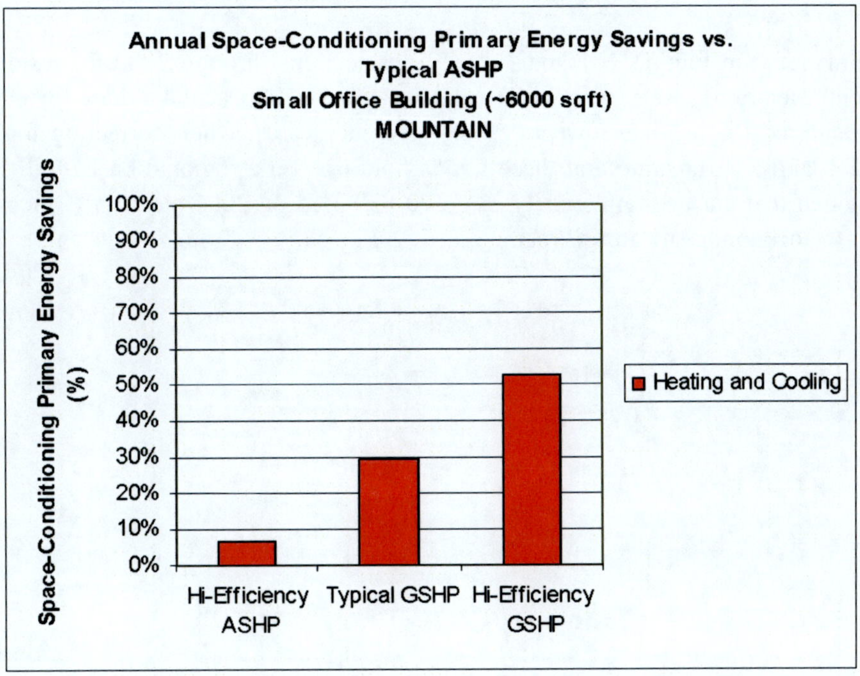
Annual Space-Conditioning Primary Energy Savings vs.
Typical ASHP
Small Office Building (~6000 sqft)
MOUNTAIN
Space-Conditioning Primary Energy Savings (%)
100%
90%
80%
70%
60%
50%
40%
30%
20%
10%
0%
Hi-Efficiency ASHP
Typical GSHP
Hi-Efficiency GSHP
Heating and Cooling

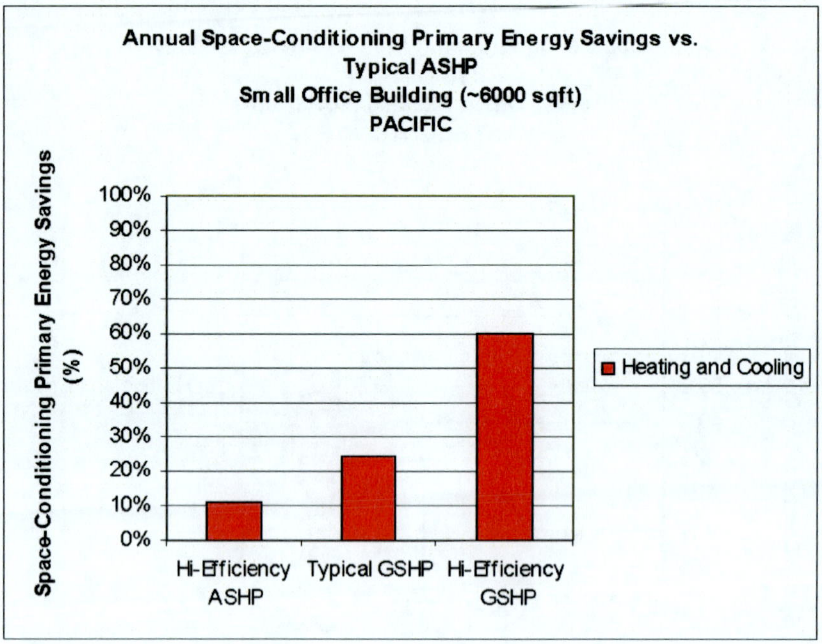

APPENDIX F: ELECTRICITY PRICE PROJECTIONS

As illustrated in Figures F-1 and F-2, EIA projections through 2030 for residential and commercial electricity prices vary relatively little from 2006 actual prices for each of the cases considered (high price, low price, and reference case), when correcting for inflation. Figures F-3 and F-4 illustrate that these trends hold true on a regional basis, too. Therefore, we concluded that our economic analyses could be based on 2006 electricity prices and still represent the economics for future years.

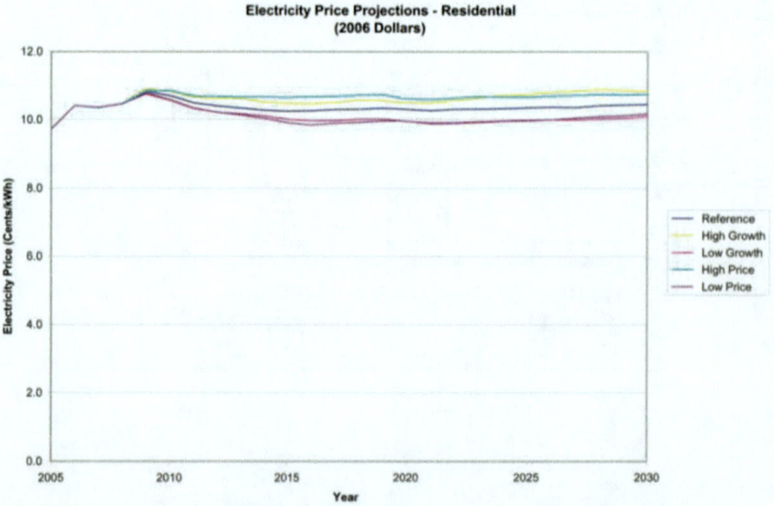

Figure F-1. Residential Electricity Price Projections

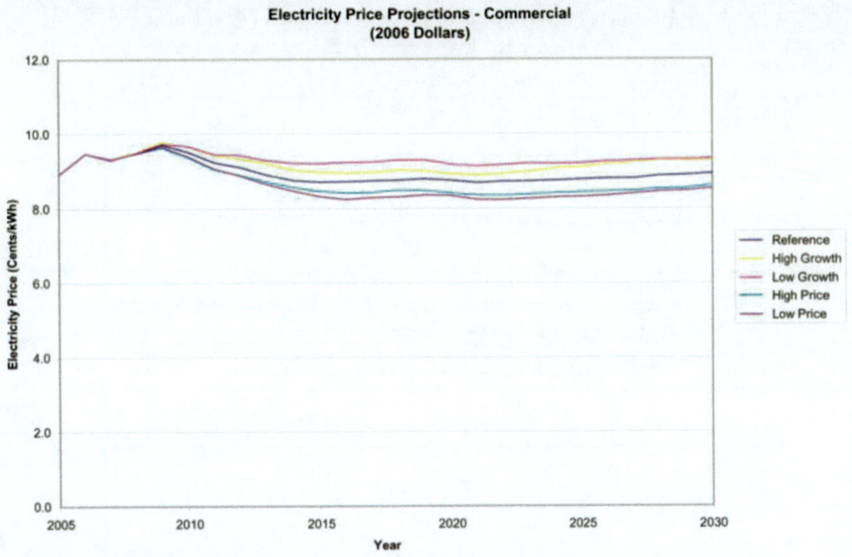

Figure F-2. Commercial Electricity Price Projections

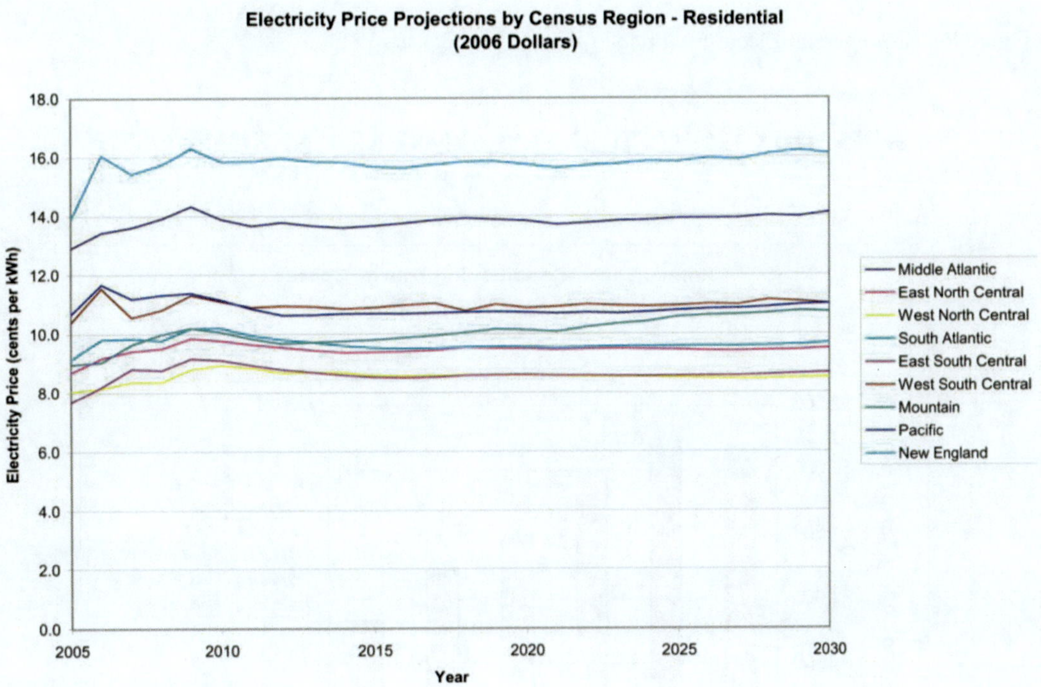

Figure F-3. Residential Electricity Price Projections by Census Region

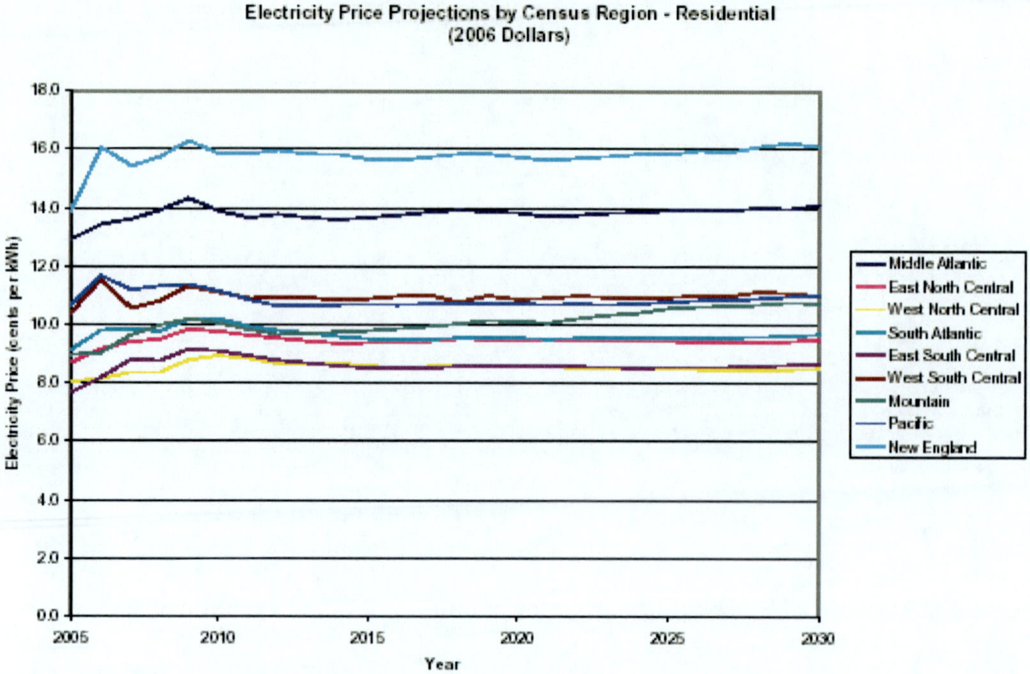

Figure F-4. Commercial Electricity Price Projections by Census Region

APPENDIX G: RESIDENTIAL ANNUAL ENERGY COSTS

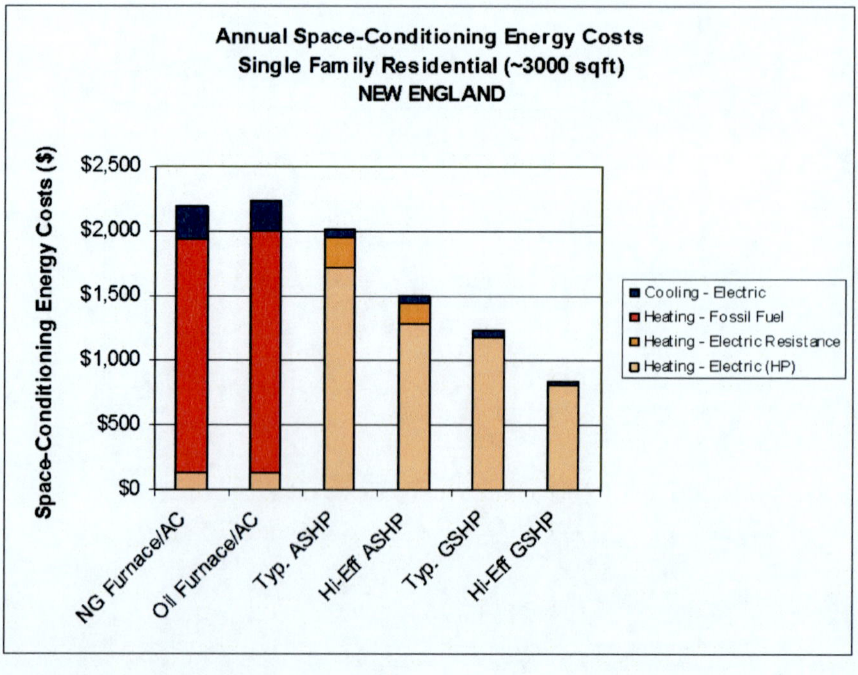

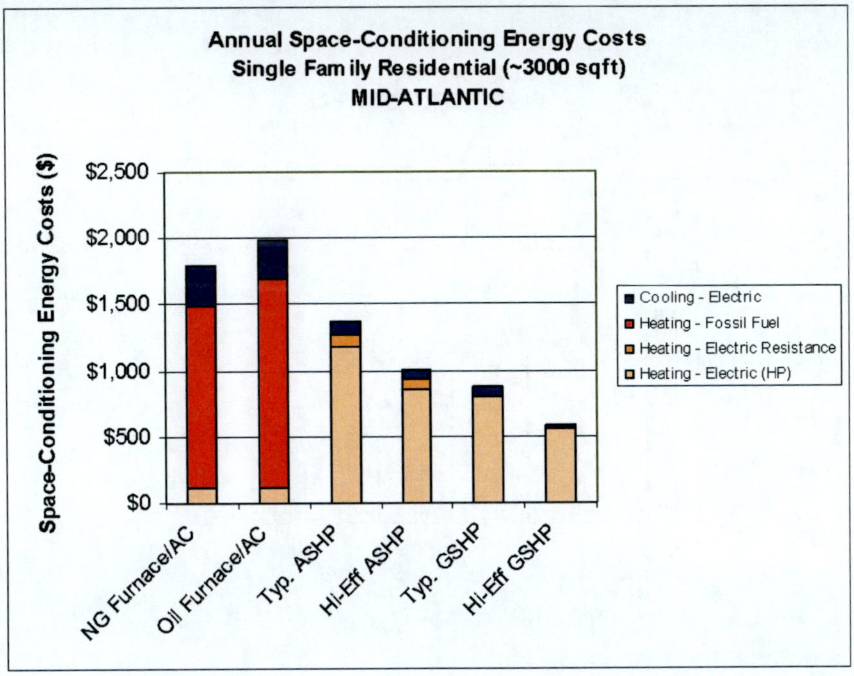
Annual Space-Conditioning Energy Costs
Single Family Residential (~3000 sqft)
MID-ATLANTIC
Space-Conditioning Energy Costs ($)
$2,500
$2,000
$1,500
$1,000
$500
$0
NG Furnace/AC
Oil Furnace/AC
Typ. ASHP
Hi-Eff ASHP
Typ. GSHP
Hi-Eff GSHP
Cooling - Electric
Heating - Fossil Fuel
Heating - Electric Resistance
Heating - Electric (HP)

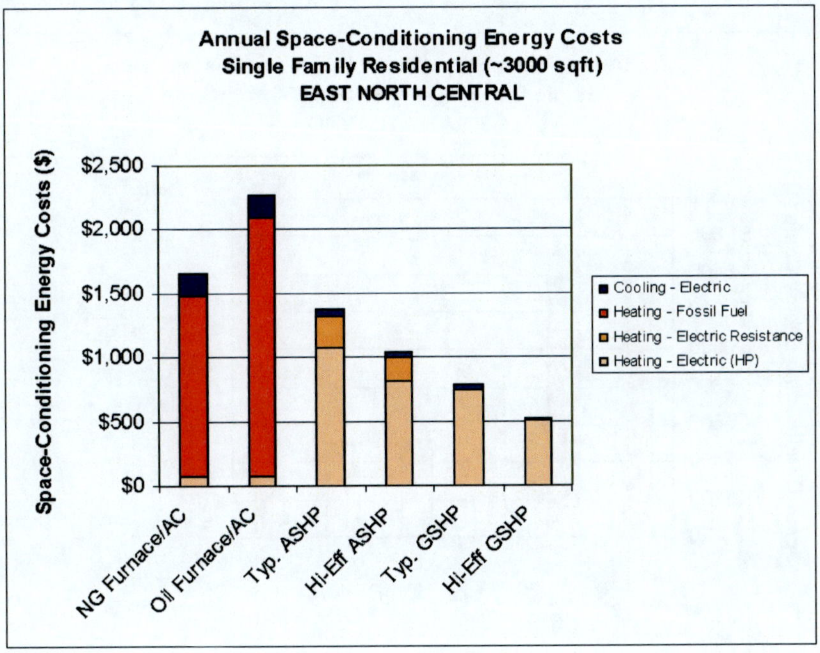
Annual Space-Conditioning Energy Costs
Single Family Residential (~3000 sqft)
EAST NORTH CENTRAL
Space-Conditioning Energy Costs ($)
$2,500
$2,000
$1,500
$1,000
$500
$0
NG Furnace/AC
Oil Furnace/AC
Typ. ASHP
Hi-Eff ASHP
Typ. GSHP
Hi-Eff GSHP
Cooling - Electric
Heating - Fossil Fuel
Heating - Electric Resistance
Heating - Electric (HP)

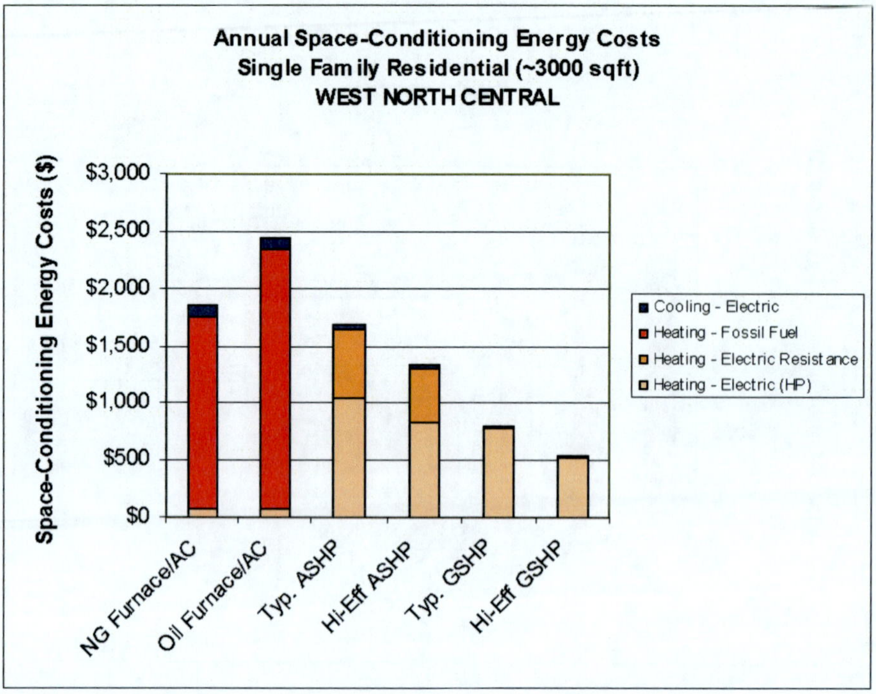
Annual Space-Conditioning Energy Costs
Single Family Residential (~3000 sqft)
WEST NORTH CENTRAL
Space-Conditioning Energy Costs ($)
$3,000
$2,500
$2,000
$1,500
$1,000
$500
$0
NG Furnace/AC
Oil Furnace/AC
Typ. ASHP
Hi-Eff ASHP
Typ. GSHP
Hi-Eff GSHP
Cooling - Electric
Heating - Fossil Fuel
Heating - Electric Resistance
Heating - Electric (HP)

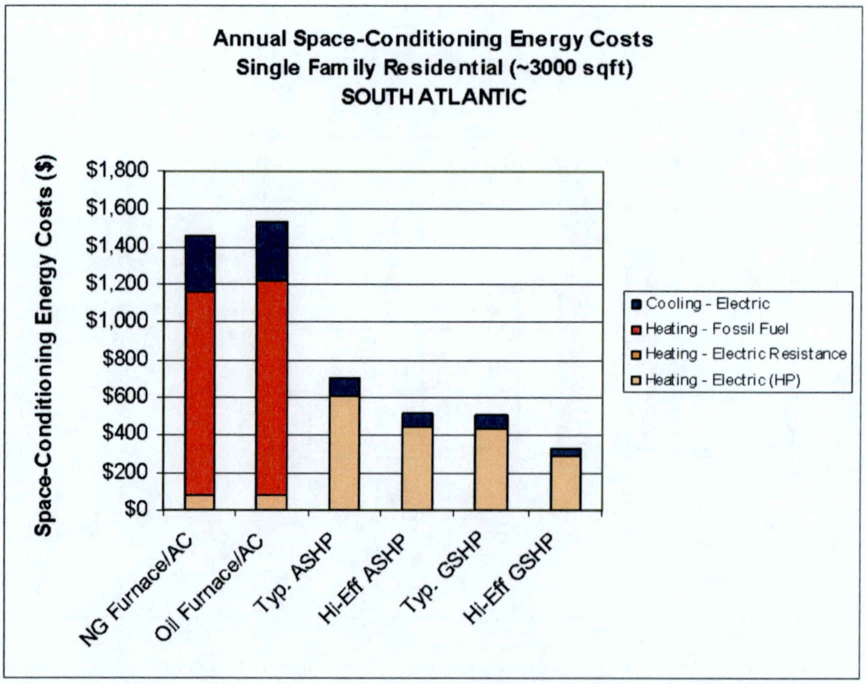
Annual Space-Conditioning Energy Costs
Single Family Residential (~3000 sqft)
SOUTH ATLANTIC
Space-Conditioning Energy Costs ($)
$1,800
$1,600
$1,400
$1,200
$1,000
$800
$600
$400
$200
$0
NG Furnace/AC
Oil Furnace/AC
Typ. ASHP
Hi-Eff ASHP
Typ. GSHP
Hi-Eff GSHP
Cooling - Electric
Heating - Fossil Fuel
Heating - Electric Resistance
Heating - Electric (HP)

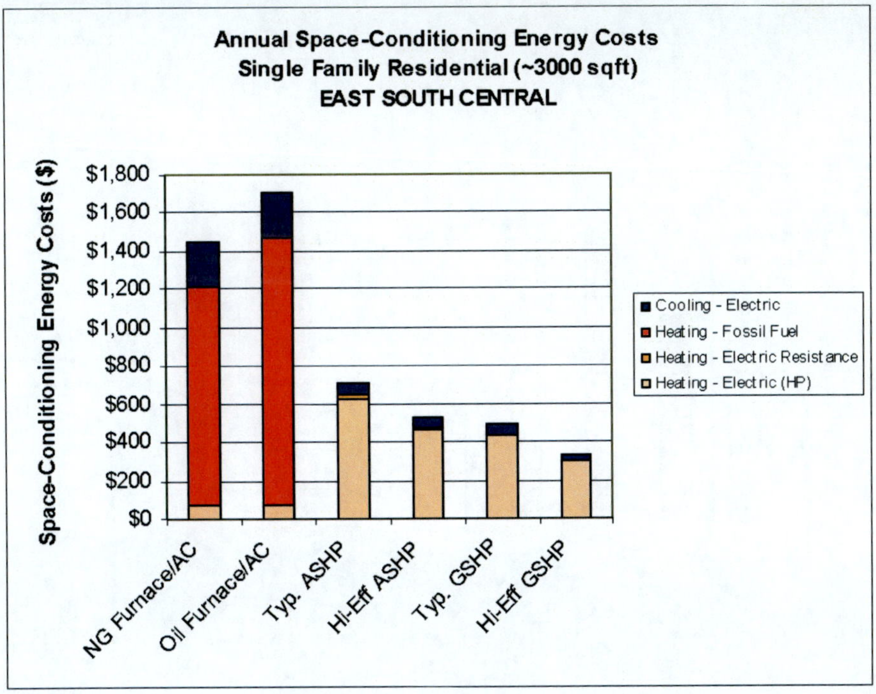
Annual Space-Conditioning Energy Costs
Single Family Residential (~3000 sqft)
EAST SOUTH CENTRAL
Space-Conditioning Energy Costs ($)
$1,800
$1,600
$1,400
$1,200
$1,000
$800
$600
$400
$200
$0
NG Furnace/AC
Oil Furnace/AC
Typ. ASHP
Hi-Eff ASHP
Typ. GSHP
Hi-Eff GSHP
Cooling - Electric
Heating - Fossil Fuel
Heating - Electric Resistance
Heating - Electric (HP)

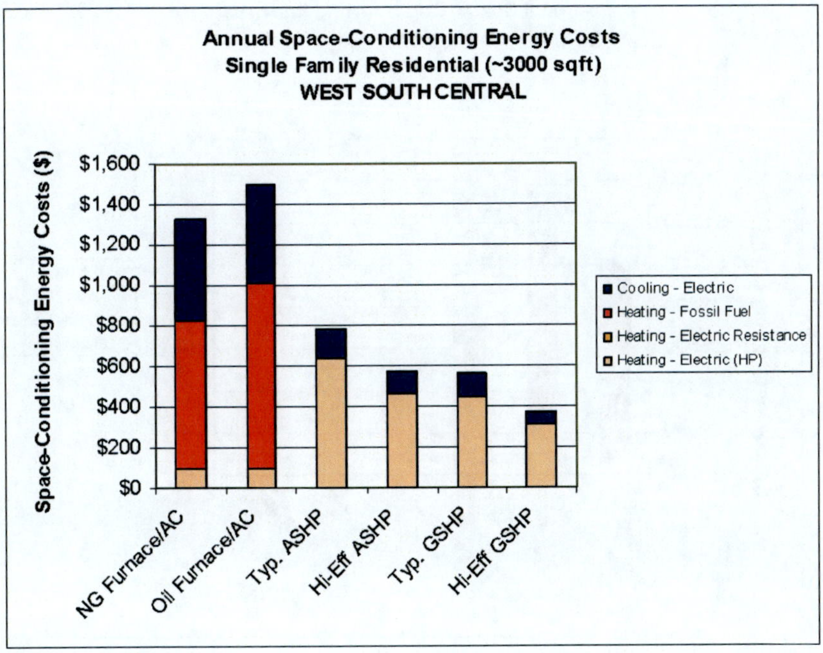
Annual Space-Conditioning Energy Costs
Single Family Residential (~3000 sqft)
WEST SOUTH CENTRAL
Space-Conditioning Energy Costs ($)
$1,600
$1,400
$1,200
$1,000
$800
$600
$400
$200
$0
NG Furnace/AC
Oil Furnace/AC
Typ. ASHP
Hi-Eff ASHP
Typ. GSHP
Hi-Eff GSHP
Cooling - Electric
Heating - Fossil Fuel
Heating - Electric Resistance
Heating - Electric (HP)

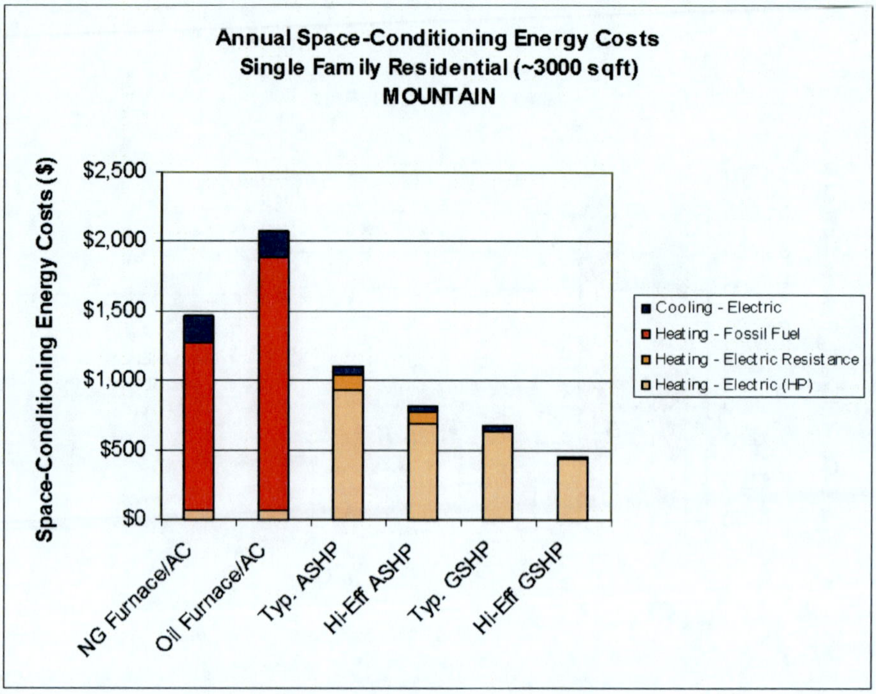
Annual Space-Conditioning Energy Costs
Single Family Residential (~3000 sqft)
MOUNTAIN
Space-Conditioning Energy Costs ($)
$2,500
$2,000
$1,500
$1,000
$500
$0
NG Furnace/AC
Oil Furnace/AC
Typ. ASHP
Hi-Eff ASHP
Typ. GSHP
Hi-Eff GSHP
Cooling - Electric
Heating - Fossil Fuel
Heating - Electric Resistance
Heating - Electric (HP)

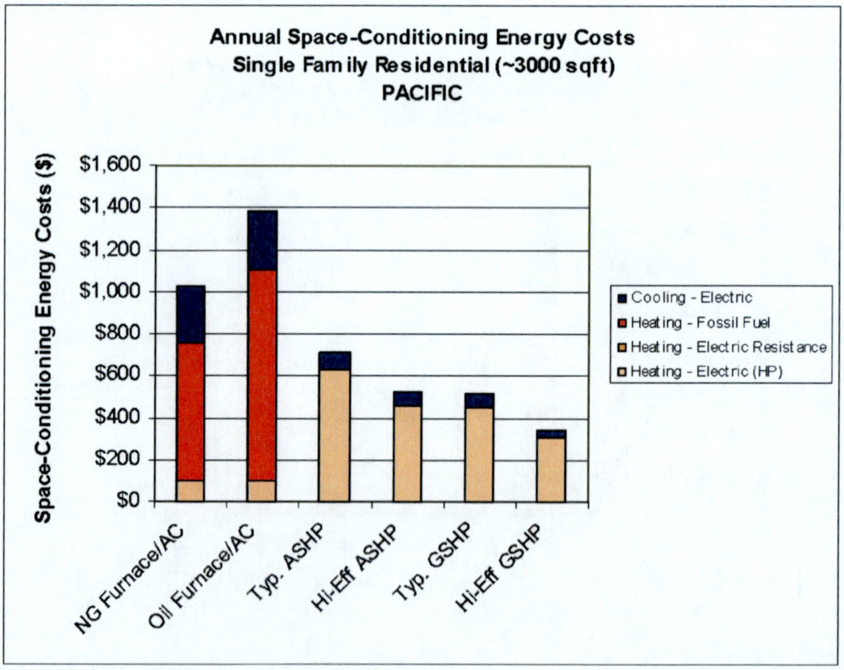
Annual Space-Conditioning Energy Costs
Single Family Residential (~3000 sqft)
PACIFIC
Space-Conditioning Energy Costs ($)
$1,600
$1,400
$1,200
$1,000
$800
$600
$400
$200
$0
NG Furnace/AC
Oil Furnace/AC
Typ. ASHP
Hi-Eff ASHP
Typ. GSHP
Hi-Eff GSHP
Cooling - Electric
Heating - Fossil Fuel
Heating - Electric Resistance
Heating - Electric (HP)

APPENDIX H: COMMERCIAL ANNUAL ENERGY COSTS

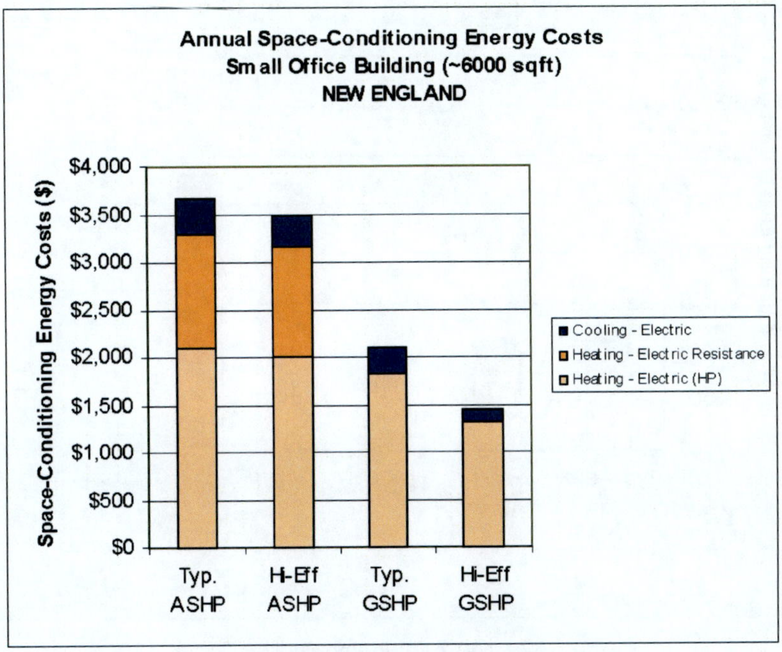

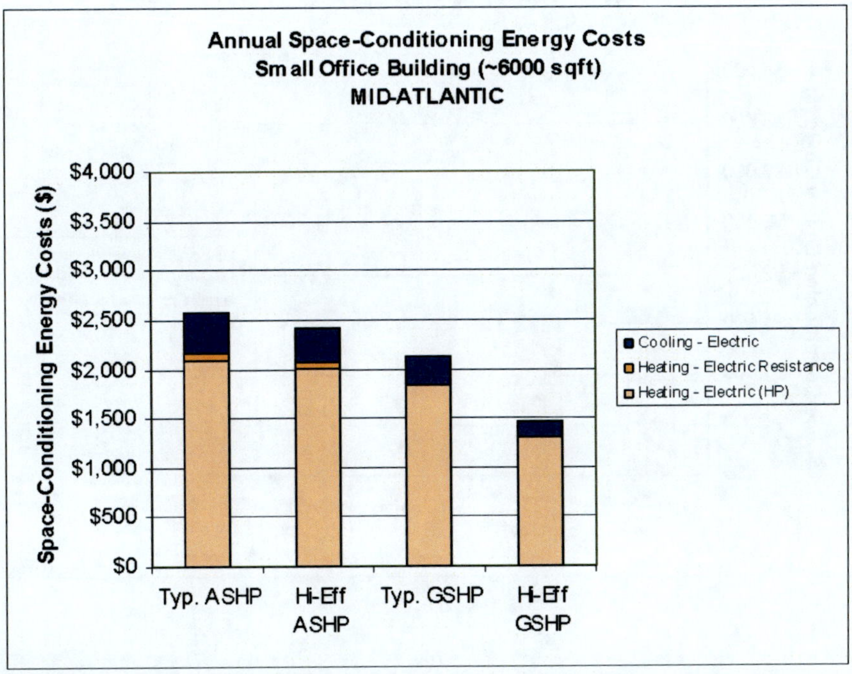

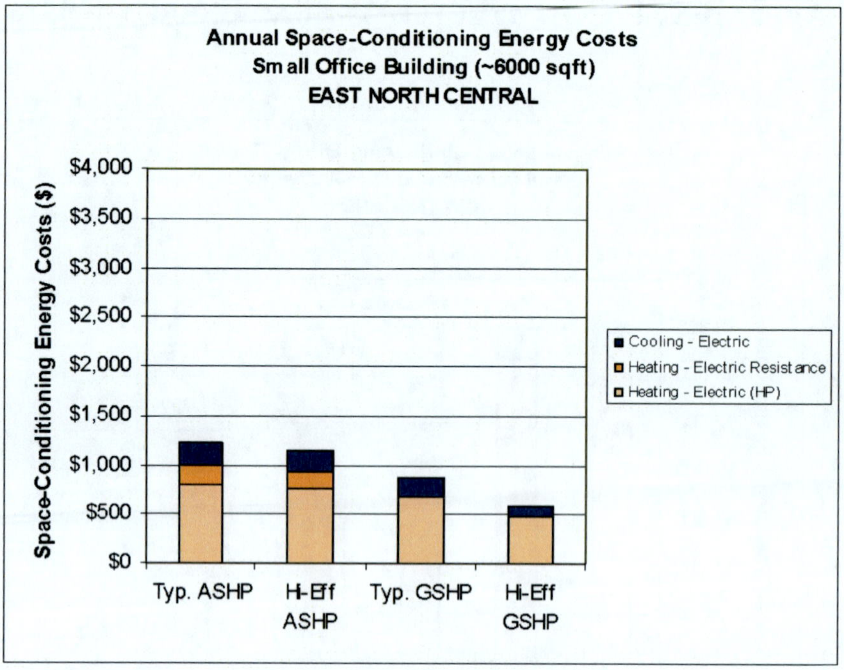
Annual Space-Conditioning Energy Costs
Small Office Building (~6000 sqft)
EAST NORTH CENTRAL
Space-Conditioning Energy Costs ($)
$4,000
$3,500
$3,000
$2,500
$2,000
$1,500
$1,000
$500
$0
Typ. ASHP
Hi-Eff ASHP
Typ. GSHP
Hi-Eff GSHP
Cooling - Electric
Heating - Electric Resistance
Heating - Electric (HP)

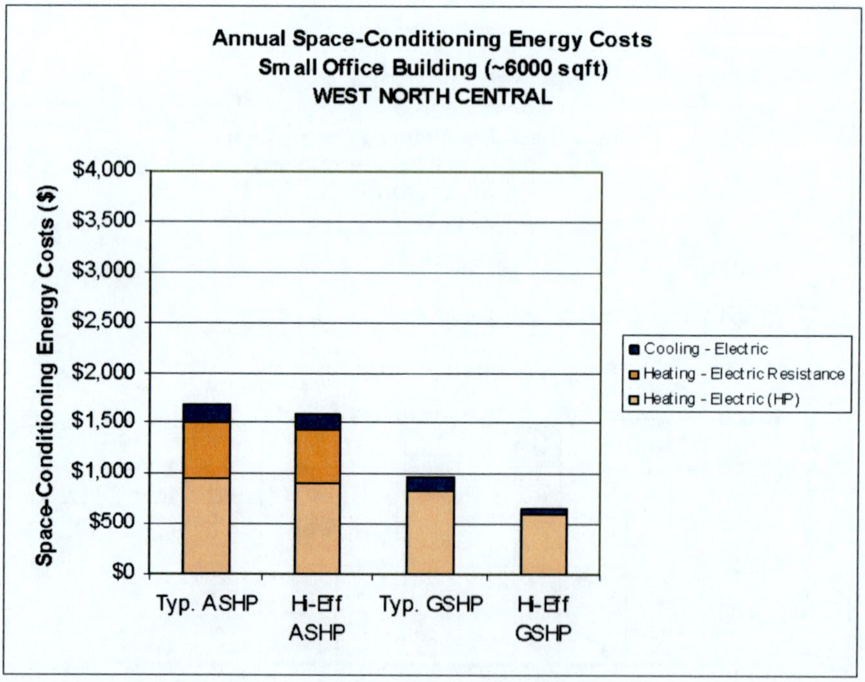
Annual Space-Conditioning Energy Costs
Small Office Building (~6000 sqft)
WEST NORTH CENTRAL
Space-Conditioning Energy Costs ($)
$4,000
$3,500
$3,000
$2,500
$2,000
$1,500
$1,000
$500
$0
Typ. ASHP
Hi-Eff ASHP
Typ. GSHP
Hi-Eff GSHP
Cooling - Electric
Heating - Electric Resistance
Heating - Electric (HP)

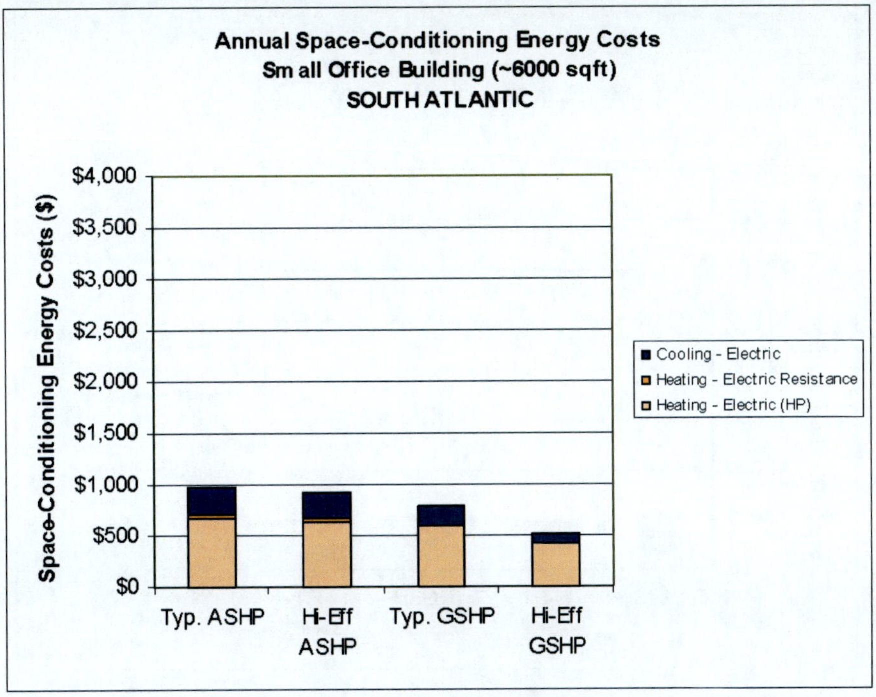
Annual Space-Conditioning Energy Costs
Small Office Building (~6000 sqft)
SOUTH ATLANTIC
Space-Conditioning Energy Costs ($)
$4,000
$3,500
$3,000
$2,500
$2,000
$1,500
$1,000
$500
$0
Typ. ASHP
Hi-Eff ASHP
Typ. GSHP
Hi-Eff GSHP
Cooling - Electric
Heating - Electric Resistance
Heating - Electric (HP)

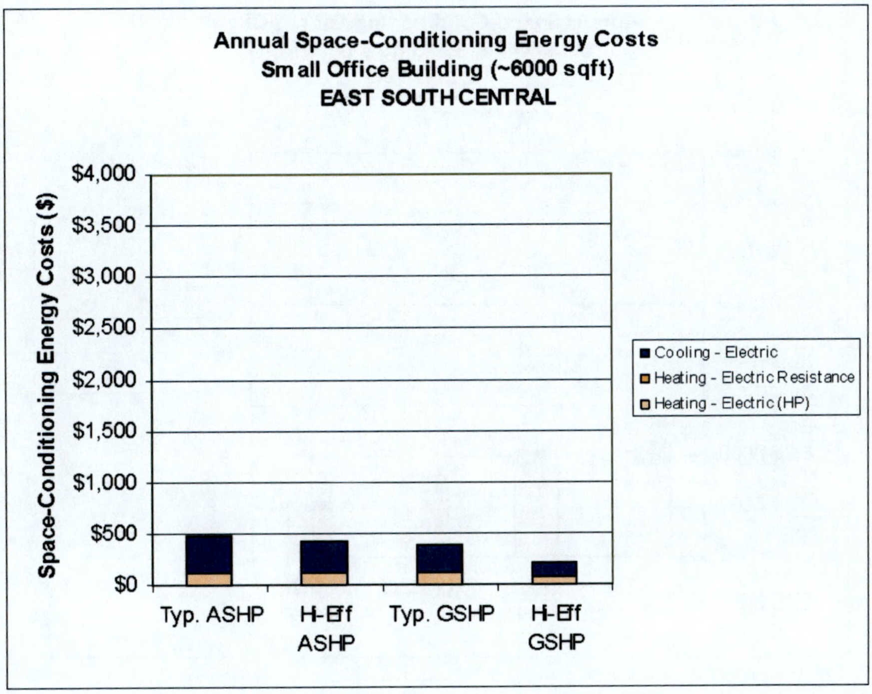
Annual Space-Conditioning Energy Costs
Small Office Building (~6000 sqft)
EAST SOUTH CENTRAL
Space-Conditioning Energy Costs ($)
$4,000
$3,500
$3,000
$2,500
$2,000
$1,500
$1,000
$500
$0
Typ. ASHP
Hi-Eff ASHP
Typ. GSHP
Hi-Eff GSHP
Cooling - Electric
Heating - Electric Resistance
Heating - Electric (HP)

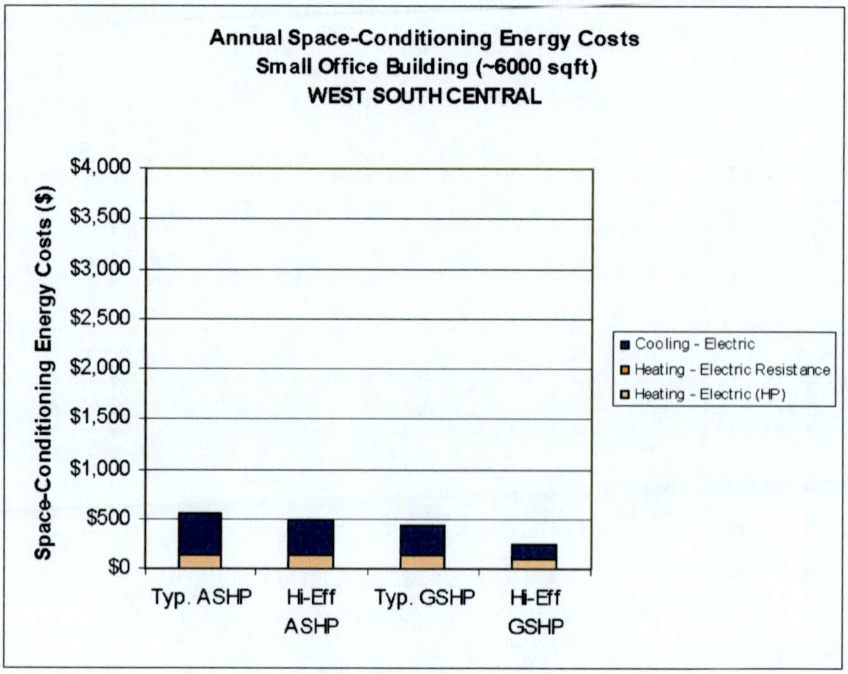
Annual Space-Conditioning Energy Costs
Small Office Building (~6000 sqft)
WEST SOUTH CENTRAL
Space-Conditioning Energy Costs ($)
$4,000
$3,500
$3,000
$2,500
$2,000
$1,500
$1,000
$500
$0
Typ. ASHP
Hi-Eff ASHP
Typ. GSHP
Hi-Eff GSHP
Cooling - Electric
Heating - Electric Resistance
Heating - Electric (HP)

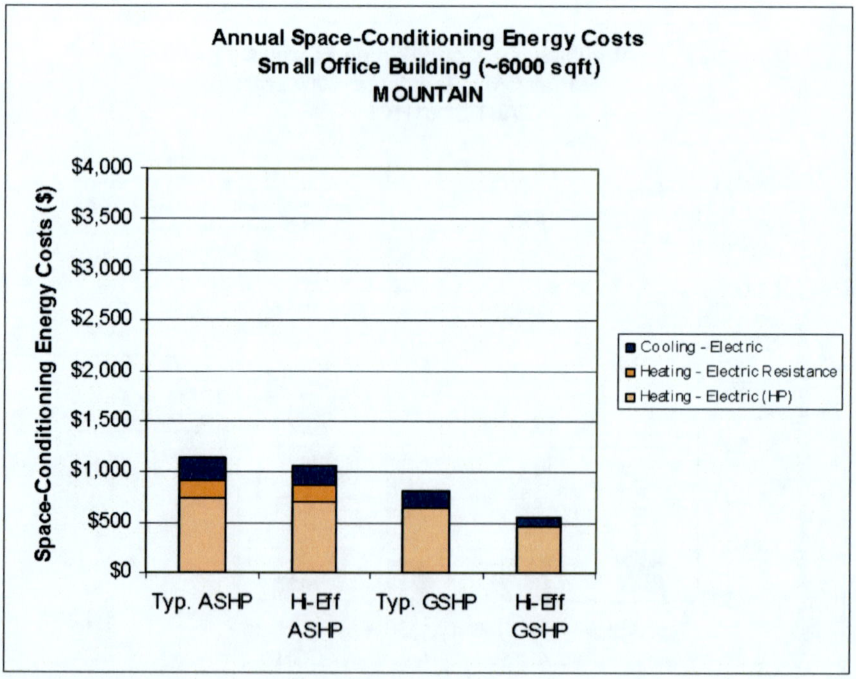
Annual Space-Conditioning Energy Costs
Small Office Building (~6000 sqft)
MOUNTAIN
Space-Conditioning Energy Costs ($)
$4,000
$3,500
$3,000
$2,500
$2,000
$1,500
$1,000
$500
$0
Typ. ASHP
Hi-Eff ASHP
Typ. GSHP
Hi-Eff GSHP
Cooling - Electric
Heating - Electric Resistance
Heating - Electric (HP)

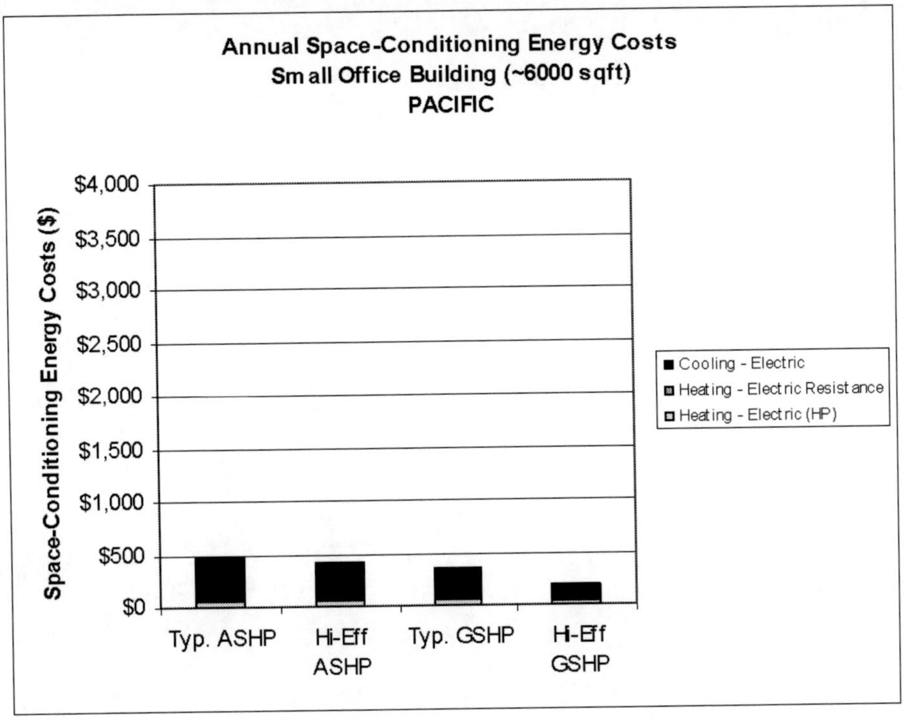

End Notes

[1] See for example http://www.mrslim.com/UploadedFiles/Resource/H2i_brochure.pdf. Products are designed achieve 100% capacity down to 5 °F outdoor and 75% capacity down to -13 °F, with a COP >1 even at those low temperatures.

[2] Austria, Estonia, Finland, France, Germany, Norway, Sweden, and Switzerland

[3] Exhaust air, air-air, air-water heat pumps, all depicted in Figure 7, fall within the air-source heat pump category. Exhaust air heat pumps are often used for the production of domestic hot water. Air-water heat pumps are typically connected to a hydronic distribution system.

[4] ARI-320 refers to ARI-rated water-source heat pumps, ARI-325 to ARI-rated groundwater-source heat pumps (open loop), and ARI-330 to ARI-rated ground-source heat pumps (closed loop).

[5] Primary energy includes the energy associated with generation (for electricity only), transmission, and distribution to the end user. For electricity, we use the national average efficiency for 2006 (31.5%). For other fuels, we neglect the transmission and distribution losses.

[6] Two homes used direct-exchange ground loops (refrigerant flows through ground loop). Report was written in 1998, so costs are dated. Also, costs for demonstration projects may not reflect typical costs.

[7] The load data are based on load profiles developed by MAISY (Jackson Associates).

[8] The load data are based on load profiles developed by Lawrence Berkeley National Lab (Huang 1999).

[9] DOE test procedures reference the following industry test procedures: ARI 320 – Water Source Heat Pumps; ARI 325 – Groundwater Source Heat Pumps; ARI 330 – Ground Source Heat Pumps

CHAPTER SOURCES

The following chapters have been previously published:

Chapter 1 - This is an edited, reformatted and augmented version of a United States Department of Energy, Bruce D. Green and R. Gerald Nix, National Renewable Energy Laboratory, Technical Report NREL/TP-840-40665, dated November 2006.

Chapter 2 – These remarks were delivered as testimony given on September 26, 2007. Susan Petty, President, AltaRock Energy, Inc., before the Sentate Committee on Energy and Natural Resources regarding Senate Bill 1543, National Geothermal Initiative Act of 2007.

Chapter 3 – This is an edited, reformatted and augmented version of a United States Department of Energy, Energy Efficiency and Renewable Energy, 2008 Geothermal Technologies Market Report, dated July 2009.

Chapter 4 – This is an edited, reformatted and augmented version of a United States Department of Energy, Energy Efficiency and Renewable Energy, Geothermal Technologies Program. Geothermal Tomorrow 2008 letter.

Chapter 5 – This is an edited, reformatted and augmented version of a United States Department of Energy, Energy Efficiency and Renewable Energy, Geothermal Technologies Progam. An Evaulation of Enhanced Geothermal Systems Technology publication, dated 2008.

Chapter 6 – This is an edited, reformatted and augmented version of William Goetzler, Robert Zogg, Heather Lisle, Javier Burgos, Navigant Consulting, Inc., submitted to United States Department of Energy, Energy Efficiency and Renewable Energy, Goethermal Technologies Program. Ground-Source Heat Pumps: Overview of Market Status, Barriers to Adoption, and Options for Overcoming Barriers report, dated February 3, 2009.

INDEX

A

B

C

H

I

N

O

P

R

S

T

U

V

W

Y